P9-CRO-089
3 1611 00225 3372

Plants in the Development of Modern Medicine

Participants in the Symposium "Plants in the Development of Modern Medicine," reading left to right:

Professor Ara Der Marderosian; Professor Bo Holmstedt; Professor Norman R. Farnsworth; Dr. Rudolf Hänsel; Professor Morris Kupchan; Dr. Albert Hofmann; Dr. Nestor Bohonos; Dr. Daniel Efron; Mr. S. Henry Wassén; Dr. T. Swain; Dr. Richard Evans Schultes; Professor John Mitchell Watt. (*Photograph by Doel Soejarto.*)

Plants in the Development of Modern Medicine

EDITED BY TONY SWAIN

Harvard University Press Cambridge, Massachusetts 1972

Library of Congress Catalog Card Number 79-169862

SBN 674-67330-1

Printed in the United States of America

Contents

Foreword

This volume has grown out of a symposium—Plants in the Development of Modern Medicine—held on May 8 to 10, 1968, in Cambridge, Massachusetts, and sponsored by the Botanical Museum of Harvard University and the American Academy of Arts and Sciences.

There has been continued interest during the past thirty years or so in the new drugs—the so called "Wonder Drugs"—from the Plant Kingdom. Meetings and symposia have been devoted to reviewing the advances made in numerous surveys of plants as possible sources of new, therapeutically active compounds. Government agencies, pharmaceutical companies, and academic institutions have all shown a vital interest and have been active in this field of investigation.

Most of these meetings and symposia have been sponsored by pharmaceutical or chemical institutions and have naturally been strongly oriented towards these two major fields. Consequently, the greatest emphasis on the papers delivered at these meetings has been on the pharmaceutical and chemical aspects of the subject, often to the neglect of other fields that play directly or tangentially a decisive role in the search for new drugs of vegetal origin. Then, too, their emphasis has more often than not been either historical or directed to the reporting of new discoveries.

The symposium from which this volume has grown differs in several respects. First: its organization stressed the interdisciplinary outlook, attempting in this way to draw into an integrated whole the contributions and potentialities of anthropological, botanical, chemical, pharmaceutical, and other pertinent fields. Second: while heeding the historical outlook and reporting recent discoveries, the symposium carried throughout its contents an emphasis on the future or potential of the Plant Kingdom as a source of new biodynamic constituents.

The symposium was made possible by an anonymous gift to the Botanical Museum of Harvard University for this purpose. The lectures were held in the Geological Museum of Harvard University, adjacent to the Botanical Museum. The American Academy of Arts and Sciences, cosponsor of the symposium, made available the house of the Academy in Jamaica Plain, Boston, for certain of the gatherings. We wish to thank the many people and institutions who through their help or collaboration made possible the successful achievement of this unique meeting and the eventual publication of this volume.

The authors of all of the papers contained in this volume have had an opportunity of extending their contributions and of bringing them up to date. Consequently, the book represents a greater coverage of the subject than that which it was possible to offer during the meeting itself.

REED C. ROLLINS Organizer for the American Academy of Arts and Sciences

RICHARD EVANS SCHULTES Organizer for the Botanical Museum of Harvard University

TONY SWAIN Editor of the Proceedings of the Symposium

Preface

In May 1968 the Botanical Museum of Harvard University and the American Academy of Arts and Sciences jointly sponsored a remarkable symposium on "Plants in the Development of Modern Medicine." It was remarkable because it brought together, as had no previous symposium, so many international leaders in so many diverse disciplines to focus directly on that single topic. This volume grew out of that symposium. The authors of the individual papers have greatly expanded and updated the material which they gave at the symposium. Their patience, wisdom, and help have made my task as Editor pleasant and worthwhile.

I am grateful to my friends, Dr. R. E. Schultes and Professor R. C. Rollins, who not only have added a Foreword, but also have extended to me the hospitality of the Botanical Museum and the Bussey Institution during my stay at Harvard.

T. S.

Introduction

PAUL C. MANGELSDORF

As the recently retired Director of the Botanical Museum which, together with the American Academy of Arts and Sciences, sponsored the Symposium which led to this volume, it was my pleasure and a very great privilege to extend to the distinguished speakers and to all other participants a brief word of welcome.

My own interest in the subject, "Plants in the Development of Modern Medicine," is largely vicarious and grew out of a course entitled "Plants and Human Affairs" that I taught at Harvard University for many years. In such a course, some attention must necessarily be given to plants as sources of medicinal drugs. When I began to prepare my first lecture on this subject in 1941, I was impressed with what then seemed to be the declining importance of medicinal drugs of plant origin. Of the ten drugs most widely prescribed in the United States at that time, only two—codeine and digitalis—were of plant origin.

A study of the pharmacopoeias confirmed my impression that medicinal drugs of plant origin were declining in importance. The first American pharmacopoeia, published in 1820, listed 650 drugs of which 455, or about 70 percent, were derived from plants. The eleventh edition of the pharmacopoeia, published in 1936, listed 570 drugs the preparation of 260 of which—or about 45 percent—were derived from plants. In that first lecture on medicinal drugs, which I delivered in 1941, I described this downward trend, told the students that it would undoubtedly continue, and predicted that by the time I would retire some 25 years later, medicinal drugs of plant origin would be of little more than historical interest.

Twenty-seven years have how elapsed since I made this bold prediction, and it turns out that I could scarcely have been more wrong.

Today there is probably more interest in drugs derived from plants than at any time in history. What has happened? Having failed so notably as a prophet, it would be presumptuous of me to now pose as a historian, but it is apparent, I think, that many factors are involved in this new interest in plant drugs. Among the more important are these:

1. The discovery of penicillin and other antibiotics has focused attention on the fungi and other lower plants as sources of physiologically active substances.

2. The brilliant success of the Indian Rauvolf drugs and their derivatives in treating various mental and other diseases has focused attention on folklore medicine as a source of new drugs. We have learned again what we should have learned from previous experience (the rediscovery of digitalis, an ancient British folk remedy, in the eighteenth century, and of ephedrine, an even more ancient Chinese remedy, in the second decade of this century), that folklore, although it may be principally nonsense, does often contain several grains of truth.

3. The discovery that certain plant drugs can produce psychotic models has resulted in a world-wide search for other plant substances possessing hallucinogenic properties.

4. Much of the research and progress today is interdisciplinary in its scope, involving not only pharmacology and botany but many branches of medicine, as well as biochemistry, anthropology, and other sciences. The extent of this interdisciplinary approach is reflected in this volume.

5. The scope of present-day research is broadly international. This too is reflected in the papers in this book in which authors from six different countries are represented.

I said to the participants of the Symposium "To all of you who have come here, from whatever part of the world, we members of the Harvard community and of the American Academy extend a warm welcome and wish you every success in your deliberations. If the program is a success, it will be because the participants have taken time from their busy lives to come here and join in this enterprise. We thank you for coming and we sincerely hope that the Symposium will prove to have been worthy of your time and effort." I think this volume shows that the Symposium undoubtedly was.

Plants in the Development of Modern Medicine

The Anthropological Outlook for Amerindian Medicinal Plants

S. HENRY WASSÉN

Gothenburg Ethnographical Museum, Sweden

To Mr. and Mrs. Gustav von Reis, generous and hospitable American friends at Hillsboro Beach, Pompano, Florida, the author gratefully dedicates this paper.

INTRODUCTION

> I have come to get you, but not without a purpose. You were placed as medicine, and it is for medicine that I seek you. Be not humiliated, oh powerful one.

The above quotation from William E. Safford's paper on "Daturas of the Old World and New" (1) explains how the Luiseño Indians in gathering *Datura meteloides* for ceremonial or medicinal purposes were treating the plant "with great deference." Before digging it up, the medicine man had to address it in the way quoted. Safford quite correctly has added that such an apologetic preliminary recalls "the customs of certain Mexican tribes in gathering the narcotic *peyote,* and those of the European herb gatherers of the Middle Ages in connection with the dreaded mandragora" (1, 2). These words of the Luiseño medicine man are a most appropriate introduction to this paper.

ANTHROPOLOGISTS VERSUS CHRONICLERS

The literal meaning of the word anthropology is "the science of man." We find this definition in hundreds of scholarly books, but we also learn, for example, from Kroeber (3) that "this literal etymological meaning is too broad and general." He took as a provisional basic definition: "Anthropology is the science of groups of men and their behavior and productions." He also stated (3) that "of all the social sciences anthropology is perhaps the most distinctively culture-conscious." Beals and Hoijer (4) express the same idea as follows: "All cultures interest the cultural anthropologist, for all contribute some evidence of men's reactions in cultural forms to the ever present problems posed by the physical environment, the attempts of men to live and work together, and the interactions of human groups with one another." The same authors continue: "Since cultural anthropology covers so wide a range of human activities, it is traditionally divided into three main branches: archaeology, ethnology, and linguistics." I have been using all of these three branches, as well as following a historical approach, in my own studies in the field of drug research among Amerindian groups.

Mandelbaum (5) has said that "each special field within anthropology has its special lesson to give," but it should be noted that Beals and Hoijer (4) stated that "anthropology was impossible until the age of European exploration brought scholars into contact with peoples diverse in physical form and culture. Even the tools of ancient man found in Europe were not understood until educated missionaries pointed out their resemblance to the tools of the American Indian." It is clear from this that the old and often quoted *cronistas* from what became Latin America could by no means be termed anthropologists but, nevertheless, as early eyewitnesses, they many times fulfilled one of the most important prerequisites for social anthropology which "deals with the behaviour of man in social situations" (6), namely, that of direct observation.

An anthropologist naturally has to look for documentation when studying any culture. In addition to all his own observations or those of his collaborators in the field, and those from his own or other collections, all kinds of written or other recorded information, printed or unprinted, are part of his documentary material. From my first years as a student of anthropology in Göteborg I recall how our professor, Erland Nordenskiöld, when lecturing on the cultures of the South American Indians, used to begin with an appraisal of the literature available up to his time and which he included in his six bibliographical maps in Volume 2 of his "Comparative Ethnographical Studies" (7). I must admit that we young students learned a great deal from such an approach, which was possible only through having access to a comparatively good library. And we must all admit that there are wonderful and detailed descriptions, for those days, in many of the very early reports of the *conquistadores,* the missionaries, and so on, which are reflected in the many works of the chroniclers.

I must say that neither this paper nor my contribution to the symposium "Ethnopharmacologic Search for Psychoactive Drugs" in San Francisco, 1967 (8) could have been written at all had I had to depend only on the works of the modern anthropologists. I stated in my 1967 paper (8) that "during his second voyage, 1493-1496, Columbus not only commissioned the Friar Ramon Pane to undertake what we now call anthropological field work among the aboriginal population of Española, but he himself made valuable observations presented in his narrative of the second voyage."

All of us should also express deep respect to the memory of the sixteenth-century Spanish soldier and chronicler Pedro de Cieza de León, who filled his work *La crónica del Perú,* with most valuable observations of the Indian cultures in Colombia. In his dedicatory chapter to his Emperor, he outlined his need for writing down his observations, using in one place the words: "*. . . pues muchas veces cuando los otros soldados descansaban cansaba yo escribiendo*", that is, "Many times when the other soldiers were resting, I rested by writing" (9).

FIELD WORK AND LIBRARY STUDIES

The necessity for both field work and library studies in any serious work regarding an anthropological approach to medicinal plants has been touched on above. In the *Notes and Queries of the Royal Anthropological Institute* (6) we read that "the belief, so widely held, that ill-health, accident, and death do not occur from natural causes alone, brings much of medicine under the heading of ritual and belief. The investigator should, therefore, observe, in addition to the practical methods of treatment by drugs, massage, manipulation, inoculation, etc., the ritual connected with these practices." For the *treatment,* it is said (6) that "all ritual connected with the collection or preparation of the *materia medica* should be observed. The ritual behavior and any payment between patient, or his intermediary, and expert should be noted, and the psychological attitude of the patient to the expert."

This is all very well, but I believe more stress should have been laid on *collecting*—where possible and available—all the medicines used to bring about a certain cure. Since plant medicines are used and referred to often by tribal healers, I am certain that valuable data could be secured by anthropologists in many cases even if they are untrained in botany. According to Schultes (10), "today, *ethnobotany* is defined usually as the study of the relationships which exist between peoples of primitive society and their plant environment." In many cases the information obtained by the anthropologists comes close to this definition as does much of the information found in the hundreds of local publications (as, for example, in South America) dealing with plant names used by local populations although these give few exact descriptions of value to the botanist. The medical evaluation naturally must be made by professionals after the plants referred to in some more or less obscure source have been found, identified, and properly investigated from chemical, toxicological, and pharmacological points of view; in other words, finally accepted or rejected as drugs. In spite of the fact that very few anthropologists have an opportunity of working in direct collaboration with botanists or ethnobotanists in the field and consequently must themselves take the initiative of observing the use of plants, and if possible, also collect them, I am certain that the information which such anthropologists present already has been, and will continue to be, of outstanding interdisciplinary importance.

We anthropologists should, however, think of the words of Schultes (10) that "in discussing the role of ethnobotany in our search for new drug plants, we must constantly bear in mind the widespread exaggeration of the usefulness of ethnobotanical data." But we might also bear in mind his assertion that "many of our basic ethnobotanical observations have been made by botanists and anthropologists as by-products, so to speak, of their major programs of research. Usually men of broad training and interests, they were often quite unaware of the contributions which they were making to ethnobotany" (10). From modern anthropology, we have indeed an excellent example of a complete study about traditional medicine and certain other folk beliefs and practices presented as "a by-product" of a housing study. I am thinking of Isabel Kelly's *Folk Practices in North Mexico,* published in 1965 (11) but based on her field study from 1953. During research for "a major housing program in the small agrarian community of El Guije, near Torreón" in the state of Coahuila, that is, the desert area of northern Mexico, this anthropologist recorded her data "on Saturdays and Sundays when, for official reasons, it was not practical to work on the housing survey" (11). Herbarium specimens were collected, "chiefly because of their alleged medicinal properties." After taxonomic determinations had been made by specialists, names of the specimens were published as an appendix to Dr. Kelly's work (11). All in all, it resulted in a most interesting book.

In the discussion at the Psychoactive Drug symposium in San Francisco in 1967, Dr. Nathan S. Kline asked: "How would you educate anthropologists in terms of going out into the field? Would there be some way, perhaps, of even preparing a field guide for them? Perhaps it should be prepared by a botanist in terms of what should be looked for from the botanical point of view, and perhaps from a clinical point of view." I told Dr. Kline afterward that I had had the same idea when, a couple of years previously, in my paper "Some General Viewpoints in the Study of Native Drugs" (12), I had written as follows:

> As naturally very few teachers in anthropology have a real interest in or knowledge of drug lore, I have a feeling that an interest in what would benefit pharmacology could be furthered if the anthropologists could be provided with concise information on the most important facts and problems relevant to this broad field. Such information should if possible be written and printed separately for the greater cultural regions of the world so that a reader, not himself a pharmacologist but with the good will to help to procure useful material for research, would not run the risk of losing himself in too many details.

I then referred to some works which I myself consider of outstanding importance for certain geographical regions. I am not going to repeat this list here but simply refer to the paper quoted (n. 12, p. 115). Such a list would now be longer and more detailed, but this is a task for those more familiar with the field than I am.

During the preparation of this paper I decided to prepare a list of some of the works related to America which were available and which I feel are of possible importance if we want to see how varied the field is. Their titles and short comments on them are given in the Bibliography following this paper. These books and papers have been selected somewhat haphazardly, but we notice immediately that the literature in the field forms a broad spectrum, from such pearls as Sahagun's "*Historia de las Cosas de Nueva España*" (= The Florentine Codex, translated by Dibble and Anderson) and certain modern investigations in the field (also including some important bibliographical papers) to a number of small and rather obscure works which might be placed in the category of folk medicine.

To show the importance of "botanizing" by using libraries in the study of Amerindian plants and their use, I wish to draw attention to two items both of which happen to come from Brazil. Remembering that Schultes (13; cf. 8,14) has warned that we cannot be absolutely sure that the snuff used by the Mura and Maué was prepared from *Anadenanthera peregrina* var. *peregrina,* since botanical considerations must be kept in mind, I refer to the rare book by Barbosa Rodrigues from 1875 (15) in which he describes his exploration of the rivers Urubú and Jatapú. He deals with the preparation of the snuff among the Mura, who he points out as taking the *paricá* in a different and much more barbaric way than the Maué. From what he says (n. 15, p. 25) it is clear that he believed the ingredients to include ripe seeds of *paricá* which came from the *Mimosa acacioides* and that ashes from *Theobroma sylvestre* were added. He described the use of this concoction either as a snuff or an enema (see Figs. 7 and 8). Whether Rodrigues' botanical detail is correct is another question; we know that the role of *Mimosa acacioides (Piptadenia peregrina = Anadenanthera peregrina* whose beans are used) has been accentuated in many nineteenth-century descriptions of the use of hallucinogenic snuff in South America. Since the work by Barbosa Rodrigues is rare, a full translation of the passage is given here. It is a description of the habits of the Mura which was not quoted by Nimuendajú in his paper on "the Mura and Piraha." (16) The translation follows:

As the Maué, they [the Mura] prepare and take *paricá* in a different and more savage way. In older days the celebration took place when some Indian had reached maturity, but today there is a celebration in June each year in some *malocas* not yet christianized or civilized. Hunting and excesses followed by drinking-bouts with *cachiry* and *caysuma* precede the paricá feast (*A'uin*) while a special house or shelter is built (*birupica*). When the *puracé*—as the natives call it—begins, the women prepare the beverages and the *beijú* (flat cakes) of manioc and *payauaru.* [Cf. Nimuendajú (16): "The Maué are very fond of a drink made from dried cakes of flour, the arobá or paiauarú of Neo-Brazilians"]. This, as well as the preparation of *paricá,* is reserved for the oldest (women) in the tribe. The dance is exclusively for the men. Before a description of the feast is given, the method for preparing *paricá* will be described so that the variation in the Maué Indian way of doing this can be seen. Ripe seeds of paricá, *Mimosa acacioides,* are collected and crushed in a wooden mortar (*tuá-ain*). During this process, and because of the oil they contain, a kind of dough is formed and therefore one adds ashes from the bark of wild cacao, *Theobroma sylvestre.* Having been kneaded well, 3-inch long cakes are formed which are put on a forked pin and dried by the fire. When dry, the material is again crushed in another mortar called *quên-uê* and the powder is kept in a box of snail shell, *tiusepôe.* The feast starts with flagellation, by two Indians whipping each other with dried branches of a tree or thongs of the hide of tapir, sea-cow or deer. Formerly, a stone or some other hard object was tied to the end. Each man flagellates the other in turn. This ceremony sometimes goes on for six days depending on the number of young men who have attained majority. All those who have been flagellated take *paricá* either as snuff or dissolved in water as a clyster. During the dancing, the old (women) who have prepared the *paricá* either fill a bamboo tube with powder and give it to one of the dancers who places one end of the tube in one of the nostrils of a comrade and blows in the other end, or gives him a syringe of rubber filled with *paricá* dissolved in cold water to administer. Taken either way, the effect is terrible and so violent that many die of suffocation or fall unconscious, while still others, resisting, continue the dance. Under the special shelter there are bows and bunches of arrows hanging for those who have suffocated and cannot continue to dance but stroll in the place with them. Usually, the clyster causes a violent intoxication.

In another printed source from Brazil, the famous general and Indian friend, Rondon (see Bibliography) mentions the now extinct Tupí group, Kepkiriwat, as snuffing through tubes. The Kepkiriwat were southern neighbors of the western Nambikwara and lived on the headwaters of the southern tributaries of the Machado river, to the right of River Guaporé (17). Lévi-Strauss (18) states that the Nambikwara were inveterate "fumeurs de cigarette" while their Tupí neighbors Kepkiriwat and the Mundé (Mondé) snuffed the tobacco through "tubes insufflateurs." In October 1916 General Rondon spoke in the "*Camara dos Deputados*" in Rio de Janeiro and he also mentioned the Indians *Kepikiri-uat,* whom he had visited many times (19) (see Bibliography). "They do not smoke, but they take snuff through a very ingenious apparatus. It consists of a tube of arrow grass (*taquarinha, Gynerium* sp.), two spans in length which in one end has a small container of coco. When the container has been loaded with tobacco powder, the person who is going to take the snuff places it to the nostrils, while another person blows from the free end so that the snuff goes into the nasal cavities of the 'tobacco-man,' who inhales deeply. This apparatus, which cannot be denied its innovative value, is not as strong as the Nambikwara cigar which in Kepkiriwat is called *nharimã-cap.*"

We have here an eyewitness who was describing a kind of shake-hands ceremony among the Kepkiriwat for his listeners in the Congress of Brazil. If Rondon really observed a tube with a special container on it, his observation is of interest and I shall return to it later. There is, however, the possibility that he was observing one of the "tubes that terminate in a hollow nut, often shaped like a bird's head" (17).

An anthropologist interested in drug research must sometimes be prepared to engage in real detective work if he comes upon details of special interest during his search in the literature. I think I am fully entitled to refer to the case of 1786 and the Brazilian naturalist Alexandre Rodrigues Ferreira. In an earlier paper (20) I touched on the story and have since found more information, which I would like to give now.

In 1967 I received from my friends in the *Instituto Nacional de Pesqui-*

sas da Amazônia a copy of *Cadernos da Amazônas,* containing a paper on the life and work of Alexandre Rodrigues Ferreira written by Gloria Marly Duarte Nunez de Carvalho Fontes (21), which complements the work by Valle Cabral (22). Ferreira, who became the pioneer naturalist of the Amazon region, was born in Bahia on April 27, 1756, and in 1770 he was sent to Portugal for enrollment in the University of Coimbra. It was a time of great scientific activity in Portugal, during which, to mention one example, a museum of natural history was founded. Ferreira was nominated *demonstrador* in Natural History at the Philosophical Faculty of Coimbra University in 1777, passed his examination in Natural History in 1778, and was soon after recommended and accepted to take part in a scientific program of research in Brazil, his age at that time being 22 years. He got his doctor's degree in 1778 and became in May 1780 a member of the Royal Academy of Sciences at Lisbon. At that time he was living in that city and, except for writing, he dedicated his time to work in the "classification and analysis of natural history specimens in the Real Museu d'Ajuda (21), which is of importance for what is to follow. This museum was founded by Portugal's famous Marques de Pombal (1699-1782), at the suggestion of Miguel Franzini, to replace some kind of a curio-cabinet which was destroyed during the great earthquake of 1755. In 1836 the Real Museu d'Ajuda was transformed into the Acadêmia Real das Ciências and thus became the forerunner of today's Museu Nacional de Historia Natural–Museu Bocage in Lisbon which belongs to the Faculdade de Ciencias de Lisbon. Also the Jardin Botanico da Ajuda was planned during Pombal's time, and Rodrigues Ferreira was one of its first directors, following the first one, Domingos Vandelli. This Botanical Garden (which had a small museum) was later annexed to the Escola Politécnica.

During studies in Portugal in July 1968, I found in the Bocage Museum several dispatch lists of collections sent to Portugal from Brazil during the eighteenth century. In these lists we often meet the phrase "com destino a o Real Muzeo Nacional da Corte, i cidade de Lisboa, a saber–", followed by an inventory. The important manuscript material, also in Ferreira's hand, now found in the Museu Bocage at Lisbon, where it was freely placed at my disposal by Dr. Maria Morais Nogueira, is there thanks to the historic origin of the museum. Shortly after Ferreira's death on April 23, 1815, the Viscount of Santarem, by an order of July 5 the same year, arranged that all of the late naturalist's papers should be placed in the Real Museu da Ajuda. The man who carried out this order was Felix de Avellar Brotero, and he wrote a catalogue with the following title "*Catalogo geral dos papeis pertencentes á Viagem do Sr. D.*^{or} *Alex.*^{e} *Rõiz Ferreira dos Estados do Brasil, que me forão entregues por ordem do Ill.*^{mo} *e Ex.*^{mo} *Sr. Visconde de Santarem*". This catalogue, which has been published by Valle Cabral (22), is very important, as it clearly shows that many of the original *Memorias* written by Ferreira now found among the *Códices* of the National Library in Rio de Janeiro were, at the time of Ferreira's death, still in Lisbon.

In 1783 Ferreira sailed to Brazil to fulfill a Royal Decree regarding a so-called Philosophical Expedition ("Expedicão Philosophica"). He went on several long journeys until his return to Pará in January 1792. His return trip to Portugal took place in 1793. In spite of being promoted to Acting Director of the Real Cabinete de História Natural and other posts, his life in Portugal until his death on April 23, 1815, was not a very happy one. Among other things, his collections, carefully labeled and classified and sent to the Real Museu d'Ajuda had been very badly treated during his absence. Most of the material had been mixed up, the labels were gone, and, still worse, he could not get financial help to publish his manuscripts—his so-called *Memórias.* A final blow was the order, which was issued by the French during the Napoleonic invasion of Portugal, that a lot of the Natural History Museum specimens should be sent to France. Fontes quotes the order in her footnote on page 21 of her paper on Ferreira's life and work (21). It was signed on June 3, 1808, by "le Duc d'Abrantes," and the man who had to permit the French representative, M. de Géoffrey, to take what he wanted in Lisbon was the director of the Real Museu d'Ajuda, Junot a Domingos Vandelli.

From this historic framework we return to Brazil, where on February 13, 1786, in Barcellos at the Rio Negro Ferreira signed a very interesting manuscript, one of his many *memórias.* This deals with the instruments used by the Maué Indians for taking *paricá;* and Ferreira had sent a collection of such instruments in his first dispatch from Rio Negro. The full translation of this *memória* is given in Appendix A. The original manuscript, which is kept in the National Library of Brazil in Rio de Janeiro, has the title: "Memória sobre Os Instrumentos de que usa o Gentio p.^{a} tomar o tabaco Paricá = os quaes forão remettidos no Caixão No. 7 da primeira remessa do Rio Negro". The original has, in *Anais de Biblioteca Nacional,* vol. 72, been described as "4p. 32×21.5 cm. Códice, Sem. o nome do autor." It seems to have been published in the "Rev. Nacional de Educacão," no. 8, pp. 74-76, Rio de Janeiro, May 1933, but this edition has not been available and it does not appear among the ten items written by Alexandre Rodrigues Ferreira listed in the "*Amazônia - Bibliografía,*" 1614/1962 published in Rio de Janeiro in 1963 by the *Instituto Brasileiro de Bibliografía e Documentacão.* The manuscript has, however, been listed by Valle Cabral (22) as follows: "Memoria /sobre/ os Instrumentos, de que usa o Gentio p.ª tomar o tabaco - Paricá - os quaes forão re-/mettidos no Caixão N.º 7 da primeira remes-/sa do Rio Negro./—Original, com correccões escriptas da propria mão de Ferreira. Cód. CXVI/16-17/2ff. não num. 26×12. No alto da primeira folha se-le: 'N.º 40.-Drummond. (Coll. Lag.)." Drummond is said to have been an ambassador of Brazil to Portugal, evidently with a vivid interest in the manuscripts by Ferreira. We can perhaps assume that he took some part in the return of the originals to Ferreira's native land. For a study of the original, I have had the most welcome help of the Ambassador of Sweden in Rio de Janeiro, Count Gustaf Bonde, who had acquired a photocopy of the manuscript for me. The transcript of the text was made by Mr. Hildon Hermes da Fonseca at the Embassy of Brazil in Copenhagen. I am most thankful to both.

As Ferreira made the earliest detailed description of snuffing paraphernalia of the Maué Indians (Appendix A) and also sent a collection of the material to the Ajuda Museum in Portugal, it is of course interesting to read his manuscript and to try to follow what happened to the collection. The wooden snuff trays from "o Gentio Magué," or the Maué Indians, are said to be in the form of animals (Appendix A). One owned by an Indian was said by its owner to be characterized as an alligator (*jacaré*). Mother-of-pearl, used as an ornamentation on the trays, is called *itaã,* and covers the place on the tray where the snuff is placed (*paricá-rendana* which Ferreira translates with "the place where one puts the *paricá*"). Double tubes of bird bone are mentioned and also the names of several birds used to make them. The snuff tubes had a fruit at one end for insertion into the nostrils. The snuffing took place ceremonially at drinking-bouts (grandes Bacchanaes) and special *paricá*-houses were built. The feast started with the ceremonial whipping between the men, and in the description we can clearly recognize one of the *yurupari* feasts for the demons of vegetation known from several tribes in Amazonas. These feasts went on for a considerable time and were given when certain palm fruits and other comestibles were available so that there was plenty to eat and drink. Ferreira finally speaks of the narcotic property of the *paricá* ("a virtude narcotica do Parica") and about the violent character of the feasts.

Illustrations from Ferreira's *Viagem filosófica* are also kept in the National Library at Rio de Janeiro. As a matter of fact, Ferreira, when starting his voyage to Brazil from Lisbon early in the morning of September 1, 1783, took with him two draughtsmen. Their names appear, together with that of a botanically trained gardener, in the title of one of the original manuscripts kept in the Museu Bocage at Lisbon which reads: "Roteiro das viagens que fez pelas capitanias do Pará, Rio Negro, Mato-Grosso e Cuyabá Alexandre Rodrigues Ferreira quem acompanharãn os Desenhadores Joseph Joachim Freire, Joachim Joseph Codina e o jardaneiro botanico Agostino Joachim do Cabo." To this title Ferreira has added in Latin: Per mare, per terras, tot adire pericula jussus" ("Commanded to meet so many perils by sea and land"). One of the illustrations (reproduced here as Fig. 1) shows "instrumentos de música, ornatos e utensilios domésticos dos gentios." It has been published in the paper by Gloria Fontes (21), and its catalogue number in the Biblioteca Nacional is 21, 1, 1 n.º 89. One of the details shows two Indians helping each other with snuff-blowing in a long tube (Fig. 2). Below this

Fig. 1 Illustration from Ferreira's *Viagem filosófica* showing snuffing paraphernalia and other ethnographical specimens. Catalog No. 21, 1, 1 no. 89 of the National Library of Rio de Janeiro.

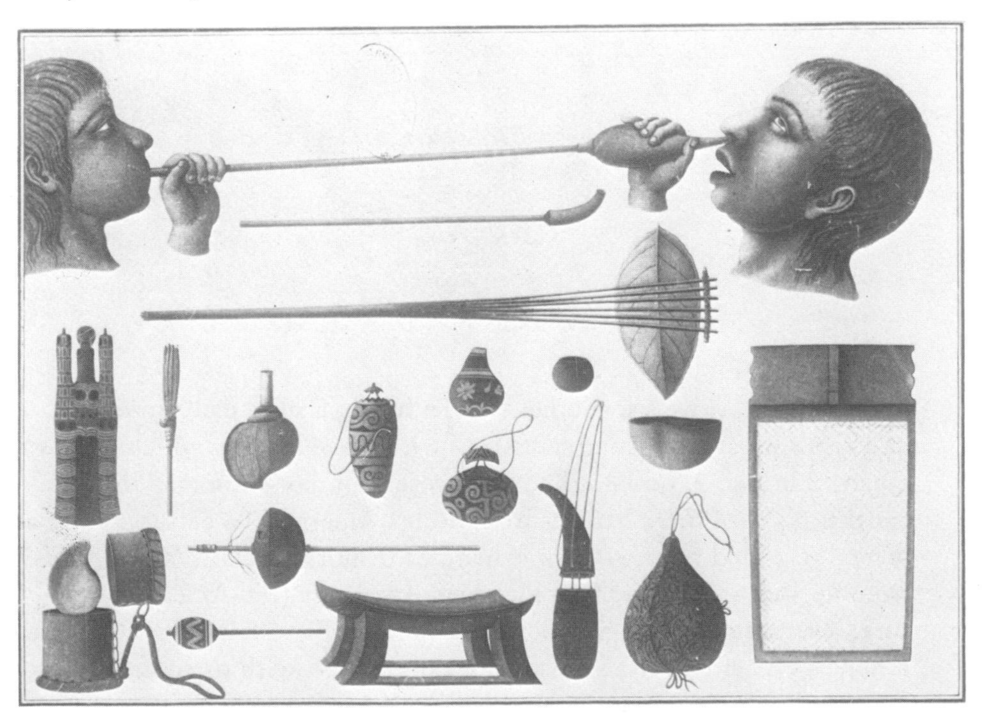

Fig. 2 Snuffing Brazilian Indians. Drawing from a photograph of a detail in Fig. 1.

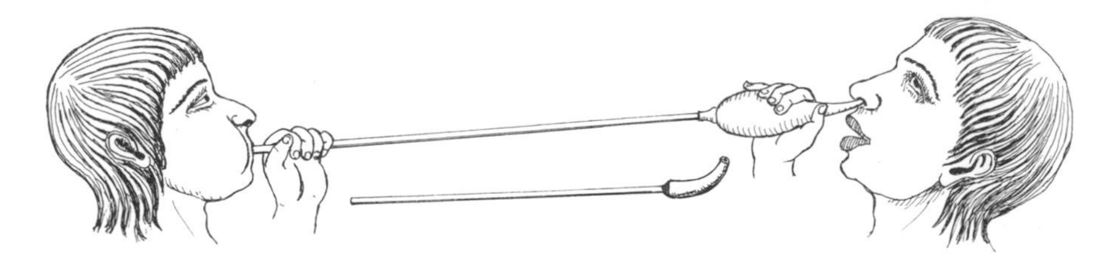

Fig. 3 Snuff tray of wood. Detail from the illustration shown in Fig. 1.

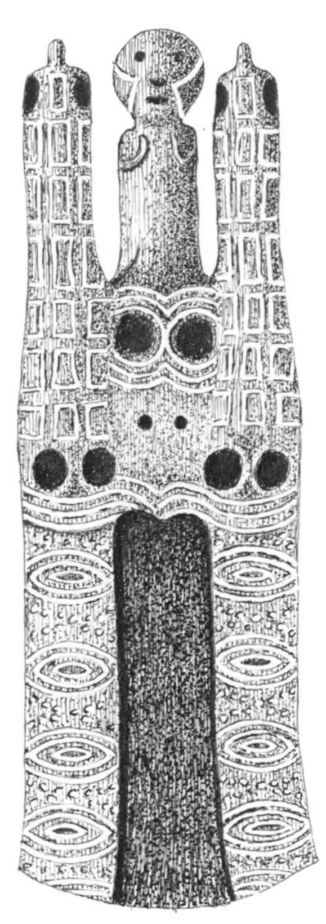

detail (in Fig. 1) we see another tube, a brush, a snail shell container, and also a most interesting snuff tray with three handles which is shown enlarged in Fig. 3. I do not know at present of any *ethnographically* used snuff trays with *three* handles from South America. To be sure, there are several recorded trays with three handles from the *archaeological* finds from northern Chile and Northwestern Argentina (the Atacameño culture; several figures are given in ref. (8) and in Fig. 28 in ref. (20). If, as it appears from Ferreira's illustration, snuff trays with *three* handles existed in his time—for example, among the Maué or other tribes in the Amazon region as late as the end of the eighteenth century—it strongly supports my idea, expressed at the San Francisco symposium in 1967 (8), that we find "a chain of evidence for an early Amazon cultural influence in the Atacaman region." This view has clearly been supported also by, among others, my colleague Otto Zerries of Munich. Zerries states (23) in his summary: "In the ethnological as well as in the archaeological field numerous parallels can be drawn between the Southern Andes and the Tropical Forests of South America." Having in his text discussed my own thoughts on an early Amazon influence in the Atacaman region, Dr. Zerries continues:

> As the emphasis on the dispersion and importance of these phenomena lies in the Tropical Forest Region, it is to be assumed that this region exerted its influence in the Southern Andes. Taking into account general ethnological considerations the same is valid for the ritual snuffpowder complex, which was formerly current among the Atacameño and Diaguita. This complex, which has only been brought to light through the archaeological finds of its paraphernalia—snuff tablets and tubes—is, however, often found in the Tropical Forests, especially in the Northwest, among the recent Indian tribes. There are remarkable parallels between the handles of the snuff tablets in the shape of animal heads found among the Atacameño on one hand, and the Tupi tribe of Mauhé on the lower Tapajoz on the other.

Zerries ends his summary by saying: "Nevertheless the question as to origin of the snuffpowder complex cannot yet be conclusively resolved, because so far, archaeological finds in Costa Rica, West Indies, the Lower Amazon Basin, South Brazil and Uruguay are most probably more recent than those in the Southern Andean Region." To this must be added that we need the results of further studies on the botanical distribution of the plant species concerned, as well as the archaeological results of analyses of snuff powder found in sites, details of which I will return to later in this paper.

Dr. Plutarco Naranjo has recently published two archaeological pottery pipes, or inhalers, found in the region of Ambato, in the central, Inter-Andean part of Ecuador (24). These specimens from the Ethnographical Museum of the Universidad Central at Quito (a third similar specimen is said to exist in the Museo del Colegio Bolivar at Ambato) are of utmost importance, as they are crowned with two and three human figures respectively (see Fig. 4). Naranjo informs us, in the legend to Fig. 18 (of ref. 24), in which he has published the two museum specimens from Quito, that only one of the heads (evidently that with a special cup at the top) joins into the central duct. According to him, the

Fig. 4 Pottery tubes or inhalers with human figures. Archaeological finds from the region of Ambato, Ecuador. Length 29 cm. After Plutarco Naranjo (24).

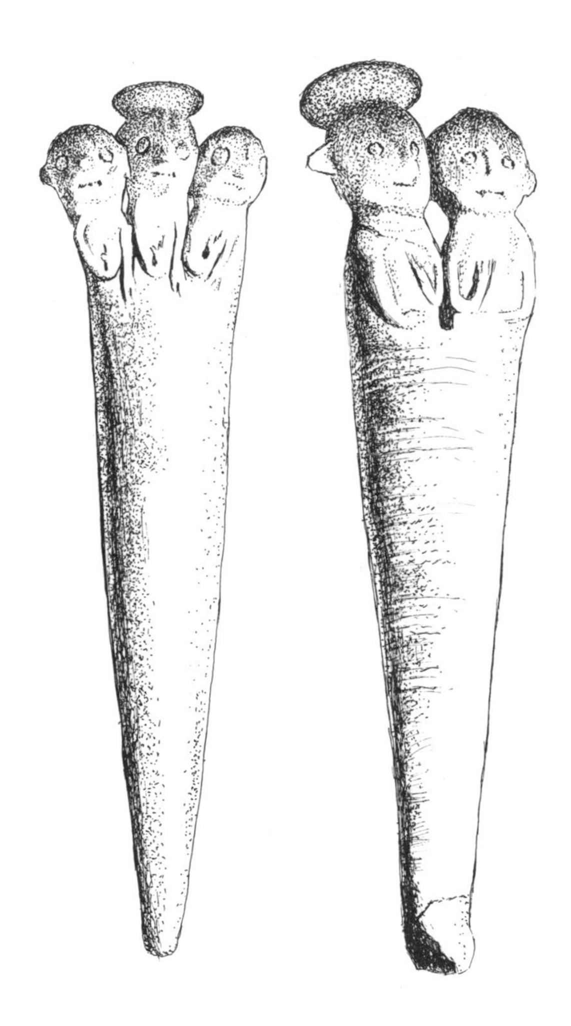

powder was blown from the part with the heads, the pointed part of the instrument having been applied to one of the nostrils (of the recipient). He considers the representation of two or more human (or animal) figures on specimens of this kind an expression of the phenomenon of impersonation which occurs during the hallucinatory phase (24). Most interesting also is the fact that he considers this type of inhaler as another example of cultural exchange between the Amazon region and the Andean highland, as the region of Ambato, since very remote times, has been in contact with Amazonian tribes through the Valley of Pastaza, one of the tributaries of the Amazon.

If we examine the illustration of the snuff tray found in Ferreira's manuscript in Rio de Janeiro (Fig. 3), we can clearly see that the middle figure of the three forming the handle is a human sculpture more or less of the same type as the two figures of the handle in the archaeological specimen shown in Fig. 31, of Wassén (20), namely, a wooden tray coming from Calama, Antofagasta, Chile, and now in the collections of the Field Museum of Natural History. The other two handles of the Ferreira specimen seem to represent animal sculptures, each of the same type. To judge from the scale-like body decoration, they are caymans or possibly snakes, especially as we observe the protruding tongue element which is typical of old Maué specimens referred to in Figs. 19 and 20 in my earlier paper (20). It is also interesting that in the Ferreira specimen we can see a conventionalized head or face below the middle figure forming the group of handles, and we may compare this with the human head seen in the same position on the Rio de Janeiro specimen 2913 from 1873 (Fig. 5), published earlier as Fig. 19 in ref. (20). In several other details of ornamentation, Ferreira's specimen comes close to this 1873 snuff tray from Rio (Fig. 5).

Having read Ferreira's *Memória* from Barcellos written in 1786, it is only natural that I should be curious to see an ethnographical snuff tray with three handles from the Amazon region. I have therefore tried to find out if Ferreira's "box No. 7" ever got to Lisbon. It appears it did, according to information from Mr. Thorsten Andersson, Director of the Lisnave shipbuilding yard, Lisbon. This Swedish friend, working in Portugal, became interested in the problem and had one of his assistants, Sr. António Luis Gomes, undertake a systematic search in Lisbon on the basis of all the available information I could give. The box itself has not been found, however, nor have the specimens mentioned in Ferreira's *Memória.* In a report of March 8, 1968, Sr. Luis Gomes told me that he

Fig. 5 Wooden *paricá* tray from the Maué. Handle is in the form of a snake. White ornaments, cutouts and inlays of mother-of-pearl. Length 38 cm. Catalog No. 2913, Museu Nacional, Rio de Janeiro. Collection "Commissão de Madeira," 1873. Photograph by Carl Schuster.

had been visiting the following institutions: Sociedade de Geografia de Lisboa, Museu Etnológico Português (Dr. Leite de Vasconcelos), Palácio Nacional da Ajuda, Junta Investigacão do Ultramar, and the Library of the Faculdade de Letras da Universidade de Lisboa. The Library of the Palacio Nacional de Ajuda was also approached. Not one of these institutions could give any direct information about the box or its contents, and Dr. Ernesto Vega de Oliveira, Deputy Director of Centro de Estudos de Antropologia Cultural of the Junta de Investigacão de Ultramar spoke of it as "an extraordinary documentation unhappily lost." A further negative reply came from the Fundacão da Casa de Braganca, in answer to a suggestion that "box No. 7" was presented to the Royal Family, and for this reason kept in the Palacio Ducal de Vicosa.

Following this research, in July 1968 I tried to find out the fate of Ferreira's collection during my stay in Lisbon, in Figueira da Foz (Museu Municipal, where a small collection from Brazil is found), in Coimbra (University Library) and the Casa da Insua, the mansion of the Albuquerque family in the village of Penalva do Castelo or Castendo, where written documentation from Ferreira is found in the Library's "Correspondencia official para Luiz d'Albuquerque 1786." According to information received by Sr. Antonió Luis Gomes from Dr. Alberto Iria, Director of Arquivo Historico Ultramarino at Lisbon, Governor Luis de Albuquerque de Mello Pereira e Caceres, Senhor da Insua, was born on October 21, 1739, and died in Lisbon on the 7th of July 1797. On July 3, 1771, King Don Jose I signed the *carta-patente,* nominating Luis de Albuquerque *Governador* and *Capitão-General* of Matto Grosso and Cuyaba. His official duty ended on July 1, 1790. In the Governor's papers dating from 1786, I found a document from Barcellos of April 15 of that year, written by Sr. Joao Pereira Caldas "para o Dr. Naturalista Alexandre Rodr. Ferreira" in which the sender has stated that a collection had been forwarded "a Soberana Prescencia de Sua Magestade." This is very important, as it confirms that Ferreira's shipments were directed to the Sovereign in Portugal, at that time Donna Maria I (b. 17 December 1734, d. in Rio de Janeiro 20 March 1816) and her *rei-consorte* Don Pedro III. This is also confirmed by a letter from Dr. Ferreira himself to *Sua Magestade.* This letter, dated at Barcellos on the 30th of March, 1786, was found by me in the documents kept in the Museo Bocage at Lisbon. Since it gives a summary of Ferreira's various shipments from Brazil to Lisbon, I am quoting the full text here, at the same time referring to Valle Cabral's list of Ferreira's various reports (22). Dr. Ferreira wrote as follows:

I have here and up to this time let Your Excellency read the history of my trip from this village of Barcellos to the first rapids of the Uaupes River. To this should be added the report of products observed and collected and now being sent to the Royal Cabinet, including the eighteen packages which comprise the shipment from this river. Grouped together, seven *Memorias,* with different titles, will be sent: (1) On the turtles which have been prepared and forwarded in box no. 1 to [illegible]; (2) on the *peixe-boi* (American sea-cow)—material which was prepared and sent from the village of Santarem in six boxes in the last shipment from the Capitania do Pará, and now being dispatched in box no. 9; (3) on the gourds which are made by the Indian women of Monte Alegre and Santarem, to be added to the samples which I forwarded in the first shipment; (4) on the pottery made by the Indian women at Barcellos, to be added to the samples sent in boxes 1, 5, and 8; (5) on the braided and painted trays made by the Indian women of the village of Santarem; (6) on the lighter, or box for tinder; and (7) and last, on the equipment used by the natives when they take *paricá.**

It is to be regretted that as yet nothing of the ethnographical material sent by Ferreira to Portugal has been found. We can agree fully with the words of Valle Cabral (22, p. 105): "Que o d.or Alexandre Rodrigues Ferreira foi sabio consciencioso e infatigavel não ha contestacâo alguma." He did not limit his consignment of paraphernalia for snuffing only to the Maué. Document No. 50 of the papers regarding Dr. Ferreira in the Bocage Museum at Lisbon gives clear evidence that Dr. Ferreira asked Sr. Joao Pereira Caldas to forward material also from the Mura Indians. This material ("*varias curiosidades do Gentio*") was collected by another person so that Ferreira himself could save his time. This document, dated the 14th of December 1786, gives a list of what specimens were sent to the "Real Gabinete de Historia Natural" from the "Gentio Múra." A later document, dated January 3, 1787, mentions still more specimens to be sent, among them "tres caixas de tabaco-Parica" (three boxes of

*Tenho ate aqui dado aler a V.Exa a Historia da minha viagem, desde esta Villa de Barcellos, ate aprimeira Cachoeiro do Rio Uaupes. Seguese ajuntar a ella a rellacão dos Productos observados, erecolhidos, que sao os que agora rernetto p.a o Real Gabinete, incluidos nos dezoito volumes, que constituem aremessa deste Rio. Vao juntas sette Memorias de differentes titulos, a saber *a primeira* = Sobre as tartarugas, que sao preparadas, e remettidas nos Caixoens N.1 ate . . . (difficult to read). *Segunda*-Sobre os peixes Bôy, que forao preparados, eremmettidos da Villa de Santarem nos seis caixoens da ultima remessa da capitania do Para, edoque agora seremette nos Caixao N.9.—*Terceira*-Sobre as cuias, que fazem as Indias de Monte Alegre, e de Santarem, para ser appensa as amostras que remetti no Caixao N.1 da primeira remessa. *Quarta*-Sobre a Louca, que fazem as Indias de Barcellos, para ser appensa as amostras delle que forao remettidas nos caixoens Nos.1, 5 e 8. *Quinta*-Sobre as salvas de palhinha pintada pelas Indias da Villa de Santarem. *Sexta*-Sobre o isqueiro ou caixa de guardar aisca para o fogo. *Septima* e ultima-*Sobre os Instrumentos de que usa o Gentio para tomar o tabaco parica.*

parica) and "dous bocaes por onde tomao [for *tomam*] o d.º Parica" (two tubes for the taking of this *parica*).

One guess would be that the snuff tubes, etc., were taken to France in 1808, but the document regarding this consignment as outlined by Gloria Fontes (21) mentions only mammals, birds, and fishes, and we can therefore disregard this possibility. Another possibility has been opened, however. According to my informant in Lisbon, an important ethnographical collection was sent from Lisbon to Madrid in 1892 on the occasion of the Fourth Centenary of the discovery of America. This material was to be exhibited in Madrid, but since then there is no known clue to its whereabouts. Some interesting items, however, are given in the printed catalogue of the specimens sent from the Royal Academy of Sciences at Lisbon to Madrid for the special exhibition. The catalogue was written by Sr. A. C. Teixeira de Aragao (25). There is no specific mention of Ferreira or of other collectors in this catalogue, but one gets the general impression that collections forwarded by Ferreira were included. There is special mention of the Rio Negro and other geographical regions visited by Ferreira as well as a statement that the collections were brought together during the eighteenth century. Since this is important, I prefer to quote from page 4 of the original (25):

> Os objectos de arte e industria dos indigenas americanos, que a Academia Real das Sciencias de Lisboa envia á exposicão de Madrid . . . pertencem . . . ao seu museu, e foram pela maior parte adquiridos no seculo XVIII nas margens do Amazonas, ilha de Marajo, grutas de Maraca, Rio Negro, . . .*

Secondly, several *snuff tubes* are mentioned in the catalogue as *tubos de cana.* This goes for numbers 379, 380, and 380*a.* Number 380*a* has a special and interesting description and accordingly fits very well with the tube shown in Ferreira's illustration (Fig. 2). Specimen 380*a* is said to be "another tube also protected with cotton strips. The container is bigger, pyramid-shaped and is glued with pitch. Length 0.73 m." (Outro tubo tamben enleado com fios de algodao. O cabaco e maior, com a forma pyramidal e esta preso com pez. Comprimento 0^{m}. 73.)

That such a tube has been understood to be a snuffing tube is clear from the direct continuation of the text in the catalogue. It says in translation:

*"The American native arts and craft objects which the Royal Academy of Sciences at Lisbon sent to the Exhibition in Madrid belong to its museum. They were for the most part acquired in the eighteenth century from the Amazon River, the Island of Marajó, the caves of Maraca, from the Rio Negro, . . ."

> These pipes are used by the Mura Indians for *paricá*—tobacco. *Paricá* is the fruit of a tree which the natives powder finely after roasting. It is much used by the tribes of the Amazon. After being powdered very finely this tobacco is placed in the gourd, which is then applied to the nostril. Another person blows hard in the other end. The effect thus produced by the tobacco is very violent, at times causing loss of consciousness and always provoking heavy discharge of mucus. Apart from pipes of cane, the thin stems which support the fruit of palm trees (*Maraja* and others) are also used. Sometimes two tubes are joined together—generally two leg bones of a hawk—then one end is placed in the nostril, permitting more moderate inhalation. *Pango,* an African tobacco, is also substituted for *Paricá.*

In analyzing this description, the question of the container on the tube (Fig. 2) first puzzled me. In my earlier paper (ref. 20, p. 107, fn. 2) I commented on the drawing in Ferreira's manuscript showing two Indians snuffing in a long tube with a "container," and speculated that the bulbous detail on the end of the tube was a misrepresentation of the nut used to fit one of the nostrils of the receiver of the snuff. Ferreira speaks of double tubes of bird bone shown (e.g. ref. 14, Figs. 9, 17, 22, 23) with two small nutshells from the Yu-hue palm (coquilhos da Palmeira -Yu-hue) attached to their upper ends. He did not mention a "container" on the snuffing tube, but the author of the catalogue of the collection which was sent to Spain in 1892 (25) must have seen a specimen with "a pyramid-shaped" *cabaço* (container), "glued with pitch." In the description quoted in Portuguese it is said that the fine powder is put in the container, which is placed in one of the nostrils; another person is then supposed to blow the powder into the nose of the receiver. We will be able to learn more exactly about this detail, it seems to me, only after finding a snuff tube of the kind shown in Ferreira's drawing. The container idea gets clear support in Ferreira's drawing, in which the receiving Indian is seen holding his hand around the container while he keeps the end of the tube in his nose (Fig. 2). As already mentioned, Rondon (see Bibliography) also speaks about a container for tobacco snuff on the *Gynerium* tubes used by the Kepkiriwat. The snuffing is said to have a violent effect which always causes sneezing.

At the end of the description in the Portuguese catalogue we suddenly came upon the information that *pango,* or "African tobacco," could be used as a substitute for *paricá.* This must refer to a *Cannabis* preparation. In the *Grande Dicionario de Lingua Portuguesa* by Antonio de Morais Silva (vol. VII, p. 733, Lisbon, 1954) the word *pango* stands for "a myrtaceous plant, also called *liamba* (*Cannabis sativa indica*), which is used by certain natives in Africa for smoking." Except for the erroneous botanical family (*Cannabis sativa L.* belongs to the Moraceae), the information is substantially correct. In Africa the names *liamba, riamba,* and also *diamba,* for *Cannabis sativa* are found among Negro tribes in Angola, and the varied first consonant is only a somewhat different way to pronounce the soft initial letter. Other words used for the *Cannabis* preparation are *aliamba, birra, dirigio, haxixe, maconha,* and *soruma.* In the new 1964 edition of Lewin's *Phantastica* (27), a cult of "Riamba" is mentioned from the Kassai. The Portuguese catalogue text seems to indicate the snuffing of hashish (Lewin [27] cites the Wanyamwesi from East Africa as smoking hemp and snuffing hashish). As far as I know, any use of *Cannabis sativa indica* must have been a post-Columbian introduction to South America; and if the word *pango* has been introduced to the Portuguese language, it could possibly have to do with the word *bhang,* used in India as well as in Tanganyika for a *Cannabis* preparation (see ref. 27, p. 865, and ref. 28, p. 761). According to Pereira (29), the new drug *dirijo,* or *diamba,* was introduced to the Maué by the *civilizados.* Dr. Stig Wikander, professor of comparative linguistics at Uppsala University, has informed me (personal communication, 1968) that *panka* was a word used in Siberia for the poison from the fly agaric. This word appears in Persian as *bang,* for hemp, and so had an interesting history before its discovery in South America via India as *pango.*

To turn for the moment from the case of Ferreira: evidently, Portugal has always had considerable trouble with the old overseas ethnographical collections. Teixeira (25) has quoted a critical statement in French written in 1880 by M. Emile Cartailhac about "Le Musee Colonial," at that time "installé à l'Arsenal" and to a large extent belonging to the Royal Academy of Sciences. According to Teixeira (ref. 25, p. 5, fn. 4), the Academy's collections were lent to the Ministerio do Ultramar in 1867 for a Colonial Section of a museum in Paris the same year. When the collection was returned to Lisbon, the Academy "emphatically demanded its restitution through several official notes, but without results" (reclamou com instancia por varios officios a su restituicao, mas sem resultado!).

In 1893 a catalogue in three volumes was published in Madrid based on the contributions from various countries to the Exhibition in 1892. In this *Catálogo General* (30) Volume III (p. 4), we find that Portugal's exhibition in Madrid had a section for American ethnography, and on page 7 a reference is made to Teixeira's catalogue already mentioned. There is not much help, however, in the problems discussed above.

WHAT IS NEW?

It is an established fact, underlined by such prominent specialists as Dr. Erwin Ackerknecht (31) and Professor Claude Lévi-Strauss, that (I am quoting the latter) "few primitive people have acquired as complete a knowledge of the physical and chemical properties of their botanical environment as the South American Indian" (32). The same author considers it "probable that only a fraction of the herbs used by modern Indians are presently known and exploited" (ref. 29, p. 484). The long and useful list of medicinal plants for various diseases published by Lévi-Strauss in his paper, "The Use of Wild Plants in Tropical South America" (32), was considered by the author as "only partial and fragmentary," but still it contains a most varied illustration of the Indians' search for medicinal plants. The summary by Lévi-Strauss is, as far as I know, the best of its kind made by any of the anthropologists working on South American cultures. We find in it the botanical names of plants used as emetics and purgatives, as remedies against gastric disturbances, for wound healing and as astringent remedies. There are also medicines to stop bleeding, to cure eye pains, several febrifuges and sedatives, as well as specific drugs against diarrhea and dysentery. If we add specific antidotes for snake bites, plants used as aphrodisiacs and contraceptives, and plants to cure hernia, pulmonary afflictions, blisters, scurvy, hemorrhoids, and catarrh, we still are left with several species of *Datura* and *Thevetia,* which are used for their anesthetic properties, and several balsams. We may wonder if all of these species and those botanically related have yet been properly tested by drug researchers. If anthropologists are properly informed about progress in the field of "drug-hunting," they might possibly report still more of the Indian secrets in this broad field. Up to the moment, as far as I am aware, reports from anthropologists of valuable plant medicines have been relatively insignificant.

Ackerknecht has stated (31) that the "knowledge of the South American Indian's pharmacopoeia brought some of the most momentous changes in our own." After mentioning such drugs as ipecacuanha, which is used against diarrhea and dysentery, and coca, tobacco, curare, and cascara sagrada, the Peruvian and Tolu balsams, and the now obsolete guaiac, jalap, sarsaparilla, and sassafras, he exemplifies the vivid interest of the South American Indians in plant medicines with the information that "the Carajá even cultivate healing herbs in gardens" (31). This probably takes place in many tribes. I know for certain that among the Cuna Indians of Panama, where the *inatulet* category of medicine men (from *ina,* "medicine" and *tule,* "man, person") have gardens in which they cultivate all kinds of rare medicinal plants (33). When the Cuna Indian, Rubén Pérez Kantule, was studying with us in Göteborg in 1931, several medicine men asked him to bring back from Sweden "seeds which they planned to sow in their cultivated plots" (ref. 32, p. 507). During our studies of the Cuna Indian culture we found that these Indians "try to protect their plants from extermination" (33) and also that a medicine man must undergo a thorough examination in order to be recognized. "Bits of bark of all different kinds of trees, seeds, etc., are mixed together by the teacher and it is up to the pupil to identify each one and to tell how it is used. If the pupil passes the examination the teacher makes public the fact that he is now a competent medicine man" (33).

The famous Nele Kantule of Ustupu, San Blas, Panama (a *nele* [orig. *lele*] man or woman who has power to see spirits; *kantule,* probably from *kammu-tule,* "fluter," name of an official at the feasts), was "exceedingly well versed in botanical knowledge. If the seeds are brought to him of any plant whatever, he will readily describe the plants to which the seeds belong. It is he who discovered that a very small dose of the sap of the manzanilla tree is an exceedingly valuable aperient" (33). "At the age of ten he wanted to go with his father in order to find medicines in the forest, on the islands, and by the rivers" (34). I have taken these examples from the Cuna Indians, a tribe which I have studied extensively, but I am quite convinced that a similar intense interest in plant medicines could be found among most of the Amerindians. From the old region of the Cuna Indians, that is, Darien, we have records that healing with herbs was used as early as 1640 in the description by Padre Adrian de Santo Thomas from the so-called Páparos, a tribal group which I have considered as Cuna or very closely related to these Indians (35). Father Adrian, who arrived in Darien in 1637, says that the *lere* (medicine man) cured a sick person "with some herbs which he seeks" (con algunas yerbas que busca) (36).

I cannot resist pointing out once again the importance of the Indians' enormous ability to observe and learn in the realm of Nature. An extraordinarily interesting lecture on this subject was delivered during the 38th Congress of Americanists in Germany in August 1968 by the Argentine ethnobotanist, Dr. Raúl Martínez-Crovetto. In his paper (37) the author showed graphically that the floristic knowledge of the Indians in northeastern Argentina studied by him is worthy of admiration (or, in the author's own words, "los conocimientos teóricos y utilitarios son de una magnitud digna de admiración"). Four of these tribes used the following numbers of plants (the total number of plants known to the same tribes is much higher): Guarani-bmiá 438 (64 for medicinal use); Toba-takshik 244 (medicinal use, 61); Mocoví 227 (medicinal use, 55); and, finally, the Vilela 130 (33 for medicinal use). In the same paper, Martínez-Crovetto published an interesting analysis of the principles for naming plants among the same group of Indians. This "architecture of phytognomy" has enabled the author to establish certain laws for the construction of plant names among the four tribal groups mentioned above (37).

Here it should be added that the German linguist Kramer, after an extensive study of the Cuna Indian literature, found that these Indians classify all plants using criteria such as foliage, branches, root system, locality, size, and so on, in what Kramer defines as practically a Linnean system (38). This observation is another example of Amerindian knowledge of the realm of plants.

To the south of the Cuna are the Chocó Indians in Panama and Colombia. The Chocó grow small gardens in discarded canoes which they place in the forks of two poles. During my field work in Chocó in 1934 I observed that the Indians were growing bulbous plants and flowers in these "canoe gardens" (see ref. 39, Fig. 6, p. 55). Nordenskiöld published in 1928 (40) a photograph of a "canoe garden" at Cocalito, noting that the Indians cultivate a little of everything around their houses, including medicinal plants. Of these plants, Nordenskiöld "once received a liana said to have the wonderful quality of producing the gift of seeing in the distance if one eats a little of it" (40). In 1935 I wrote the following: "On R. Docordó I even observed the cultivation of flowers within a rectangular wooden frame placed in a corner of the main floor of the house. Evidently this is a very ancient practice among these Indians, seeing that 'handmade gardens' are mentioned as early as in the account of Garciá Montaño's journey down the San Juan (in March 1593), published by Pedro Simon" (39). Montaño observed "*jardines hechos á mano,*" with plants beautiful to look at. The Indians (Waunana) explained their use in the following way: One herb was taken in the mouth during their drinking-bouts so that they should keep sober longer. Another was used for curing the wounds from arrows and splints. Still another kind of herb was used for washing prisoners so that they should lose their ferocity and also the memory of their native country; thus, a mixture of medicinal and magic use.

It is the same Chocó Indians, long studied by Swedish anthropologists, who use an arrow-poison, *pakurú-neará.* According to earlier determinations, this poison may possibly have come from *Ogcodeia ternstroemiiflora,* which, according to the late Professor Santesson, contains a non-nitrogenous glucoside that is a powerful cardiac poison (see ref. 41, pp. 79-80, and references quoted here). Now (in 1969) Dr. C. C. Berg of the Botanical Museum and Herbarium of the State University of Utrecht, Netherlands, after having examined leaves and bark of the plant collected by me in 1934, declares that "the leaves belong to a species of the genus *Naucleopsis*" (letter of December 11, 1969). Dr. Berg, a specialist in this group of plants, continues: "I never met with a specimen of this genus collected in Chocó. The only collection of *Naucleopsis* known from the Pacific coastal region of South America is the type collection of *N. chiguila* from northwestern Ecuador. This material does not belong to that species. The characters of the twigs and leaves match those of *N. amara* Ducke. Although this species is known at present only from the Amazon Basin, it is very probable that the present material belongs to it. But it is impossible to determine with certainty without an inflorescence or fruit."

If the *pakurú* arrow poison (samples of which have been sent by me to Dr. N. G. Bisset, Department of Pharmacy, Chelsea College of Science and Technology, University of London, for further chemical examination) should come from the species *Naucleopsis amara,* we can perhaps assume that the tree was brought to the Chocó by the actual so-called Chocó Indians, who, to judge from ethnographical evidence, originally came to western Colombia and Ecuador from the Amazon region in the East. Schultes has suggested in a similar way that specimens of *Banisteriopsis caapi,* recently found in Chocó, were brought there by the Indians from the East.

The same Indians have also learned to use another arrow-poison, namely, the strongest cardiotoxin known, the three-dimensional batrachotoxinin A, which they get from the *kokoi* frogs (*Dendrobates* or *Phyllobates* species). For the historical background to this observation (I was collecting such frogs as early as 1934), I refer to my earlier publications (39, 41, 42), as well as to the recent reports by Brown (43), Tokuyama (44), and other experts on the *kokoi*-poison.

RECENT WORK ON COHOBA AND EPENÁ

During the symposium "Ethnopharmacologic Search for Psychoactive Drugs" in January 1967, I concentrated on an anthropological survey of the use of South American snuffs, using both ethnographical and archaeological data for explaining the use and distribution of such drugs as *cohoba* (*Anadenanthera* sp.) and *epéna* (*Virola* sp.), etc. Since these drugs of evident ethnopharmacologic interest were treated in some detail at that time, one may ask if there is now anything new to report. I think there is, and will deal with new developments in the last part of this paper.

Mr. Georg Seitz (45) has a photograph of a Waica Indian *häkula*-dancer. In my own later paper (20) I used another of Seitz's photographs (Fig. 40) showing a Waica *häkula* performer under the influence of the *Virola* drug *epená*. The drug-influenced Waica is said to invoke the spirits of the mountains and the waterfalls, and this is supported by the missionary Franz Knobloch, who in 1967 published a monographic study of the Aharaibu—one of the Xirianá tribes—also written as Waharibo (46). Knobloch, who uses the term *epéna* for the hallucinogenic powder and identifies *hekurá* with supernatural beings, says that the *hekurá* live on the mountaintops and in the waterfalls (46). Becher (47) connected the gigantic animal and plant spirits, *hekurá,* who have their abodes on special mountain ridges (*die auf bestimmten Bergzügen wohnen*), with the name-giving of three-year old children: the father receives the names from these spirits after a snuff-taking ceremony and suggests that the life-destiny of a child is governed by the respective plant or animal. In her wonderfully illustrated popular work on the Waica in 1968, Dr. Inga Goetz only mentions the use of *ebena* in passing. We learn that the fruit, which is used for one end of the snuff tube, comes from the *cucurito* palm; and the author assures us that the women never take *epéna,* and that not all of the men use the powder. Some of the *hecurá* spirits are said to enter into the breast of a snuffing man while other of these spirits remain close to him (48).

The *häkula*-conception, however, seems to hold something much more complex, and we have only been able to grasp the whole explanation after the publication of the paper by Holmstedt and Lindgren (49). These authors stated that "under the influence of the drug, the Indians identify themselves with the gigantic spirits of animals and plants (*hekurá, hekula*), and also have the impression that they themselves personify the *hekurá* (Surára tribe)."

In the works by Wilbert (50) and Barandiarán (51), we find the word *häkula* as *híkola* or *hekorá* among the Sanemá, one of the Yanoáma groups. (Chagnon [52] uses the tribal denomination Yanomamö and calls the snuff *ebene.*) Biocca (53) has dealt extensively with the culture and grouping of the Yanoáma (the tribal denomination, Yanoáma), and describes the snuff as *epená* (or *ebená*), and the spirits mentioned here as *hekurá.* In Volume II of ref. (53), he has a detailed description of the preparation and use of "epená, polvere allucinogena," and has collaborated with Marini-Bettolo and Delle Monache (54) in examining the active principle of the snuff. (I should mention that Biocca [53] lists a myristicaceous plant [*Virola* sp.] among the contraceptives used among the Tucano.)

In Biocca's more popular work (55), Helena Valero, a woman of Spanish blood who for many years was a captive among the Yanoáma Indians, in the chapter "le droghe allucinanti" (ref. 55, pp. 146-151) describes many details of great interest in the study of the use of hallucinogenic snuff among the Waica. From her description we obtain a most vivid picture of the internal duelling among these Indians with fists, cudgels, or, nowadays, with stolen axes. The fighting, which goes on because of—as the Indians explain it—their desire to still their fits of anger, is preceded by the taking of *epená.* According to Helena Valero, the *tucciano* ("chief") calls out: "Let us now take *epená,* and when properly intoxicated we shall really talk together." Chagnon (56) has given a full explanation of the social background of "The Feast" among the Yanoamá: the invitations, preparations, arrival of the invited visitors, and so on. It is while waiting for the guests that the "adult men in the village take hallucinatory drugs" (56).

The effect of the drug is dramatically referred to in Helena Valero's story. One of the young boys who had been taking the drug started exclaiming that his mother was dead, when as a matter of fact she stood next to him. Another was asking: "Father, why do you run away?," when his father had absolutely no intention of doing so. The most interesting detail in Valero's description of the symptoms is, however, the exclamation by another boy that the whole site was whirling round and that his father was standing upside down. The latter description immediately recalls the words of Ramon Pane regarding the use of *cogioba* among the Taino of Española during the days of Columbus: "Consider what a state their brains are in, because they say the cabins seem to them to be turned upside down and that men are walking with their feet in the air" (8).

The phenomenon of macropsia (see below) is also noted in Helena Valero's narrative. One of the oldest cried out: "You have giant faces and enormous teeth." During the duelling with cudgels there is a form of challenge: "I have called upon you to see what a man you really are. If you are a man, we shall see if we can be friends again when our fury passes away." I cannot resist recalling here nearly the same form of challenge during the *balseria* ceremonies of the Guaymí Indians in Chiriqui, Panama. It is a game of stick dancing mentioned as early as the seventeenth century by the Dominican father, Adrian de Santo Thomás, and excellently described by Alphonse (57).

The game consists of dancing while balsa wood sticks 5 feet in length are thrown at the dancer, who performs with his back turned to his opponent. When a man chooses to hold a balseria, he assumes the name of *koböbu*. The *koböbu* comes out with empty hands and turns his back to the man challenged and begins his rhythmic dance, singing the while: "Brani! Brani!" (meaning "Man you say you are"). The challenged (*etebali*), coming out with a stick aimed at the other's legs and dancing rhythmically back and forth, replies in a challenging tone, "Man I am," and he lets drive with both hands as hard as he is able. When the sticks for one team are through, the opposing side falls in and repeats the same procedure; thus both teams belabor one another for a whole day. At the end of the contest hundreds of bruised legs are seen.

This information from Alphonso has been chosen to show the basic similarities between the duels among the *epená*-snuffing Yanoáma in South America and the Guaymí. In both places the challenger offers food to the challenged party, and the fighters depart as friends, both among the Waica and the Guaymí.

Together with the chief's daughter, Helena Valero once succeeded in snuffing *epená,* otherwise absolutely forbidden for the women. The powder, which made her and her companion giddy, was a preparation from bark.

With reference to the botanical discussion on whether the Waica use the beans of *Anadenanthera peregrina* or not, and the evident use of a myristicaceous snuff among them (58), it is of great interest to learn from Helena Valero that "the Namoeteri did not have plants with fruits but obtained the beans from other Indians." Chief Fusiwe planted such beans in a clearing, and Helena Valero recalls how one day after perhaps four years, he came to her carrying a small basket with such beans in it. He said to her: "Look! This epená that I planted has grown well and it already bears fruit. It is not long ago since I planted them. The plants are still small, but full of seed." And she continues, "I went with him to see the plants which were filled with fruits." Schultes and Holmstedt (58) have discussed the problem of the occurrence of *Anadenanthera peregrina* among certain Waica groups, pointing out that seeds are obviously imported, and ending the discussion as follows: "The question remains, nonetheless: why are the Rio Marauiá Waiká cultivating *A. peregrina?*" If Helena Valero refers to the seeds or beans of *Anadenanthera peregrina,* which seems probable, her answer indicates that the *epená* prepared from the seeds is considered to be stronger than that prepared from bark (i.e. *Virola*).

Schultes and Holmstedt (58) have recently been interested also in the problem of the ash of the bark of *Elizabetha princeps,* which is added to *Virola justicia* snuff. "We have every reason to believe that this ash may be an inert ingredient, although it may serve as a means of drying, to free the alkaloids more easily from the resin, to keep the snuff from deteriorating rapidly when stored in the bamboo tubes, or merely for mechanical purposes." Helena Valero tells us (ref. 55, p. 152) about the use of seeds which always are somewhat humid and that ashes are therefore added "to dry the mass, but they never take too much of the ashes, for in that case, as they say, they get a sore nose."

Among the reasons for taking *epená* Helena Valero mentions that the Indians use it to avoid spirits pursuing them when they are traveling or hunting, and, in the nighttime, to chase away forest-spirits, "as the priests do when blessing" (55).

According to Barandiarán (51) *híkola* or *hékola* should be derived from *hea* or *he,* "head or (something) at the top" (cf. ref. 47, p. 22). It should be translated as "vital power" or "vital energy," a power found in men but not in women and children. According to Barandiarán (51), a gigantic demon, "the long-tailed one," was overcome by a shaman and from the body of the rotten giant, *híkola* penetrated to plants, animals and man as an indestructible power. We can, as anthropologists, thanks to Holmstedt and Lindgren (49), now understand the Indians' explanation that only men have this vital power, and also their idea of identifying themselves with gigantic spirits of animals and plants. Since only men take the *epená* drug and as "the disturbance of the visual appreciation of the size of objects" has been shown to exist in the forms of *macropsia* and *micropsia* among those taking the drug (49), the belief in giants among the *epená*-using Yanoáma groups has a logical and natural explanation. The explanation by Chagnon (52) that the "drug allegedly produces colored visions, especially around the periphery of the visual field, and permits the user to enter into contact with his particular *hekurá, miniature* [italics supplied] demons that dwell under rock and on mountains" should be observed, as we evidently find here an example of a *micropsia*-caused phenomenon. Apparently it can occur also in intoxications with *Banisteriopsis* (49).

The descriptions of both macroptic and microptic illusions as a result of drug-taking among the Yanoáma show once again the importance of a close collaboration between anthropologists and doctors of medicine. In cases like the above, they should both be capable of understanding and explaining macroptic illusions as described by the German anthropologist Becher (47) and analyzed by Holmstedt and Lindgren (49). The immediate importance of a broad and detailed collaboration between scholars from different branches of science who are, if possible, already in the field, is also clearly shown by the report of Schultes and Holmstedt (58) on their participation in Phase C of the "Alpha Helix" Amazon Expedition of 1967. Such close collaborations should be the ideal for all to aim at.

I would also like to refer to the joint paper of Agurell et al. on the chemical constituents of the snuff plants described here (59). Polykrates, who has recently published a monographic work on two Yanoáma tribes (60), has a special chapter on the *epená*. Unfortunately, although he has experienced the drug himself, he has not seen the analytical work of Schultes and Holmstedt (58).

With regard to analytical contributions which attempt to explain the use or the effects of snuff drugs among the South American Indians, I think it proper to refer first to the new and rather difficult work by Claude Lévi-Strauss (61), *L'Origine des manières de table*. In this book he tries to explain the ritual use of narcotic snuff. He refers to the Tukuna legend, "The Errors of Cimidyuë" (62), which deals with an adventurous wife, and places it in the same group as the Tukuna myth, "Monmaneki the Hunter and His Wives," (62) in spite of the many superficial differences between the two texts. On her way back to her father's house from the jungle (where she had been left by her treacherous husband) "the big blue butterfly (*Morpho menelaus*)," reciting a magic formula "transformed Cimidyuë into a red dragonfly and flew with her to the other side" (of the Solimões). It was the side of her father's house, and finally the father and some helpers managed to capture the woman. The father brought his daughter home "and gave her an emetic, so that she vomited violently and was thus restored to reason" (62).

Seler, in his analytical work on animal figures in the Mexican and Mayan codices (63), has analyzed the partly sinister role played by the butterfly, or *papalotl,* among both the Nahua and the Maya. "Flowers, butterflies, and water were sacred symbols of the Nahuas" (64) and *Itzpapalotl,* the Obsidian Butterfly, was "a demonic goddess of unpredictable fate represented as beautiful but with death symbols on her face (65). The sinister role of the butterfly seems also to be reflected in South America.

Gerardo Reichel-Dolmatoff (66) has in passing remarked that the *morpho* is considered an evil omen to the Tucanoan Desana, and in analyzing the part played by this butterfly, Lévi-Strauss points out that the butterflies, and especially the *morpho,* have an evil connotation in the whole region of Guiana and Amazonas, among the Tukuna as "the mistress of the kapok tree" (58): "The spirits of certain trees such as the kapok . . . have the tendency to wound menstruating women with 'arrows,' even by day; this is why women do not like to stray far from the house yard while menstruating." In spite of this, the *morpho* becomes in the Tukuna myth a "guide and protector of the heroine" (61). Lévi-Strauss, who in his work also has a picture of the *paricá* tray from the Natterer collection in Vienna (Coll. No. 1377, see [14], Fig. 12), makes the very interesting remark that the insects seen on the handle of this Maué tray should not be looked upon as copulating dragonflies, but instead—exactly as we find it in the Tukuna legend—a butterfly towing a dragonfly. If we return to the role of helper which the *morpho* butterfly plays in the legend about the Tukuna woman who got lost, Lévi-Strauss does not find it inconceivable that this "retournement de la valeur semantique de papillon fut lié à l'usage rituel des narcotiques ou stupéfiants prisés sous forme de poudre, notament du paricá" (ref. 61, p. 98).

Lévi-Strauss has also published a picture of the Tukuna snuff tray found in Oslo University's Ethnographical Museum (see [14], Fig. 41). He admits the correctness of my interpretation of the hook-nosed figure in the middle of the shaft as a prego monkey (*Cebus fatuellus*) but he also sees a butterfly or dragonfly with folded wings. Even if we content ourselves with seeing only a dragonfly in the upper part of the handle of the Tukuna tray, the parallel with the insect motif on the Maué snuff tray is interesting. In continuation of his analysis, Lévi-Strauss (61) recalls that the majority of *paricá* trays from the Amazon region represents an animal which has been identified as a snake or cayman, an animal which plays the part of ferryman instead of the butterfly in the myth version among several other tribes. According to Lévi-Strauss, these myths have one thing in common: they relate to the peregrinations of a hero or a heroine among treacherous and evil animals, often gigantic ones. *To get in friendly contact with these animal spirits should, according to Lévi-Strauss, be the principal aim of the rites based on the use of narcotic powders.* We can fully grasp the importance of such spiritual contact only if we note from Reichel-Dolmatoff's description of the symbolism of the Desana (66) that "the jaguar, as the principal representative of the sun, symbolizes nature's fertilizing energy. He is a protector of the *maloca* and the forest, and for his color he is associated with the fire and for his roaring with the lightning. He is closely associated with the *payé* (shaman, or medicine man) because what the latter is for the society, the jaguar is for the whole nature. The jaguar and the payé are not identical but compensatory."

As an addition to the historical-ethnographical notes on the taking of *paricá* in South America which I presented to the San Francisco symposium in 1967, and as an addition to the information on the Guahibo Indians given by Spruce and published in my earlier paper (ref. 14, p. 104), I can now, with the permission of the Royal Botanic Gardens at Kew, Surrey, describe the apparatus for making and taking Niopo snuff procured from Guahibo Indians at the cataracts of Maypures which was collected and sent to Kew by Richard Spruce in 1855. This apparatus (Fig.

Fig. 6 *Yopo* snuffing paraphernalia procured from the Guahibo in 1855 by Richard Spruce. Crown copyright. Reproduced by permission of the Controller of Her Majesty's Stationery Office and of the Director, Royal Botanical Gardens, Kew.

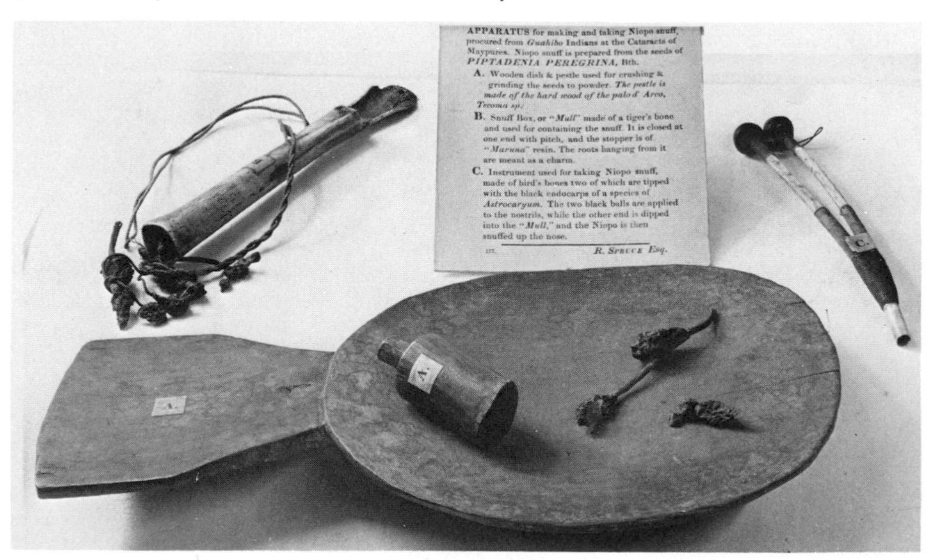

6), is still in the Kew collections (No. 177) and consists of a round wooden dish and a pestle for crushing and grinding the seeds of what Spruce termed *Piptadenia peregrina,* Benth., as well as a snuff box made of a jaguar's bone and, finally, a Y-shaped snuff tube made of bird's bone. Complete information on this set of specimens as well as Spruce's note on No. 166, the *caapi,* and its effects is given in Appendix B. I know that Schultes is specially interested in Spruce's suggestion "that *Haemadictyon* might be used with *Banisteriopsis,*" and he suggested to the Director of the Royal Botanic Gardens that an analysis should be made of some of Spruce's collections as "it would be a most extraordinary feat if, after a century, we could get positive tests for alkaloids." According to information from Professor Holmstedt, such material later was received from Kew and gave a positive result, indeed a remarkable result.

For my earlier paper I prepared a map showing the distribution of psychotomimetic snuffing in South America (ref. 8, Fig. 25). It is evident that the majority of tribes known to be using *paricá* and similar drugs live in Brazil. Eduardo Galvão, in his contribution, "Indigenous Culture Areas of Brazil" (67), drew up a map with eleven culture areas within Brazil. Of these, the following are specified in his text as places where *paricá,* etc. are used (67).

I. NORTH AMAZON

Subarea A, Brazilian Guiana:	"Use of *paricá* (Piptadenia) for ritual purposes"
Subarea C, Rio Negro:	"Use of *paricá* (Piptadenia), ipadu (coca), and kaapi (ayahuasca), besides tobacco"

II. JURUA-PURUS:	"Use of tobacco, *paricá,* and chicha"
III. GUAPORE:	"Use of snuff (*paricá* and tobacco) for shamanistic purposes"
V. UPPER XINGÚ:	"Use of tobacco"
VI. TOCANTINS-XINGÚ:	"Use of tobacco but absence of tobacco cultivation among the majority of these tribes. Other narcotics or stimulants unknown"
VIII. PARAGUAI:	"Use of tobacco and fermented drinks"
IX. PARANÁ:	"Use of tobacco and of chicha"
X. TIETÉ-URUGUAI:	"Use of tobacco and fermented drinks"
XI. NORTHEAST:	"Use of smoke and of an intoxicating drink, made from the sap of the "*jurema*" tree." *Jurema* is a Tupi word for the *Acacia jurema* Mart. (fam. *Mimosaceae*).

As "the Maué of the northwest of this area" are mentioned on p. 191 (67) in culture area IV, *Tapajós-Madeira,* this area must have been omitted by mistake as one where *paricá* is used. Area VII is termed *Pindaré-Gurupi,* a forest zone, bounded by the Pindaré and Gurupi rivers. "Fermented alcoholic beverages made with manioc and other fruit" are mentioned (ref. 67, p. 198). It would be interesting to know whether in this area or in any of the others where only tobacco is mentioned (V, VI, VIII, IX, X, and XI) there are botanical reasons for the fact that no psychotomimetic snuffs are used.

A considerable part of my earlier paper (8) dealt with archaeological snuffing paraphernalia in the form of trays and tubes. My dream has been to obtain archaeological snuffing paraphernalia also containing associated powder so that an analysis of the latter could be made by Professor Holmstedt of the Karolinska Institutet in Stockholm. From the data of Dr. Junius Bird of the American Museum of Natural History, I described a whalebone snuff "tablet" and its corresponding tube from near Huaca Prieta, Chicama Valley, Peru, as the oldest then known (as of 1966). The age for this find was given as c. 1200 B.C. (ref. 8, Fig. 17). At that time we unfortunately did not know that the archaeologist Dr. Frédéric Engel had published in 1963 the results from some mounds located in the drainage of the Omas River, 110 km south of Lima, near an Indian community called Asia (68). It was a German archaeologist, Dr. Henning Bischof, who participated in the work at this pre-ceramic settlement on the Central Coast of Peru who kindly drew my attention to Dr. Engel's find in Grave 47. According to Engel (64), "the radiocarbon date for Unit 1 of Asia reads 1225 ± 25 B.C.," which means that it fits very well with Dr. Bird's find of a whalebone snuff tray from near the Huaca Prieta. Grave 47, which yielded *not only a whalebone tray and a tube but also associated* (and afterward lost) *powder,* has been described by Engel as follows:

> An oval pit, 0.90 m. east-west × 0.50 m. north-south × 0.90 m. deep, originating just under the surface, had been dug into the first layer of shell refuse, down into the soil. The pit was filled with loose ashes, refuse, and earth, and contained a bundle covered with a few loosely piled small crude earth lumps, and wedged in with a few stones around the waist. The eastern end of the bundle came close to the surface of the mound, while the western end (the head) lay deeper down, near the ground. Inside the bundle was the skeleton of an adult, tightly flexed on its left side, the handle joined at the chin or face, the skull lying on the left cheek, head west. This skeleton was in good condition, with abundant remains of flesh and skin. With it was a bowl-shaped gourd containing a gray powder and closed with a wooden stopper, a pointed bird bone, a carved whalebone tray, a wood tube, a few cut mussel shells with no distinctive shape, probably blanks, and a mammal-bone implement.

The reported finds of two whalebone snuff trays and tubes from an early epoch on the Peruvian coast are most important, as they show that the use of some kind of snuff was spread along the coastal zone at that time. In order to attempt to get an analysis of the powder found by Engel, I contacted him (1968) and he informed me that the whole collection of his work at Asia is now in the Anthropology Museum of Peru in Pueblo Libre, Lima, but the search for the powder from Grave 47 has been negative (information of Dr. Jorge Muelle, Director of the Museum in Pueblo Libre, to Mr. G. Rudbäck, Lima, 1968). My hope is now that more powder associated with snuffing paraphernalia will be found during archaeological work on the Peruvian coast and be brought to my attention by my Peruvian colleagues.

As the climatic conditions in northern Chile offer good possibilities for the study of archaeological snuffing paraphernalia and associated powder, I am glad to report that Dr. Lautaro Nuñez, Director of the Department of Archaeology and Museums at the University of Chile in Antofagasta, was kind enough to send me three samples marked P. 1820, P. 1822, and P. 2300, in October 1966. These samples when received were immediately forwarded to Professor Holmstedt for analysis. Unfortunately, the small box in which the material was packed had been crushed in the mail from Chile to Sweden, so that the experts at Karolinska Institute received too limited a supply of material. According to Dr. Lautaro Nuñez (letter of October 1966), the box contained "a small bag together with a tray for inhaling narcotics, and also two small bags containing powder" (una pequena bolsa associada a una tableta de insuflar narcoticos, y dos bolsas pequenas, de las cuales dos tenian polvos). It was a sample of this which was sent. "The bag contained vegetable matter which, because of being macerated and combined with a (snuff) tray, should offer a good opportunity for analysis. The context refers to a pre-Columbian grave on the coast near Iquique (Bajo Molle)" (La bolsa tenia restos vegetales, que por estar macerados, junto a una tableta ofrece una buena posibilidad de analisis. *El contexto corresponde a una tumba preincaica, de la costa cerca Iquique* [*Bajo Molle*].) Dr. Lautaro Nuñez sent me also a manuscript in Spanish in which he refers to the find as having come from the site Patillos-1 in the Province of Tarapaca, Northern Chile. According to the drawings attached to this manuscript, the corresponding snuff tray is a very fine one made of algarrobo wood (*Prosopis*) with a figure of a high-ranking official carrying a knife and a severed human head. For comparative reasons Lautaro Nuñez has placed the Patillos-1 culture in a period from A.D. 700 to 1450 (69).

Thanks to Dr. Julio C. Montané, chief of the section of anthropology of the Natural History Museum in Santiago, I received four samples of powder from these archaeological sites, all from Caspana (Lat. 22° 40′ S., Long. 68° 14′ W.). These samples have been submitted to Professor Holmstedt. According to information received from Dr. Montané (letters of March and July 1968), two of the samples come from small boxes (*cajitas*) and two others from two tubes, all of this material belonging to the *complejo del rapé.* I also submitted the "small cylindrical lump of *Piptadenia* mixed with ashes of *Cecropia*" from the 100-year-old Silva Castro collection in the Ethnographical Museum in Stockholm to be analyzed by Professor Holmstedt (Coll. No. 1865.1.41, shown in Fig. 27 in ref. [14]). Between 20 and 850 mg of each sample were first extracted with methanol, and the fraction which should contain alkaloids prepared using conventional techniques. The components of this fraction, and those of a sample of the original methanol extract, were examined by thin-layer chromatography; Dragendorff's reagent, and a reagent containing iodoplatinate being used for the development of spots. Unfortunately, none of the samples mentioned above, either from archaeological investigations in Chile or from the Silva Castro collection, produced positive results.

Dr. Eskil Hultin, at the Institute of Organic Chemistry and Biochemistry, University of Stockholm, in a letter of October 1968, sent me the following comment on the possibility of deciding by chemical analysis whether or not ancient snuff material originally contained psychoactive alkaloids:

> The alkaloids of primary interest would be those belonging to the indole group. These alkaloids are fairly sensitive to deterioration, and especially so if they occur in a powdered preparation and have not been stored under particularly favorable conditions. The analysis of such breakdown products of indole alkaloids has not hitherto attracted much interest so that methods have not been worked out by which one may find out from such products in an old preparation which alkaloids, if any, were originally present. In the regular analysis for alkaloids, the quaternary alkaloids (which cannot be separated from other components of a sample as easily as the tertiary alkaloids) are very often disregarded, and consequently our present knowledge of this group of alkaloids is much less developed than the knowledge of tertiary alkaloids. It is possible that at least some quaternary alkaloids do not deteriorate in the same way, or as easily, as do the tertiary alkaloids. Considering this, it is not surprising that normal tests for alkaloids in old snuff material may give negative results; methods at present available do not appear to permit any definite conclusion whether or not the material originally contained indole alkaloids.
>
> It may be added here that a satisfactory answer to some ethnological questions may possibly be obtained from analyses for "accessory" substances which are fairly

stable chemically and whose presence or absence is closely related to that of the substances of primary interest.

Once the need for special chemical methods for the study of the composition of archaeological snuff material has arisen, it can be expected that, with time, new and suitable methods will be developed, provided the material in question is collected and properly preserved, preferably over dry inert gas in sealed glass tubes, and thus saved for future investigations.

Dr. Hultin has suggested a joint anthropological-archaeological-chemical discussion about the analyses needed and the methods which should be developed. I think that, in this paper at least, I have answered the first question (see also additional notes by Wassén and Lüning in ref. 69).

Finally, I want to point out that the recent research on psychotomimetics in Latin America has aroused the interest of other anthropologists. Dr. Peter Furst of the UCLA-Latin American Center, one of the participants of the Dumbarton Oaks conference on the Olmec in 1967, in his paper "The Olmec Were-Jaguar Motif in the Light of Ethnographic Reality" (70), quotes my data on the old snuff tray from Huaca Prieta, Chicama Valley, Peru, and says that

these, and the many effigy snuffing implements of more recent date discovered in archaeological sites in Chile, Argentina, Peru, Uruguay, and the Amazon basin provide a remarkable thread of continuity, both in form and iconography, which leads from the prehistoric lowland tropical forest cultures to the Andean civilizations, and from them directly into the historic period and the contemporary ethnographic scene. This archaeological evidence and related ethnographic data leave no doubt that the numerous psycho-active plants known and ritually used by South American Indians have important bearing on the shaman-jaguar complex as well as the phenomenon of shamanic flight or celestial ascent (70).

The same author (ref. 70, fn. 7) also writes that "a bifurcated snuffer with two tubes ending in a small bowl, from Jalisco, is believed to have been found in a shaft and chamber tomb of ca. A.D. 100-250 . . . A redware horned figurine from Colima holding a snuffing tube to his nose and Colima redware snuffing tubes are also known to me." Based on the possibility of the Olmec jade "spoons" having been "receptacles for psychotomimetic snuff," Furst seems inclined to assume some kind of snuff-taking among the Olmec, with the reservation, however, that "this is speculative as we do not know whether the Olmec shamans used snuff or other narcotics; however, in view of the great antiquity of snuffing and the widespread use of psychotropic plants in South and Central America, as well as in Mexico, it would be surprising if they did not" (70).

As regards Dr. Furst's knowingly speculative suggestion about possible psychotomimetic drugs among the Olmec, even if all this is highly conjectural, it is tempting to quote the following answer of February 1969 from Schultes to a letter of mine regarding our present botanical knowledge of interesting plants in the Mexican Gulf Coast region.

> We are finding so many plants with tryptamines—the active principle of many of the snuffs of South America—that it is very possible that in the Mexican Gulf Coast area the Indians could have found a plant which, prepared in the form of a snuff, could intoxicate as does the snuff of the Waikas. One of these is *Psychotria*, a species of which in South America has now been found to have *N,N*-dimethyltryptamine. *Psychotria* occurs up as far as Vera Cruz and it is possible that other species have this principle. Furthermore, Holmstedt believes that he has found this same chemical in our species of *Justicia* which is added to *Virola* snuff by the Waikas. Other species of *Justicia* occur as far north as Vera Cruz and may possibly also have this chemical constituent" (71).

If—and it is still conjectural—we were able to prove that any of the *Psychotria* or *Justicia* species found as far north as Vera Cruz have the same macropsia-causing effects as the plants used by the Waica, we could perhaps explain the colossal stone head monuments from San Lorenzo, Veracruz, as inspired by the drugs (55).

To this I would like to add that Dr. Alfonso Caso, in his paper "*Religión o religiones Mesoamericanas*," presented during the 38th Congress of Americanists (August 1968), expressly stated that "the magical practices of telling the fate by means of beans, the ingestion of hallucinogenic drugs from plants, the conception of *tona* and *nahual*, and innumerable coincidences in the ritual are generally accepted all over Mesoamerica."* We may conclude that many of these traits in one or other form also have a strong bearing on the Indian cultures south of Mesoamerica.

One may finally ask whether the use of narcotics—of the types known from Mexico, for example—has been reflected in Amerindian art. Alfonso Toro (72) gave the extreme reply when he suggested that the old Mexican art was "an abnormal art, a product of a feverish imagination intoxicated by the sacred plants" (un arte anormal, producto de una imaginación calenturienta intoxicada por las plantas sagradas). For the presentation of the maguey in Mexican art we have the work by Gonçalves de Lima (73); and Martí and Prokosch Kurath (74) answered the question with "Yes, since in Mexico, ecstatic dancing was induced by intoxicants or narcotics,

*". . . Las prácticas mágicas para decir la suerte con frijoles, la ingestión de drogas alucinógenas sacadas de vegetales, el concepto de la *tona* y el del *nahual* e innumerables coincidencias en el ritual, son generalmente aceptadas en toda Mesoamérica."

Fig. 7 Rubber syringe with a bird-bone tube for *paricá* clysters. Length ca. 10.5 cm. Caripuna Indians, R. Madeira, Brazil. Collection No. 1050, Johann Natterer in the Vienna Ethnographical Museum. Hugo Th. Horwitz reported briefly about this specimen in *Arch. Ges. Med.* (Leipzig), *13* (1921), 181-182. Photograph by Museum f. Völkerkunde, Vienna.

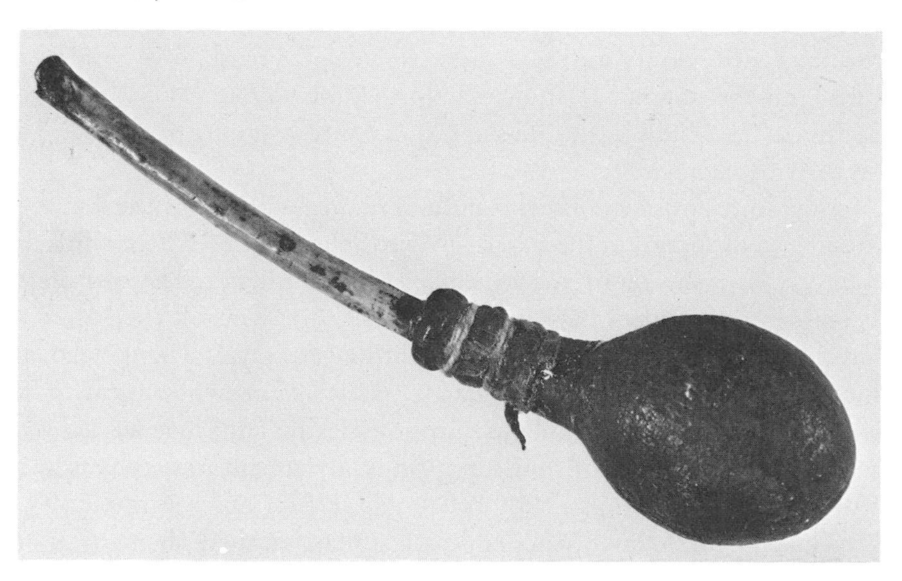

Fig. 8 Three rubber bottles in the Ethnographical Museum, Vienna. *Left* and *right,* tobacco snuff bottles" (Collection Nos. 53.552 and 53.551, the latter with a plug of light wood). Originally in the collection of Johann Natterer in Brazil, where he worked from 1817 to 1836. Natterer mentions snuffing and intoxicating clysters as particularly used by the Maué, Mura, Caripuna, and other tribes. The middle bottle is made of *Ficus elastica* and has a stopper of vegetable ivory. Collection No. 10.451, originally from the Tirolean Ambras Castle and the collection of Archduke Ferdinand of Tirol. Photograph by the Museum f. Völkerkunde, Vienna.

as in remote regions to this day." According to these authors, "narcotic visions" could explain "distorted and ecstatic postures." Figure 59 in Martí and Kurath's *Dances of Anáhuac* (ref. 74), illustrates "nude males in various antinatural positions, one with a double mask, one upside down" (cf. the statements about the *upside-down effect* of *cohoba* from the Taino, West Indies, and the same *epêna* effect among the actual Waica.

Upside down, often seemingly smiling, figures with wings (bird men) are sometimes found on the Paracas Necropolis mantles. These winged figures are carrying plants and have staves in their hands. The carrying of staves is a detail they have in common with the wing-cloaked attendants to the central figure on the monolithic gateway at Tiahuanaco. Since the Paracas Necropolis and Early Classic Tiahuanaco appear to be overlapping or comtemporaneous chronologically, and since we know that there was early use of snuff drugs in Peruvian cultures (tablets and snuff tubes dating from ca. 1200 B.C.), and later a representation of Tiahuanaco with "incised wooden snuff tablets" in the Calama oasis of North Chile (75), we should perhaps not totally exclude the possibility of drug effects having been reflected in the art motifs mentioned above. Naturalistic bird symbolism and identification can evidently still be observed among drug-taking South American Indians.

Appendix A *Alexandre Rodrigues Ferreira's Report from Barcellos (Rio Negro) on the Paricá Snuffing among the Maué Indians.*

Transcript and translation of the original manuscript dated February 13, 1786. (See pp. 9-10 for comments on this text.)

Memorandum regarding the instruments used by the savages for taking paricá-snuff and sent in box 7 in the first shipment from Rio Negro.[1]

When the Maué savages want to take *paricá*-snuff in their own way they need all of the following equipment: A mortar (*induá*) with its pestle (*induá-mena*), a small brush (*tapixúna*), a snail shell (*yapuruxitá*), a wooden tray and two joined bones from the wings of a bird. (See the following explanation.)

One of the two halves from a divided shell of the "*castanha do Maranhão*"[2] serves as mortar. In this, they rub to a fine powder the seeds from the *paricá*-tree after the seeds have been roasted. This is the kind of snuff they like best of all. The small brush is made of the tail hair of an anteater (*tamandua*). It is used for cleaning the mortar and to spread the snuff over the hollow surface of the board.

The shell (*Helix terrestris*) in accordance with its use is called *paricá-rerú*, which means *paricá*-box. With another piece of the same kind of shell, they shut its opening, glueing it with resin of *ananý*, and the snuff box is thus ready without any difficulty. In order to fill it and empty its contents onto the snuff tray they open the helical vertex of the shell and into this opening glue the mouthpiece from the neck of a gourd.

The tray is usually in the form of some animal. One I have been shown was according to its Indian owner, shaped like an alligator (*jacaré*). The carving of the figure and the rest of the handiwork is done using teeth from the agouti (Cutia) or other animals, as such teeth are their scooping-out tools, chisels and planes. They make eyes inlaid in the cavities representing them with mother-of-pearl from the snail shell *itaa*. The end of the hollowed-out shovel-formed specimen is called *paricá-rendana* which means "the place where to pour the *paricá*."

The two wing bones are selected from birds which have the longest, such as *tujujús*, *maguary* and *ayayás*, and after removing the marrow, one of the bones is tied to the other by means of a fine cotton thread. By the insertion of two splints from the *paxiúba* palm they prevent the bones coming too close together at the mid-point and upwards to make a hindrance to the separation or interstice of the nostrils.[3] In order to make the bones fit well into the nostrils, they glue two nuts from the *yu-hue* palm to their upper ends having taken out the core and opened the holes after peeling off the outer bark.

Let us now see how the *paricá* is taken.

After the portion to be taken has been emptied unto the snuffing board the powder is evenly spread by the handle of the brush.[4] The one who is going to inhale grasps the handle of the tray which represents the neck of the alligator (*jacaré*) with his left hand. Turning the hollow part towards himself he puts the upper parts of the two bones to his nostrils with the right hand and the lower ends to the cavity of the tray. In this way the snuff can be taken up both tubes. The savages use it during extensive *paricá*-drunken orgies ("*grandes Bacchanaes, chamadas do paricá*"). For this they have a big house called the *paricá* house. It is constructed for this purpose alone and has no partitions.

The ceremonial drinking-bout begins with a most cruel flagellation. They reciprocally flog each other with thongs of leather from manatee (*peixe-boy*), tapir (*anta*) or deer. If these are missing a cord of *pita* well twisted and about a six-foot length will do. At the end, they tie a stone or anything which is solid and gives wounds. They flog each other in pairs. The one receiving the scourging stands up with open arms and the scourger whips him as he pleases. Soon afterwards the flagellant passes to be flagellated and in this way each pair gets its turn. They continue with this flagellation ceremony during eight days and during this time the old (women) prepare *paricá* and beverages of fruit and *beijú*.[5] Those who took part in the whipping also partake of this. The narcotic power of *paricá*, the way of absorbing and the superabundance of drinks work with such a violence that those who do not die, sometimes are suffocated by snuff, fall half dead and continue in this way until the intoxication has passed. When one drinking-bout is over, a second one follows and according to the rules for the feast the drinking should go on as long as the pitchers last.

Barcellos, February 13, 1786.

NOTES TO APPENDIX A

1. In Brazil the word *tabaco* often is used for snuff even if there is no tobacco at all in the powder.
2. Formerly also "castanha do Pará," now "noz do Brasil" (14).
3. These technical details can be clearly observed in the double snuffing tubes published in Wassén (14, Figs. 9, 17, 20, 21, and 22).
4. "*Que representa huma contrabuxa*" has been omitted in the translation as the word *contrabuxa* is not given in any dictionaries at my disposal. *S. H. W-n.*
5. Cakes of manioc.

Appendix B *Information by Richard Spruce on a paricá tray from the Guahibo Indians, No. 177 in the Museum entry book of the Royal Botanical Gardens, Kew, Richmond, Surrey, and also his note on No. 166, the caapi and its effects.*

(Material received from Sir George Taylor, Director, Royal Botanic Gardens)

1. Spruce 65:1855. No. 177 (see Fig. 6).

Apparatus for making and taking Niopo snuff, procured from Guahibo Indians at the Cataracts of Maypures. (The *Niopo* of Venezuela is the same as the *Paricá* of Brazil, and is used on the Upper Orinoco, Guaviare, Vichada, Meta, Sipapo, etc. There is no doubt of its being prepared from *Acacia niopo,* Humb. which is perhaps not different from *Piptadenia peregrina,* Benth. My specimens of the Paricá tree from the Barra are referred to the latter species by Mr. Bentham. I did not see the tree from which the Guahibos obtained their Niopo, and which they told me was planted in their canncos near the head waters of the river Tupáro; but the Paricá I have seen on the Amazon and all the way up the Rio Negro, planted near the villages, belongs to but one species, which, on passing the Venezuelan frontier takes the name of Niopo.)

In preparing the snuff, the roasted seeds of the Niopo are placed in a shallow wooden platter, which is held on the knees by means of a broad handle grasped in the left hand, then crushed by a small pestle of the hard wood of the Pao d'Arco (Tecomae sp.) which is held between the fingers and thumb of the right hand.

The snuff is kept in a "mull" made of a tigers bone, closed at one end with pitch and at the other stopped with a cork of marúna. It hangs from the neck, and has attached to it the tubiferous rhizomes of some Cyperaceae. (*Hypoporum nutans,* Nees (?)) which are slightly odoriferous. These, or the tubers of some allied species, are used throughout the Amazon, Rio Negro, Uaupés, etc. among the Indians of the forest. With a piece of Piripirióca (the name given to them in Lingoa Geral) about the Person, one is safe from the bad wish and evil eye.

The instrument for taking the snuff is made of birds' bones, and differs somewhat from that used by the Catauixi Indians (see Jour. Bot. v. 5, p. 246). Tow tubes end upwards in little black balls (the endocarps of some species of *Astrocaryum*) which are applied to the nostrils, while the single tube in which they unite at the lower end is dipped into the "mull," and thus the Niopo is snuffed up the nose.

I enclose a piece of Caápi, from which the Indian who was grinding the Niopo every now and then tore a strip with his teeth and chewed with evident satisfaction. It had been slightly toasted over the fire. "With a chew of Caápi and a pinch of Niopo," said he to me, in imperfect Spanish, "one feels so good—no hunger—no thirst—no tired!". A piece of Caápi is generally suspended along with the snuff box, but the snuffer or snuff-taker is stuck in the thick bushy hair of the head.

2. Spruce 6:1854. No. 166.

Portions of the stems of a Malpighiaceous twiner, apparently an undescribed *Banisteria* (2712 to Benth.), called by the Indians Caá-pí; and of the roots and leaves of a Haemadictyon, called *Caapí-piniona* (i.e. painted caapí) the leaves being veined with red. From these ingredients, the *Banisteria* entering much more largely than the *Haemadictyon,* is prepared an intoxicating drink known to all the natives on the Uaupés by the name of *Caapí.*

In the Dabocurés (or festas) of the Uaupé Indians, the young men who figure in the dances drink of the Caapí 5 or 6 times during the night, the dose being a small cuya, the size of a very small teacup, twice filled. In two minutes after drinking it, its effects begin to be apparent. The Indian turns deadly pale, trembles in every limb, and horror is in his aspect; suddenly contrary symptons succeed–he bursts into a perspiration and seems pop-eyed with reckless fury–seizes whatever arms are at hand, his murucú, cutlass, or bow and arrows, and rushes to the doorway, where he inflicts deadly wounds on the ground or doorposts, calling out "Thus would I do to such a one (naming some one against whom he has a grudge) were he within my reach." In the space of 10 minutes the effects pass off, and the Indian becomes calm, but appears much exhausted.

References

1. William E. Safford, "Daturas of the Old World and New: an account of their narcotic properties and their use in oracular and initiatory studies" (Washington, D.C., Annual Report of the Board of Regents of the Smithsonian Institution, 1920), 537-567.

2. William E. Safford, "An Aztec narcotic," *J. Hered. 6* (1915), 291-311.

3. A. L. Kroeber, *Anthropology*, rev. ed. (New York, 1948).

4. Ralph L. Beals and Harry Hoijer, *An Introduction to Anthropology*, 2nd printing (New York, 1953).

5. David G. Mandelbaum, "The transmission of anthropological culture," in Mandelbaum, Lasker, and Albert, eds., *The Teaching of Anthropology* (Berkeley and Los Angeles, Calif., 1963), pp. 1-21.

6. *Notes and Queries on Anthropology*. 6th ed. Revised and rewritten by a committee of the Royal Anthropological Institute of Great Britain and Ireland (London, 1964).

7. Erland Nordenskiöld, *The Changes in the Material Culture of Two Indian Tribes under the Influence of New Surroundings*, Comparative Ethnographical Studies, 2 (1920), Göteborg.

8. S. Henry Wassén, "Anthropological survey of the use of South American snuffs," in *Ethnopharmacologic Search for Psychoactive Drugs*, U.S. Department of Health, Education and Welfare, Publ. no. 1645 (Washington, D.C., Government Printing Office, 1967).

9. Piedro de Cieza de León, *La crónica del Perú* (Madrid, 1553, 1922).

10. Richard Evans Schultes, "The role of the ethnobotanist in the search for new medicinal plants," *Lloydia 25* (1962), 257-266.

11. Isabel Kelly, *Folk Practices in North Mexico*. Latin American Monographs, vol. 2 (Austin, University of Texas Press, 1965).

12. S. Henry Wassén, "Some general viewpoints in the study of native drugs, especially from the West Indies and South America," *Ethnos* (Statens Etnografiska Museum, Stockholm) *29* (1964), 97-120.

13. Richard Evans Schultes, "Ein halbes Jahrhundert Ethnobotanik amerikanischer Halluzinogene," *Planta Medica 13* (1965), 125-157.

14. S. Henry Wassén, "The use of some specific kinds of South American Indian snuff and related paraphernalia," *Etnologiska Studier* (Göteborg) *28* (1965), 1-116.

15. João Barbosa Rodrigues, *Exploracão dos Rios Urubú e Jatapú* (Rio de Janeiro, Brazil, 1875).

16. Curt Nimuendajú, "The Maué and Arapium" and "The Mura and Piraha," *Handbook of South American Indians* (Washington, D.C., Bureau of American Ethnology, 1948), Bulletin 143, vol. 3, pp. 245-254, 255-269.

17. Claude Lévi-Strauss, "Tribes of the right bank of the Guaporé River," *Handbook of South American Indians* (Washington, D.C., Bureau of American Ethnology, 1948), Bulletin no. 143, vol. 3, pp. 371-379.

18. Claude Lévi-Strauss, *Tristes topiques* (Paris, 1955).
19. Candido Mariano da Silva Rondon, "Conferencias realizadas nos dias 5, 7 e 9 de outubre de 1915 pelo Sr. coronel Candido Mariano da Silva Rondon no theatro Phenix do Rio de Janeiro sobre os trabalhos da commissão Telegraphica e da Expedicão Roosevelt," *Diario do Congresso Nacional* (Rio de Janeiro, Brazil, Republica dos Estados Unidos do Brazil, Anno XXVi, 1916).
20. S. Henry Wassén, "Om några indianska droger och speciellt om snus samt tillbehör" (Göteborg, Sweden, Etnografiska Museet, Årstryck, 1963-66), pp. 97-140.
21. Glória Marly Duarte Nunes de Carvalho Fontes, "Alexandre Rodrigues Ferreira (Aspectos de su vida e obra)," *Cadernos da Amazonia* (Manaus, Brazil, Instituto de Pesquisas da Amazonia, 1966), vol. 10.
22. Alfredo do Valle Cabral, "Alexandre Rodrigues Ferreira. Noticia das obras manuscriptas e inéditas relativas á viagem philosophica do Dr. Alexandre Rodrigues Ferreira, pelas capitanias do Grão-Pará, Rio Negro, Matto Grosso e Cuyabá-(1783-1792)," *Annaes da Bibliotheca Nacional do Rio de Janeiro I* (1876-1878), 1, 2, 3.
23. Otto Zerries, "Beiträge zur kulturgeschichtlichen Beziehung der Südanden zum Tropischen Waldland Südamerikas," *Tribus* (Stuttgart) *17* (1968), 129-142.
24. Plutarco Naranjo, "Etnofarmacología de las plantas psicotrópicas de América," *Terapia, Revista de información médica* (Laboratorios Life, Quito) *24:1* (1969), 5-62.
25. A. C. Teixeira de Aragão, "Catálogo dos objectos de arte e industria dos indigenas da America que pelas festas commemorativas do 4.° centenario da sua descoberta a Academia Real das Sciencias de Lisboa envia á Exposicão, Madrid" (Madrid, Spain, 1892).
26. Conde de Ficalho, "*Plantas úteis de Africa Portuguesa*" (Lisbon, Portugal, Agencia Geral das Colonias, 1947).
27. Louis Lewin, *Gifte und Vergiftungen. Fünfte unveränderte Ausgabe des Lehrbuches der Toxikologie* (Berlin, Darmstadt, Lizensausgabe des Verlages George Stilke, n.d.), and *Phantastica. Narcotic and Stimulating Drugs, Their Use and Abuse* (London, 1964; reprinted with a new foreword by B. Holmstedt).
28. John Mitchell Watt and Maria Gerdina Breyer-Brandwijk, *The Medicinal and Poisonous Plants of Southern and Eastern Africa,* 2nd ed. (Edinburgh and London, 1962).
29. Nunes Pereira, *Os Indios Maúes. Coleção "Rex,"* (Rio de Janeiro, 1924).
30. *Catálogo general de la Exposición Histórico-Americana de Madrid,* 3 vols. (Madrid, Spain, 1892-93).
31. Erwin H. Ackerknecht, "Medical practices," *Handbook of South American Indians* (Washington, D.C., Bureau of American Ethnology, 1949) Bulletin no. 143, vol. 5, pp. 621-643.
32. Claude Lévi-Strauss, "The use of wild plants in tropical South America," *Handbook of South American Indians* (Washington, D.C., Bureau of American Ethnology, 1950), Bulletin no. 143, vol. 6, pp. 465-486.
33. Erland Nordenskiöld, *An Historical and Ethnological Survey of the Cuna Indians,* ed. Henry Wassén (1938). Comparative Ethnographical Studies, 10, published by the Gothenburg Ethnographical Museum, Göteborg.
34. S. Henry Wassén, "Original documents from the Cuna Indians of San Blas, Panama," *Etnologiska Studier* (Göteborg) *6* (1938), 1-178.
35. S. Henry Wassén, "De la identificación de los indios páparos del Darien." *Revista del Centro de Investigaciones Antropológicas de la Universidad Nacional,* vol. I (1962), 1-11. Hombre y Cultura, Panama.

36. Juan Requejo Salcedo, "Relación histórica y Geográfica de la Provincia de Panamá." *Colección de libros y documentos referentes á la Historia de América* (Madrid, Spain, 1640, 1908). Relaciones Históricas y geográficas de América Central, vol. VIII.

37. Raúl Martinez-Crovetto, "Introducción a la etnobotanica aborigen del nordeste argentino," *Etnobiologica 11* (1968). Facultad de Agronomia y Veterinaria, Univ. Nacional del Nordeste, Corrientes, Argentina.

38. W. Fritz Kramer, "Die Schriftsysteme der Cuna. Literatur und Schrift eines Naturvolkes." *Studium Generale 20* (Springer-Verlag, Berlin-Heidelberg-New York, 1967), 574-584, and "The Literature among the Cuna Indians," *Etnologiska Studier, 30,* Göteborg, 1970.

39. Henry Wassén, "Notes on southern groups of Chocó Indians in Colombia," *Etnologiska Studier* (Göteborg) *1* (1935), 35-182.

40. Erland Nordenskiöld, "Indianerna på Panamanäset" (Stockholm, 1928).

41. S. Henry Wassén, "On Dendrobates-Frog-Poison Material among Emperá (Chocó)-speaking Indians in Western Caldas, Colombia. Appendix: Short Emperá Word List" (1957). Etnografiska Museet, Göteborg, Årstryck för 1955 och 1956, pp. 73-94.

42. S. Henry Wassén, "Colombianska pilgiftsgrodor toxikologiskt högintressanta" (Nytt och Nyttigt, en medicinsk tidskrift utgiven av Hässle, no. 4, 1963). Göteborg.

43. Weldon G. Brown, "X-ray study reveals venom congener structure. Steroidal alkaloid structure accommodates choline precursor in congener of strongest cardiotoxin known." *J. Amer. Chem. Soc. 90* (1968), 7.

44. T. Tokuyama, J. Daly, B. Witkop, Isabella L. Karle, "The structure of batrachotoxinin A, a novel steroidal alkaloid from the Colombian arrow poison frog, *Phyllobates aurotaenia.*" *J. Amer. Chem. Soc. 90* (1968), 7.

45. Georg J. Seitz, "Einige Bemerkungen zur Anwendung und Wirkungsweise des *Epena*-Schnupfpulvers der Waika-Indianer." *Etnologiska Studier* (Göteborg) *28* (1965), 117-132.

46. Franz Knobloch, "Die Aharaibu-Indianer in Nordwest-Brasilien," *Collectanea Instituti Anthropos* (St. Augustin, Germany, Anthropos-Institut, 1967), vol. I.

47. Hans Becher, "Yanonámi. Neue wissenschaftliche Erkenntnisse über die älteste Bevölkerungsgruppe Amazoniens." *Vortragsreihe der niedersächsischen Landesregierung zur Förderung der wissenschaftlichen Forschung in Niedersachsen* (Göttingen) *23* (1962).

48. Inga Steinvorth de Goetz, *Uriji jami! Impresiones de viajes orinoquenses por aire, agua y tierra* (Caracas, 1968).

49. Bo Holmstedt and Jan-Erik Lindgren, "Chemical constituents and pharmacology of South American snuffs," in *Ethnopharmacologic Search for Psychoactive Drugs.* Proceedings of a Symposium held in San Francisco, California, Jan. 28-30, 1967. Workshop Series in Pharmacology, N.I.M.H., no. 2. Health Service Publ. no. 1645 (Washington, D.C., Government Printing Office, 1967), pp. 339-373.

50. Johannes Wilbert, *Indios de la Región Orinoco-Ventuari.* Monografía no. 8 (Caracas, Venezuela, Fundación La Salle de Ciencias Naturales, 1963).

51. Daniel de Barandiarán, "Mundo espiritual y shamanismo sanemá," *Antropológica* (Caracas) *15* (1965), 1-28.

52. Napoleon A. Chagnon, *Yanomamö. The Fierce People* (New York, Holt, Rinehart and Winston, 1968).

53. Ettore Biocca, "Viaggi tra gli Indi. Alto Rio Negro - Alto Orinoco." *Appunti*

di un biologo, vol. I: Tukâno-Tariâna-Baniwa-Makú; vol. II: Gli Indi Yanoáma; vol. III: Gli Indi Yanoáma, Appendici; vol. IV, Dischi. Rome (1965-66).

54. G. B. Marini-Bettolo, F. Delle Monache, and E. Biocca, "Sulle sostanze allucinogene dell' Amazzonia. Nota II. Osservazioni sull' epená degli Yanoáma del bacino del Rio Negro e dell' Alto Orinoco" *Ann. Chim. 54* (1964), 1179-1186.

55. Ettore Biocca, "Yanoáma. Dal racconto di una donna rapita dagli Indi." Leonardo da Vinci, Bari (1965).

56. Napoleon A. Chagnon, "The feast," *Natural History* (Journal of the American Museum of Natural History) 77 (1968), 4, 34-41.

57. Ephraim S. Alphonse, "Guaymí grammar and dictionary with some ethnological notes," *Bur. Amer. Ethnol. Bull.* (Washington, D.C.) *162* (1956).

58. Richard Evans Schultes and Bo Holmstedt, "The vegetal ingredients of the myristicaceous snuffs of the Northwest Amazon." (De Plantis Toxicariis e Mundo Novo Tropicale Commentationes II). Reprinted from *Rhodora 70* (1968), 113-160.

59. Stig Agurell, Bo Holmstedt, Jan-Erik Lindgren, and Richard Evans Schultes, "Alkaloids in certain species of *Virola* and other South American plants of ethnopharmacologic interest," *Acta chem. scand. 23* (1969), 903-916.

60. Gottfried Polykrates, "Wawanaueteri und Pukimapueteri. Zwei Yanonami-Stämme Nordwestbrasiliens." Publications of the National Museum Ethnographical Series (Copenhagen) *XIII* (1969).

61. Claude Lévi-Strauss, *L'Origine des manières de table* (Paris, Librairie Plon, 1968).

62. Curt Nimuendajú, *The Tukuna.* University of California Publications in American Archaeology and Ethnology (Berkeley, Calif., 1952), vol. 45.

63. Eduard Seler, "Die Tierbilder in den mexikanischen und den Maya-Handschriften." *Gesammelte Abhandlungen zur Amerikanischen Sprach- und Altertumskunde, 4* (Berlin, 1923 and Graz, 1961), 453-758.

64. Irene Nicholson, *Firefly in the Night. A Study of Ancient Mexican Poetry and Symbolism* (New York, Grove Press, 1959).

65. C. A. Burland, *The Gods of Mexico* (London, Eyre & Spottiswoode, 1967).

66. G. Reichel-Dolmatoff, *Desana. Simbolismo de los indios Tukano del Vaupés* (Bogota, 1968).

67. Eduardo Galvão, "Indigenous culture areas of Brazil, 1900-1959," *Indians of Brazil in the Twentieth Century* (Washington, D.C., Institute for Cross-Cultural Research, 1967), ICR Studies 2, pp. 169-205.

68. Frédéric Engel, "A Preceramic Settlement on the Central Coast of Peru: Asia, Unit 1." *Trans. Amer. Phil. Soc.* (Philadelphia) [n.s.] *53* (1963), 3.

69. A. Lautaro Núñez, "Informe arqueológico sobre una muestra de posible narcótico, del sitio Patillos-1 (Provincia de Tarapaca, Norte de Chile)." Con notas adicionales de S. Henry Wassén y Björn Lüning ("Problems in Analyzing Indian Snuffs" and "Note on the Identification of Indian Snuffs"), *Etnografiska Museet* (1969), 83-95, Göteborg, Arstryck 1967-68. Göteborg.

70. Peter T. Furst, "The Olmec Were-Jaguar Motif in the Light of Ethnographic Reality" (Washington, D.C., Dumbarton Oaks Conference on the Olmec, 1968), pp. 143-174.

71. Richard Evans Schultes, "Hallucinogens of plant origin," *Science* (New York) *163* (1969), 245-254.

72. Alfonso Toro, "Las plantas sagradas de los aztecas y su influencia sobre el arte precortesiano." *Proceedings of the Twenty-Third International Congress of Americanists* (New York, 1930), 101-121.

73. Oswaldo Gonçales de Lima, "El maguey y el pulque en los codices mexicanos" (México - Buenos Aires, 1956).
74. Samuel Marti and Gertrude Prokosch Kurath, *Dances of Anáhuac* (Chicago, Aldine Publishing Company, 1964).
75. Wendell C. Bennett and Junius B. Bird, "Andean Culture History," *American Museum of Natural History Handbook,* 2nd ed. (New York, Museum of Natural History, 1960), ser. no. 15.
76. Rodrigo Hernández Príncipe, "Mitología Andina." *Inca* (Lima, 1923), 1, 25-68.
77. J. J. von Tschudi, "Culturhistorische und sprachliche Beiträge zur Kenntnis des alten Perú." *Denkschriften der Kaiserlichen Akademie der Wissenschaften in Wien,* XXXIX (1891), Vienna.
78. Nils M. Holmer, "Indian place names in South America and the Antilles. III." *Names 9* (1961), 37-62. Berkeley, Calif.
79. Hugo Th. Horwitz, "Über eine indianische Gummispritze aus der ersten Hälfte des 19. Jahrhunderts." *Archiv f. Gesch. Med.* (Leipzig) *13* (1921), 181-182.
80. Bo Holmstedt, "Historical Survey," in *Ethnopharmacological Search for Psychoactive Drugs* (Washington, D.C., U.S. Public Health Service Publication 1645 (1967), pp. 3-32.
81. Richard Evans Schultes, "The widening panorama in medical botany." *Rhodora 65* (1963), 97-120.
82. Richard Evans Schultes, "Botanical sources of the New World narcotics," *Psychedelic Review 1* (1963), 145-166.
83. Richard Evans Schultes, "Hallucinogenic plants of the New World," *The Harvard Review* (Cambridge, Mass.) *1* (1963), 18-32.

Bibliography

Aguirre Beltrán, Gonzalo "Medicina y magia. El proceso de acultaración en la estructura colonial." México, D.F., 1963.

In Chapter 6 of this work the author discusses the *Libellus* of Martín de la Cruz and the work of Francisco Hernandez, "*Rerum medicarum Novae Hispaniae thesaurus*," considered by Schultes (10) as "one of the richest sources of medicinal information of native peoples." He also discusses the medicinal herb *Montanoa tomentosa* and *Artemisia mexicana* Willd (the "*estafiate*"), as well as tobacco, *ololiuhqui* and the *teonanacatl*. Chapter 7 deals with the *peyotl zacatequensi*. Lots of valuable information.

Allen, Paul H. "Indians of Southeastern Colombia." *The Geographical Review* (New York) *37* (1947), 567-582.

Has botanical information for the *payé* equipment from the Uaupés, the taking of *paricá* and *caapi* in connection with the *yuruparí* ceremonial.

Anderson, Arthur J. O. "Medical Practices of the Aztecs." *El Palacio* *68* (1961), 113-118. Santa Fe, New Mexico.

Contains valuable botanical information. "One is impressed by the Aztecs' knowledge of their flora and its properties. There are astringents, "sneeze-plants," irritants, emollients, local anaesthetics, poisons, intoxicants, spices, and so on. Roots, barks, piths, leaves, saps, resins, juices, and fruits are used, often given in infusions or in powders" (p. 115). For the discussion of the Aztec *tlatlauqui*, a plant for which an identification has not been made, see the same work p. 115.

Armentia, Nicolás (supposed author) "Relación histórica de las misiones franciscanas de Apolobamba, por otro nombre, Frontera de Caupolicán." La Paz, 1903.

Contains information on healing herbs used by the Tacanan speaking Araona in eastern Bolivia.

Barbosa Rodrigues, João "Exploracão dos Rios Urubú e Jatapú." Rio de Janeiro, 1875.

Extracts from this work in the text, see p. 6.

Basto Giron, Luis J. "Salud y enfermedad en el campesino peruano del siglo XVII." Instituto de Etnología y Arqueología, Lima, 1957.

Has a list of medicines of plant origin, p. 61.

Bolinder, Gustaf "Die Indianer der tropischen Schneegebirge," Stuttgart, 1925.

In the chapter on the *mamas* (medicine men) and their medicine the author has

(p. 133) a list of some medicinal plants used by the Ijca Indians of the Sierra Nevada de Santa Marta in Colombia.

"*Botanical Museum of Harvard University, A Quarter Century of Publications.*" Cambridge, Mass., 1966.

An indispensable bibliography for botanical and ethnobotanical research regarding the plants used by the Amerindians.

Botany of the Maya Area: Miscellaneous papers, I-XII (1935-1936). Carnegie Institution of Washington.

A collection of papers written by professional botanists.

Brüzzi Alves da Silva, Alcionilio "A civilização indigena do Uaupés." São Paulo, 1962.

Pages 227-232 contain information about the use of tobacco, *capsicum, caapi,* and coca.

Cardús, José, R. P. Fr. "Las Misiones Franciscanas entre los infieles de Bolivia, Descripción del estado de ellas en 1883 y 1884." Barcelona, 1886.

This work on eastern Bolivia has on pp. 331-360 a detailed list of trees and plants, also information on fish poisons, arrow poisons, etc.

Carvajal, P. A. "Plantas que curan, plantas que matan." México, 1960.

Has a presentation of popular descriptions of diseases, folk medicines, and plants used in curing.

Caso, Alfonso "Representaciones de hongos en los códices." Estudios de Cultura Nahuatl, vol. IV, pp. 27-36. México, 1963.

An applied study based on the interest awakened by the modern knowledge of Mexican hallucinogenic mushrooms.

Cobo, Bernabé. *Historia del Nuevo Mundo, I-II.* Ed. Francisco Mateos, Madrid, 1956. (Biblioteca de autores españoles, vol. 91-92.)

The drug list of this classic author, found in books 4, 5, and 6, pp. 152-284, in vol. I of the Madrid edition of 1956, gives a wealth of information. With reference to what has been said in Wassén (8) about the Peruvian *coro,* etc., an Argentine colleague, Dr. Ana M. Mariscotti, living in Marburg/Lahn, Germany, has in a letter of April 1, 1968, kindly drawn my attention to the fact that Bernabé Cobo (born 1580 and working in Peru in the beginning of the seventeenth century) mentions two kinds of tobacco. He says in Book IV, Chap. LVI (p. 185 of the edition quoted here): "Hállanse dos diferencias de *tabaco:* uno, hortense, que es el que aquí he pintado, y otro, salvaje, que nace in lugares incultos, el cual no crece tan alto ni produce tan grandes hojas, pero es de más fuerte y eficaz virtud que el hortense." This tobacco, "caliente en tercero grado," was used for various curing purposes. Cobo further says that the root of this tobacco was called *coro,* ("A la Raíz del tabaco silvestre llaman los indios del Perú, *coro,* de la cual usan para muchas enfermedades").

During the 38th International Congress of Americanists, celebrated in Germany in 1968, Dr. Raúl Martínez-Crovetto from Argentina presented a very interesting paper called "*Identificación botánica del "coro", antiguo fumatorio de los indios del Chaco.*" He had found actual Chaco Indians cultivating this plant which was secured for his herbarium. This ethnobotanist further found that this *coro* was an unclassified species, which he named *Nicotiana paá* (belonging to the genus *Petunia*), taking *paá* from the name given by the Toba and Mocoví Indians to this plant meaning something like "*raiz por excelencia.*" This information was given to me by Dr. Mariscotti (letter of August 31, 1968) who also referred me to the almost unknown but most interesting chronicler, Lic. Rodrigo Hernández Príncipe, who in 1621 wrote down most valuable information on Andean mythology in the province of Huailas. With reference to the *huaca* Yallpu Huallanca (in the village of Urcón) and a kind of smoke incense used by the old priests, he refers to *conóc* herbs which they obtained from faraway regions, saying that "*habiendo encendido los unos brasas de unas muy estimadas conóc yerbas traídas de muy lejos, le sahumaban e inciensaban el rostro y con esto entraban al sacrificio*" (76). I believe this *conóc* to have been tobacco. Bernabé Cobo also has (p. 186) the very interesting information that it was the Spaniards who propagated the custom of taking tobacco in the form of snuff through the nose. "Aunque los indios, de quien se tomó esta costumbre de tomar *tabaco,* lo usaban solamente en humo, han inventado los españoles otro modo de tomarlo más disimulado y con menos ofensión de los presentes, que es en polvo, por las narices; el cual hacen y aderezan con tantas cosas aromáticas como clavos, almizque, ámbar y otras especies olorosas, que da de sí gran fragancia." A reference can here be made to Tschudi's remarks on *sairi* in his work from 1891, p. 131 (77) "In vorspanischer Zeit wurden die grünen Blätter nebst anderen Kräutern, besonders Daturaarten und Aehnliche, von den Priestern verwendet, um sich in Extase zu versetzen."

Comparetti, André. "Observações sobre a propriedade da quina do Brasil por André Comparetti P. P. P. Traduzidas do italiano por José Ferreira da Silva. Lisboa, 1801.

A description of *Portlandia hexandria* L. Also case studies of patients being treated with this drug.

Cortés, Santiago. (n.d.) Flora de Colombia. Bogotá. (Libreria de el Mensajero).

This book outlines the flora of Colombia both geographically and also according to its therapeutic use. Contains an index of popular names on plants used in Colombia, pp. 191-301.

Coury, Charles. *La Médicine de l'Amérique Précolombienne.* (Paris, Les Editions Roger Dacosta, 1969).

This nicely illustrated compilation of 350 pages covers the whole American continent. In spite of the late printing date the author uses old botanical denominations which have been changed in the nomenclature, e.g., *Banisteria* and *Piptadenia peregrina* instead of *Banisteriopsis, Anadenanthera peregrina,* etc. Some statements seem to be open to questions. Was for instance the *ayahuasca* liana really "*bien connue des anciens paysans incas*"? (p. 225).

Curtin, L. M. S. "Some plants used by the Yuki Indians of Round Valley, Northern

California." Southwest Museum Leaflets, No. 27. Los Angeles, California, 1957.

This small publication contains lists of medicinal, poisonous, and food plants used in this region of California.

Curtin, L. S. M. "Healing herbs of the Upper Río Grande." Southwest Museum, Los Angeles, California, 1965.

An important work in the field of ethnobotany in New Mexico. As has been pointed out by Dr. M. Ries, Director, Laboratory of Anthropology, in his "By Way of Preface," the author has studied the significant relationship between "the plant-medicine of the United States Southwest and that of distant Morocco," the latter, naturally being introduced through the Spanish conquerors and their Moorish associations.

Dance, Charles Daniel. *Chapters from a Guianese Log-Book.* Georgetown, Demerara, 1881.

This book has an appendix in which "snake-bite remedies," "some of the vegetable poisons known to the Indians," and "the medicinal properties of plants" are given. In the same appendix there is also a discussion on "contributions towards a Guiana Indian pharmacopæia."

Cruz, Martín. "Libellus de Medicinalibus Indorum Herbis." Manuscrito azteca de 1552 según traducción latina de Juan Radiano. Versión española con estudios y comentarios por diversos autores. México, 1964.

After the discovery of the Latin original of this work in the Vatican Library in 1929 by Dr. Charles Upson Clark, it was considered to be of outstanding importance for our knowledge of Aztec medicinal lore and herbal cures. This Mexican edition with its explanatory chapters written by Mexican specialists is outstanding. In the discussion of the identification of the plants we find a contribution by Ángel Ma. Garibay K. on the etymology of the plant names in Nahuatl.

Dibble, Charles E. and J. O. Anderson. Florentine Codex. *General History of the Things of New Spain, Fray Bernardino de Sahagún.* Book 11: *Earthly Things.* Translated from the Aztec into English. Santa Fe, New Mexico, 1963.

Beginning on p. 129: "Seventh Chapter, which telleth of all the different herbs," we find in this critical and unsurpassed translation of Sahagún's work a real gold mine of original descriptions and terms from the mouth of the Aztec. The material has been used by modern specialists mainly concerning the question of herbs in the first paragraph, those "which perturb one, madden one," like *ololiuhqui, peyote, tlapatl* (*Datura stramonium*), etc. Naturally the descriptions of the mushrooms have been much referred to by specialists like Dr. Gordon Wasson and others. From the fifth paragraph, "which telleth of the medicinal herbs and of different herbs," we conclude that some of the well-known "maddening" species like *tlapatl, peyote,* and *teonanacatl* evidently also were used as pure medicine, e.g., as fever medicine, etc. This double function of specific drugs might well be specifically investigated by specialists, if it has not already been done.

Dimbleby, G. W. *Plants and Archaeology.* (London, John Baker Publishers Ltd., 1967).

Chapter 5 in this work has the caption "Plants used in ritual and medicine," which, however, does not add very much of interest. A list of detailed references (pp. 171-176) has been added to the general one (p. 169).

Duke, James A. *Darien Ethnobotanical Dictionary.* (Columbus, Ohio, Battelle Memorial Institute, Columbus Laboratories, January 4, 1968.)

Because of its wealth of information and the colloquial names of plants used in this anthropologically little-known region of Panama, this study of research ecologist Dr. James A. Duke, which is part of the BMI Bioenvironmental Program, must be considered as a most important tool for all interested in the Indians of Darien and the use of plants there. The author's role as ecologist has been to travel some of the little-known areas of the Cuna and Choco Indians, "studying the relationships between man and plant." This work is not only important, it is indispensable.

Eder, P. Francisco Javier. "Descripción de la Provincia de los Mojos en el Reino del Perú. Traducida del latin por el P. Fray Nicolás Armentia." La Paz, 1888. (Original edition: "Descriptio Provinciae Moxitarum in Regno Peruano, quam e scriptis posthumis Franc. Xav. Eder e Soc. Jesu," etc. Budae, 1791.)

Missionary Eder's work, of outstanding importance for the study of the Mojos region, has in Book IV a chapter on arrow poisons.

Ernst, A. "Memoria botánica sobre el embarbascar ó sea la pesca por medio de plantas venenosas." Caracas, 1881. (Del tomo I de los esbozos de Venezuela por A. A. Level.)

A paper which deals with the use of fish poisons and a list of plants used as *barbasco* in various parts of the world.

Ewbank, Thomas. *Life in Brazil; or, A Journal of a Visit to the Land of the Cocoa and the Palm.* With an Appendix containing illustrations of Ancient South American Arts (New York, 1856).

This is not a very important book of travel, but it has, in Appendix A, an illustration which shows a snuff tray of wood with a human figure as handle, evidently from the Atacameño culture. This early illustration of a snuff tray is described as "a prettily carved snuff or other mill for rubbing dry leaves to powder. It resembles current Brazilian apparatus."

Fernandez, R. P. Fr. Wenceslao, misionero dominico. "Rincones del Amazonas. Diario de un misionero." Arequipa, 1942.

Has in Chap. XII, "Flora patológica" a list of poisonous plants. Chap. XIII lists narcotic plants, and also names for plants used as fish poisons. All kinds of popular beliefs are reflected. An herb called *kudirinchipini* is said to have fruits which give flute-playing qualities, one who eats the fruits "becomes a Paganini in flute-playing."

Friedberg, Claudine (1958-1959). "Contribution à l'étude ethnobotanique des tombes précolombiennes de Lauri (Pérou)." *Journal d'agriculture tropicale et de botanique appliqué* (Paris), *6-7* (1958), 397-428; *8-9* (1959), 405-434.

This modern work discusses (pp. 416-417) the use of *Fevillea cordifolia* L. (Cucurbitaceae) as a medicinal plant.

Friedberg, Claudine. "Des *Banisteriopsis* utilisés comme drogue en Amerique du Sud. Essai d'étude critique," *Journal d'agriculture tropicale et de botanique appliqué* (Paris), *12,* nos. 9-12.

A detailed tribal study of the preparation, use, and chemical quality of the drug prepared from species of *Banisteriopsis* and known in South America as *ayahuasca, caapi, yajé,* etc.

Furst, Peter T., and Barbara G. Myerhoff. "Myth as History: The Jimson Weed Cycle of the Huichols of Mexico." *Antropológica* (Caracas) *17* (1966), 3-39.

Discusses the use and mythology of *Datura* among the Huichol, who call it *kiéri.* Also gives information about the *Datura* in Pre-hispanic Mexico.

Gillin, John (1936). "The Barama River Caribs of British Guiana." *Papers of the Peabody Museum of Archaeology and Ethnology, Harvard University, 14* (1936), 1-274, Cambridge, Mass.

Has information on healing herbs used by the Barama River Caribs.

Gines, Hno, and Felipe Matos. "Algunos datos etnobotánicos." La Región de Perijá y sus Habitantes. Publicaciones de la Universidad de Zulia, 1953, pp. 341-343.

These three pages in the Perijá volume give us a few glimpses of botanical material used by the Motilon Indians for various purposes. The same authors and Ernesto Foldats contribute in the same volume (pp. 345-553) with a "*Florula de la Cuenca del Río Negro, Perijá.*"

Goldman, Irving. "The Cubeo, Indians of the Northwest Amazon." Illinois Studies in Anthropology No. 2. Urbana, Ill., 1963.

The author, in the chapter on Religion in this modern anthropological study, has several references to plant poisons, "which go by the *lingoa geral* term *marakimbára,* and in Tucanoan by the term for "illness," "*ihé*" (p. 268).

Gonzales Sol, Rafael. "La farmacoterapia presalvaradeana en Centro América." San Salvador, El Salvador, 1943.

In this small publication there is a discussion of medicines of plant origin with explanations of the Mexican names.

Grossa, Dino J. "Una visita a los indios yaruros de Riocito." *Boletín Indígenista Venezolano* (Caracas) *10* (1966), 69-79.

This paper, which gives some information on modern Yaruro at Riocito de Apure, has a short but detailed description of a ceremony involving the preparing and taking of *yopo* powder (pp. 75-76). It is said that a root with a stronger narcotic effect and named *tcuipán* could be taken "antes de yoparse" by those wanting the stronger effect.

Guerra, Francisco. "Bibliografía de la Materia Médica Méxicana." La Prensa Médica Méxicana, Mexico, 1950.

A bibliography of 5357 works regarding the medicinal properties of Mexican drugs. Of outstanding importance for the specialist.

Guerra, Francisco. "Mexican Phantastica—A Study of the Early Ethnobotanical Sources on Hallucinogenic Drugs." *Br. J. Addict 62* (1967), 171-187. Pergamon Press Ltd.

With the facilities of the Wellcome Historical Medical Library, London, at his disposal, the author has prepared a review of the historical documents on hallucinogenic drugs used in Mexico. He considers the survey of early documents "a much more reliable method of ethnobotanical research than modern pharmacological techniques have been inclined to credit" (p. 172), and he finds historical documents on hallucinogenic drugs "far more accurate than modern treatises, because they were written when the use of Phantastica was flourishing among the aboriginal communities, and such studies were the result of first-hand observations made by some Spanish natural philosophers" (p. 172). Dr. Guerra points out (his Figs. 2 and 3) that although not previously noted, *Codex Magliabecchi* (on f. 71 and f. 72) contains descriptions which refer to "the identification and glyph of *ololiuhqui* and their use." On pp. 179-187 the author gives an appendix containing pertinent texts from the references quoted.

Gusinde, Martin. "Medicina e Higiene de los antiguos araucanos." *Publicaciones del Museo de Etnología y Anthropología de Chile* (Santiago de Chile) *1* (1917), 87-293.

Has a careful explanatory list of medicinal plants with the Araucanian names in alphabetical order. The botanical classification and the vernacular names are also given, as well as descriptions of the use of the medicines in question.

Gusinde, Martin. "Plantas medicinales que los indios Araucanos recomiendan." *Anthropos 31* (1936), 555-571, 850-873. Mödling-Vienna.

Extensive list of healing herbs used by the Araucanian Indians.

Gusinde, Martin. "Der Peyote-Kult. Entstehung und Verbreitung." *Festschrift zum 50jährigen Bestandsjubiläum des Missionshauses St. Gabriel Wien-Mödling* (1939), 401-499.

A complete description of the use of *Lophophora Williamsii* and its cult in Mexico and North America.

Hagen, V. Wolfgang von. "The Tsátchela Indians of Western Ecuador." Indian Notes and Monographs, No. 51. New York, Museum of the American Indian, 1939.

Has on pp. 57-58 an analysis of a powder prepared from the roots of *Banisteriopsis caapi.*

d'Harcourt, Raoul. "La Médicine dans l'Ancien Pérou." Paris, 1939.

The author discusses (pp. 86-90) the use of Datura, *ayahuasca* and *yajé* (*Haemadictyon amazonicum,* Benth.).

Herrera, Fortunato L. "Nomenclatura indígena de las plantas. Flora cuzcoensis." *Inca 1* (1923), 607-623.

A partly comparative study of various plant names.

Herzog, Th. "Die Pflanzenwelt der bolivischen Anden und ihres östlichen Vorlandes." (Die Vegetation der Erde, editors A. Engler and O. Drude. XV.). Leipzig, 1923.

Apart from all the professional botanical information in this book, it is of value to the anthropologist through its list of vernacular plant names presented on pp. 240-244.

Holland, William R. "Medicina maya en los altos de Chiapas." México, 1963.

Has an Appendix (C) with medicinal plant names in Tzotzil explained in Spanish together with the botanical terms.

Karsten, Rafael. "Beiträge zur Sittengeschichte der Südamerikanischen Indianer, Drei Abhandlungen." (1. "Das Pfeilgift" pp. 1-27; 2. "Berauschende und narkotische Getränke," pp. 28-72.) *Acta Academiae Aboensis, Humaniora 1* (1920), 4. Åbo.

Information based on the author's research among the Jivaro Indians.

Kelly, Isabel. "Folk practices in North Mexico. Birth Customs, Folk Medicine, and Spiritualism in the Laguna Zone." Latin American Monographs No. 2. Austin, Texas, 1965.

See text, p. 4.

Krukoff, B. A., and N. N. Moldenke. "Studies of American Menispermaceae, with special reference to species used in preparation of arrow-poisons. *Brittonia,* November 1938.

A study undertaken "largely with the idea of clearing up taxonomic and nomenclatural uncertainties," but because of its historical account of "field and botanical investigations of Curare" also of outstanding importance for the Americanists.

La Barre, Weston. *The Peyote Cult.* New and enlarged edition. Hamden, Conn. The Shoe String Press, Inc., 1964.

Pages 10-22 give a summary of "botanical and physiological aspects of peyote" as well as the ethnobotany of the genus *Lophophora williamsii.*

Latcham, R. E. "Ethnology of the Araucanos." *Journal of the Royal Anthropological Institute* (London) *39* (1909), 334-370.

The author mentions, p. 352, the "*gatuhue,* or syringe, used for clysters, composed of a bladder and bone tube," among the chief instruments used by the *machi,* "medicine-man, seer, and exorcist."

Latcham, Ricardo E. "La agricultura precolombiana en Chile y los países vecinos." Ediciones de la Universidad de Chile, 1936.

Gives only cultivated plants.

León, Juan José. "Ensayo de Botánica Médica yucateco-tabasqueña. (1861). Reimpresión. Villahermosa, Tabasco, 1947.

One of the many small folklore-type books found in Latin America. Plants are listed according to their botanical names.

Lenz, Rodolfo. "Diccionario etimolójico de las voces chilenas derivadas de lenguas indijenas americanas." Santiago de Chile, 1904.

This valuable work has on p. 425 information on *Latua venenosa:* "árbol de los brujos."

Lima e Silva, Leopoldo de. "Estudos sobre o Curare." Rio de Janeiro, 1935.

As the title indicates, deals only with studies of curare.

Lira, Jorge A. "Farmacopea tradicional indígena y practicas rituales." Lima, 1948.

Vernacular names and folklore information on medicines, plants, etc., mainly from the region of Cuzco.

Lundell, Cyrus Longworth. "The vegetation of Petén. With an Appendix. Studies of Mexican and Central American Plants. I." Carnegie Institution of Washington, Publ. 478. Washington, D.C., 1937.

Important work for the Maya region.

Madero Moreira, Mauro. "Voces, usos y costumbres del folklore médico ecuatoriano." Guayaquil, 1967.

Folklore information and explanations of vernacular names for diseases, organs of the human body, plants used as curatives, etc., in the Ecuadorian population.

Manfred, Leo. "600 plantas medicinales argentinas y sudamericanas." Rosario, 1940.

A book about plants from Argentina and other parts of South America said to have curative effects.

Martinez, Maximino. "Las plantas medicinales de México." Cuarta edición. Mexico, 1959.

Illustrated work explaining the names and giving information about Mexican medicinal plants.

Mostny, Greta. "Culturas precolombinas de Chile." Santiago, 1960.

The author deals on p. 54 with the snuff trays, tubes, etc., from the Atacameño culture. The snuff supposed to originate from *Piptadenia macrocarpa.* As an interesting detail, the author connects the monstrous figures observed on snuff tubes with the faces of wooden masks also found archaeologically.

Müller, Franz. "Drogen und Medikamente der Guaraní (Mbyá, Pai und Chiripá) Indianer im östlichen Waldgebiete von Paraguay." Festschrift P. W. Schmidt: 501-514. Vienna, 1928.

Contains a list of healing herbs used by the Guarani of Eastern Paraguay.

Munizaga A., Carlos. "Nuevas investigaciones en el Norte de Chile y sus posibles vinculaciones con países limítrofes." (Jornadas Internacionales de Arqueología y Etnografía, Buenos Aires 11 al 15 de Noviembre de 1957, I "Vinculaciones de los Aborígenes Argentinos de los países limítrofes," pp. 98-106). Buenos Aires, 1962.

The possible connection between the word *khoba* used in Argentina for *Lepidophyllum quadrangulare* and the Arawak Island word *cohoba* was dealt with in Wassén (8). When presenting that paper in San Francisco I was unaware of this work by Carlos Munizaga A. I have now found that he also has referred to *Lepidophyllum quadrangulare* as *chacha* or *koa* (*coca*) (p. 102), adding the most interesting information from the inhabitants of Socaire that *chacha* is the word for the plant as it grows in the soil, and that *koa* is the expression for "*algo sagrado*" into which the *chacha* plant is transformed when used ceremonially. This strengthens my súpposed connection with the *cohoba* word found in the northern part of the South American continent.

Nierembergius, I. E. "Historia naturae maxime peregrinae." Antwerp, 1635.

This compiler has on p. 332, "caput CXII" information about the *cohobba* ("sic herba vocant inebriantem, qua & Boitij ub surorem statim vertuntur"), which he connects with tobacco. ("Suspicor hanc herbam tabacum esse"). Lib. XIV of the work deals with "de plantis peregrinis" and lib. XV with "de plantis Indicis."

Noguera, Eduardo. "El uso de anestésicos entre los aztecas." *Revista Mexicana de Estudios Históricos 2* (1928), 162-169, México.

Discusses the problem of the *yauhtli* of the Aztecs.

Nordenskiöld, Erland. "Recettes magiques et médicales du Pérou et de la Bolivie," *Journal de la Société des Américanistes de Paris,* n.s. *4* (1907), 153-174.

During his stay in Mojos (N.E. Bolivia) the author had an opportunity of observing how the Quichua and Cholos, when building a house, buried several things which were sold in the market at La Paz. He mentions specifically (p. 157) "des graines de *Piptadenia macrocarpa,* des graines de *Ormosia sp.* et de petites figures fondues d'étain."

Oblitas Poblete, Enrique. "Cultura Callawaya." La Paz, 1963.

This work is of interest for the understanding of the history and practices of the much discussed institution of the *callawayas.* See the following title.

Oblitas Poblete, Enrique. "Plantas medicinales de Bolivia." La Paz, 1969.

A well-illustrated work which in the form of a dictionary presents the "farmacopea callawaya" in Spanish with the plant names also given in Latin and Indian languages.

Pardal, Ramon. "Medicina aborigen americana." Buenos Aires, 1937.

A work of general interest. The author deals with the following drugs discovered by Amerindians and used in contemporaneous medicine (pp. 355-366): "*La Quina,*" the cinchona bark (*Cinchona* sp.). "*Lo que se sabe bien concretamente es que sólo en 1638 un indio de Malacotas en Loxa reveló a un español la quina y sus propiedades.*" "*La Jalapa,*" the purgative jalap drug from Mexico, *Ipomoeapurga Exogonium jalapa.* "*La Ratania,*" rhatany, *Krameria triandra* (from Peru). South American leguminous shrub, the root of which is used as an astringent to stop bleeding. Also used against intestinal diseases. "*La raíz de Polígala,*" *Polygala* sp., milkworts. North American diuretic ("Seneca snakeroot, "*Polygala Senega*). "*El Podofilo,*" podophyllum, the dried rhizome of the May apple, *Podophyllum peltatum,* used for its emetic properties by North American Indians and introduced in the pharmacopoeia of the United States in 1820. "*El Quillay,*" quillay, soapbark tree, in Chile *Quillaja saponaria,* used as a substitute for soap. According to Pardal, used by Indians in Central America for the treatment of skin affections. Also name of "any of various other saponaceous barks, as of several tropical American shrubs of the mimosaceous genus *Pithecelobium*" (Amer. Coll. Dict.). "*La Ipecacuana,*" ipecac, *Cephaelis ipecacuanha* and *C. acuminata.* The Tupí-Guaraní Indians revealed the emetic and purgative effects of the dried root to Piso and Marcgrav (*Historia naturalis Brasiliae,* 1648). According to Pardal, the Tupí-Guaraní also used another plant called *ituvu* (*Hyabanthus ipecachuanha*) with a milder effect. "*La Copaiba,*" capaiba, "an oleoresin obtained from various tropical (chiefly South American) trees of the caesalpiniaceous genus *Copaiba*" (Amer. Coll. Dict.). Used by the Tupí-Guaraní for wound healing and several diseases. Synonym: *kupahi.* Known in Europe 1570 through a Portuguese Franciscan mendicant. "*El Jaborandi,*" jaborandi, a shrub, *Pilocarpus* sp. (fam. *Rutaceae*). Used by the Tupí-Guaraní as a sudorific and sialagogue. "*El Chenopodio,*" chenopod, *Chenopodium ambrosioides.* The seeds of this plant used from Mexico to Río de la Plata against intestinal parasites. The *epazotl* of the Aztecs, and *caa-ne* of the Guaraní. Described by Francisco Hernández (Rerum Medicarum Novae Hispaniae Thesaurus, 1651) as *Atriplox odorata mexicana.*

Pérez Arbelaez, Enrique. "Plantas medicinales y venenosas de Colombia." Bogotá, 1937.

As the title indicates, this work gives a description of medicinal and poisonous plants found in Colombia. The author has sometimes uncritically accepted erroneous statements by other specialists. For example, on p. 174, writing on *yagé,* the author, without criticism, quotes Dr. Barriga Vallalba as having isolated two alkaloids from *yagé,* one of which was cocaine. Pérez de Barradas (see below) read the same erroneous statement and rejected it.

Pérez de Barradas, José. "Plantas mágicas Americanas." Madrid, 1957.

A work of 342 pages filled with historical and other information of use for anthropologists and other investigators wanting data about the "magic plants" of America. The early interest which was shown by the Spaniards in Mexico for the plants used by the Indians in curing illnesses (the number of plants known to the

Mexican Indians has been given as 3,000 by Francisco Hernández in 1651) was also found in Peru. The author has quoted a letter from Viceroy Toledo to the Emperor in Spain (p. 25): "Los médicos que Vuestra Majestad mandó venir van tomando alguna experiencia de las medicinas y remedios de la cosecha de la tierra y de la virtud de las plantas con que los indios naturales curan; irse a sacando libros de ellos, como se mandó, con el color, fruta y hoja, entiendo que será de provecho, y así mandado que en su curiosidad me traigan razón de ella de las provincias, para enviarlo a Vuestra Majestad." The author (p. 97) regrets the loss of the 55 volumes on the botany of Guatemala which Blas Pineda de Polanco had written at the end of the eighteenth century. This work would have been of outstanding importance in increasing our knowledge of Maya medicine.

Poeppig, Eduard. "Reise in Chile, Peru und auf dem Amazonenstrome, während der Jahre 1827-1832." vol. II. Leipzig, 1836.

Pages 455-457 give information about "the Peruvian arrow poison."

Pompa, Gerónimo. "Medicamentos indígenas." (Distribuidora Continental, Caracas. Printed in Barcelona) (n.d.).

This book, printed in Spain, gives a popular description of various native medicines in a South American country, apparently Venezuela.

Rabanales O., Ambrosio. "Uso tropológico, en el lenguaje chileno, de nombres del reino vegetal." (Editorial Universitaria, 1950).

One of the most useful books for students interested in the plant material used by the natives of Chile.

Raffauf, Robert F., and Siri von Reis Altschul. "The Detection of Alkaloids in Herbarium Material." 1968.

This paper describes new ways for tracing plant alkaloids, following up Dr. von Reis Altschul's systematic search of herbarium materials (see below).

Reinburg, P. "Bebidas tóxicas de los indios del Noroeste del Amazonas: El Ayahuasca - El Yajé - El Huanto." (Transl. from French by Ernesto More.) Universidad Nacional Mayor de San Marcos, Serie: Ciencia Nueva, No. 4. Lima, 1965.

The author describes an experiment in which he took *ayahuasca* among the Záparo Indians in April-May 1913. He also gives comparative information from the literature, including the Indian terminology. The *huanto* drug (the Záparo name given as *issiona*) is identified with *Datura arborea* L. (in Spanish *floripondio*).

Reis Altschul, Siri von. "Herbaria, Sources of Medicinal Folklore." *Economic Botany 16* (1962), 283-287.

The author presents a new and original approach to the search for useful plants by using notes on herbarium sheets and records.

Reis Altschul, Siri von. (1967). "Vilca and Use." *Ethnopharmacologic Search for Psychoactive Drugs.* Proceedings of a Symposium held in San Francisco, California, Jan. 28-30. Workshop Series of Pharmacology, N.I.M.H. No. 2: 307-314. Publ. Health Service Publ. No. 1645. Washington, D.C. Government Printing Office, 1967.

A botanist's critical and skillful interpretation of the *vilca* problem according to the author's search in the literature for an interpretation of the known data.

Reis Altschul, Siri von. "Unusual Food Plants in Herbarium Records." *Economic Botany 22* (1968), 293-296.

A continuation of the author's searching "the combined collections of the Arnold Arboretum and Gray Herbarium of the Harvard University." Cf. Raffauf and Siri von Reis Altschul (1968).

Reko, Blas Pablo. "De los nombres botánicos aztecas." *El México Antiguo 1* (1919-1922), pp. 113-157. México.

One of the important publications for the studies of the various plants used by the Aztecs. On p. 116 Reko suggests that the island of *Cuba* got its name from *cohoba* for the powder taken by the natives at the time of the conquest. More reliable, however, is the explanation by Professor Nils M. Holmer (78), who as a linguist considers *Cuba* as "an Arawak word meaning 'land' (cf. *akoba* 'field' or 'ground' in the dialect of Guiana); this designation would rather correspond to that of the 'mainland,' in relation to the number of smaller islands among the Antilles."

Relaciones geográficas de Indias. Perú. vols. I-IV. Madrid, 1881-1897.

A wealth of information can be found in the answers to the famous questionnaire which was circulated in 1577. One of the questions to be answered, No. 26, was: "*Las yeruas ó plantas aromaticas con que se curan indios, y las virtudes medicinales ó venenosas de ellas.*"

A study of these four volumes shows that in many cases no answer at all was given to this question. Positive answers are as follows: *Tomo I:* P. 69 ("Provincia de los *Yauyos*"), an herb used as effective laxative; p. 87 ("Provincia de *Xauxa*"), herbs earlier used as ointment for wounds; p. 124 ("Ciudad de *Guamanga*"), a poisonous herb called *miu,* according to M. Jiménez de la Espada, *garbancillo* (*Astragalus garbancillo*); p. 175 ("Repartimiento de *Atunsora*"), *molle* and *chillca* (*Auchenia huanacu*); the same remedies are mentioned also on p. 192 for the "Repartimiento de *Atunrucana*"; pp. 211-212 ("Rep. de los *Rucanas Antamarcas*"), *chilca* and *molle* are again mentioned. Also *zaire,* tobacco snuff. *Ancocho* (*Baccharis latifolia*), *caruancho* and *pomachoc* (*Krameria triandra*) are also mentioned.

Tomo II: Answers to the question are found on pp. 15, 19, 23, 29, 32, 47, 76. On p. 172 information about "*yerba muy mortal,*" an arrow poison, is given from Santa Cruz de la Sierra. The answers to question 26 continue on pp. 204, 214 and 218.

Tomo III: Answer on p. 61 from Quito mentions *zarzaparilla* as a cure against syphilis; on pp. 114 and 115, mention is made of "*un médico gran herbolario*" called "doctor Heras" (from Otavalo). He had special volumes for plant medicines, etc. Other answers on pp. 160, 187 and 199.

In *Tomo IV* there is an answer to question 26 on p. 2, this coming from "la ciudad de *Zamora de los Alcaides.*"

Rochebrune, A. T. de. "Recherches d'ethnographie botanique sur la flore des sépultures péruviennes d'Ancon." *Actes de la Société Linnéenne de Bordeaux,* 4th ser., *3* (1880), 343-358. Bordeaux.

One of the early papers pointing on the need of botanical examination of archaeological plant material found in Peru. In Section II of this paper the author deals with edible and medicinal plants.

Rondón, Candido Mariano da Silva. "Comissão de Linhas telegráphicas estrategicas de Matto Grosso ao Amazonas. Annexo 5. Rio de Janeiro, 1913.

A list of healing herbs used by the Arawakan Paressí of Central Brazil.

Rondón, Candido Mariano da Silva. "Conferencias realisadas nos dias 5, 7, e 9 de outubro de 1915 pelo Sr. coronel Candido Mariano da Silva Rondon no theatro Phenix do Rio de Janeiro sobre os trabalhos da Commissão Telegraphica e da Expedicão Roosevelt." (Diario do Congresso Nacional, Anno XXVI, Domingo, 30 de janeiro de 1916, N. 214). Rio de Janeiro, 1916.

See text p. 20.

Roys, Ralph L. "The Ethno-Botany of the Maya." The Tulane University of Louisiana, Middle American Research Series Publ. No. 2. New Orleans, La., 1931.

Pages 211-326 of this outstanding volume can serve as a model of research for an ethnobotanical survey of a special region. This section of the work contains, e.g., a "Vocabulary of Maya terms relating to the growth, parts and environment of plants," as well as an "annotated list of Maya plant names" and a "table of nomenclature."

Safford, William E. "Daturas of the Old World and New: An Account of their Narcotic Properties and their Use in Oracular and Initiatory Studies." Annual Report of the Board of Regents of the Smithsonian Institution, pp. 537-567, 13 plates. Washington, D.C., 1920.

The author wrongly identified the *Datura meteloides* Dunal with "*ololiuhqui,* the magic plant of the Aztecs." His identification of the Mexican *teonanacatl,* or "sacred mushroom" with the *Peyotl Zacatecensis* was, as we now know, also premature.

Sanchez Labrador, P. José. "La medicina en el Paraguay Natural" (1771-1776). (Exposición comentada del texto original por el Dr. Anibal Ruiz Moreno.) Tucumán, 1948.

The historically important work of an observing and traveling naturalist of the early eighteenth century.

Santesson, C. G. "Einige Mexikanische Rauschdrogen." *Arkiv för botanik 29* [A], no. 12 (1939). Stockholm.

The Swedish pharmacologist, Prof. C. G. Santesson, has here presented the first

European research on *teonanácatl* material sent to him from Mexico. His paper should for historical reasons be compared with the letter of January 31, 1937, from Dr. Reko sent to Wassén in Göteborg, Sweden, published in full in Holmstedt (80). (Cf. ref. 20, pp. 98-99.) Schultes has also explained this historical connection: "When I was a graduate student, seeking material for my doctoral thesis amongst Indians isolated in the mountains of Oaxaca, Mexico, I could find no pharmacological house in the United States with the time or interest to investigate witch-doctors' plants, and I was forced to send my material to Sweden to the late Dr. C. G. Santesson who, since he was retired, could study whatever it pleased him to investigate."

Schultes, Richard Evans. "A Contribution to our Knowledge of Rivea Corymbosa, The Narcotic Ololiuqui of the Aztecs." Botanical Museum of Harvard University, Cambridge, Mass. (1941).

The epoch-making study of the *ololiqui*-problem in Mexico. For some historical connections cf. ref. 20, p. 98, and in this Appendix: Santesson: 1939.

Schultes, Richard Evans. "De plantis toxicariis e mundo novo tropicale commentationes V, *Virola* as an orally administered hallucinogen." *Botanical Museum Leaflets, Harvard University 22,* no. 6 (1969), 229-240.

The author follows in this paper his series of most important works in which he combines recent botanical and chemical data on the South American tryptaminic narcotics (as beyond *Virola, "Anadenanthera peregrina* and other species of this leguminous genus; *Mimosa hostilis; Banisteriopsis Rusbyana; Psychotria psychotriaefolia;* and possibly *Justicia pectoralis* var. *stenophylla*") with new field observations of his own (from February 1969). Thanks to this report, it has now been brought to our attention that the ancient Witoto (and evidently also other Indians as the actual Muinane and Bora) have been eating pellets made of a resin, probably from *Virola theiodora* to obtain hallucinogenic effects. On p. 236 the author points out that this newly discovered hallucinogen "bears very directly on certain pharmacological matters, and, when considered with other plants with psychotomimetic properties due to tryptamines, this new oral drug poses problems which must now be faced and, if possible, toxicologically explained."

Smith Jr. C. Earle, Eric O. Callen, Hugh C. Cutler, Walton C. Galinat, Lawrence Kaplan, Thomas W. Whitaker, and Richard A. Yarnell. "Bibliography of American Archaeological Plant Remains." *Economic Botany 20:* 4 (1966), 446-460.

Bibliographies of this kind are of outstanding importance also for the anthropologists.

Stone, Doris. "The Northern Highland Tribes: "The Lenca." *Handbook of South American Indians 4* (1948), 205-217. Washington, D.C.

The author has on p. 216 a short list of the most popular plants and herbs used for curing purposes by the Lenca of Central America. E.g.: The leaves called "hoja del aire" (*Bryophyllum pinnatum*) "are boiled and taken for colic."

Stradelli, E. "Nell' Alto Orenoco." (Bollettino della Societá Geografica Italiana, Agosto-Settem. 1888). Rome, 1888.

The author has (p. 40) an illustration of typical snuffing paraphernalia from the Guahibo. His wordlist from the same Indians at the Rio Vichada gives the word *siripú* for the snuff tube, "forchetta di osso per annasare el 'tubere.'" The word *tubere* is explained as the powder from the leaves of an unknown plant, used for snuffing.

Thurn, Everard F. im. "Among the Indians of Guiana, being sketches chiefly anthropologic from the Interior of British Guiana." London, 1883.

Pages 436-438 form a detailed index of "plants used by Indians."

Towle, Margaret A. *The Ethnobotany of pre-Columbian Peru.* Viking Fund Publications in Anthropology: 30. New York, 1961.

A modern work covering the whole field of Peruvian ethnobotany.

Valdizan, Hermilio, and Angel Maldonado. "La medicina popular peruana. Contribución al "Folk-lore" médico del Perú." vols. 1-3. Lima, 1922.

A wealth of local lore and popular medicinal ideas of the Indians and other members of the population in Peru.

Walker, Edwin F. "World Crops derived from the Indians." Southwest Museum Leaflets, no. 17. Los Angeles, California, 1953.

A small publication full of informative facts. Some medicinal items, such as quinine and cocaine are also mentioned.

Villarejo, Avencio. *Así es la Selva.* Lima, 1943.

The book is published in the form of paragraphs of which 460 deals with composition of an ayahuasca drug taken by the "iquitos."

Yacovleff, E. and F. L. Herrera. "El mundo vegetal de los antiguos peruanos." *Revista del Museo Nacional 3* (1934), 241-322, and *4* (1935), 29-102. Lima.

A most valuable systematic study covering the years 1533-1703.

Yarnell, Richard A. "Prehistoric Pueblo Use of Datura." *El Palacio 66:* 5 (1959), 176-178. Santa Fe, New Mexico.

Reports the now many archaeological finds of the genus *Datura,* "all in the greater Pueblo area."

Ypiranga Monteiro, Mário. "Antropogeografia do guaraná." Cadernos da Amazonia, 6. Instituto Nacional de Pesquisas da Amazonia. Manaus, 1965.

A monographic description of the origin, use, properties, trade, etc., of the medicinal *Paullinia*-paste from Brazil.

Magic and Witchcraft in Relation to Plants and Folk Medicine

JOHN MITCHELL WATT*

Department of Physiology, University of Queensland,
St. Lucia, Queensland, Australia

INTRODUCTION

Magic is a belief in supernatural explanations for many mundane happenings. For example, on the Continent of Africa there is a general belief that illness (other than trivial) and death are due to some malevolent activity directed purposely against the individual concerned.

The association of plants with healing, religious practices, and magic is of early origin. The ancient Greeks and Romans worshipped flowers and herbs because of their beauty as well as for their healing virtues (1). They dedicated them to their gods, decorated their altars with them, crowned their priests with them, wore wreaths themselves on gala occasions.

Plato, visiting the aged Cephalus, found the latter sitting upon a cushioned seat with a garland of flowers on his head "as he happened to have been sacrificing in the court" (1). The making of wreaths and garlands for these occasions was one of the ancient professions, the predecessors in office of our florists.

Another ancient practice is perpetuated in the bride's bouquet (2). This started simply as a spray of orange blossom which was carried by the Saracens as a symbol of fertility. The Saracens, incidentally, were Muslims. The Roman bride clutched ears of wheat as a symbol that her

*Emeritus Professor of Pharmacology and Therapeutics, University of the Witwatersrand, Johannesburg, South Africa.

husband's grain bins should always be well filled. By the time of the Renaissance, the bride carried a sheaf of corn (wheat) but by the eighteenth century, it had become a bouquet of flowers.

ASTROLOGY

The ancient astrologers taught that the seven "planets"—Sun, Moon, Mercury, Venus, Mars, Saturn, and Jupiter—had an influence over plant life and that each had its own particular trees, herbs and flowers which resembled them in a special way. Many of the data in this introduction are derived from the material in Oldmeadow's interesting book (1).

The sun "owned" the ash, famous for its healing properties and known as the tree of life. Friede (3) notes that "in Scandinavian mythology the sacred ash-tree, Yggdrasil, was the whole world," the roots reaching down to hell, the trunk traversing the earth, the leaves forming the clouds and the fruits the stars. The bay tree was the Sun's too, as shown by the following: "All the evils old Saturn (the sun) can do to the body of man, and they are not a few; yet neither witch, nor devil, thunder nor lightening, will hurt a man where a bay tree is." Marigolds, true flowers of the sun, were often called "the Sun's brides" or "the Sun's herb."

The Moon played an important part in the gathering of "simples," and farmers and gardeners, who wanted good crops, were careful to sow seeds at the new Moon and to harvest them when the Moon was on the wane. I knew a man in South Africa who would plant fruit trees and seedlings only at certain phases of the Moon. The Bushmen of Angola, incidentally, worship the Moon as the source of all life.

Astrologers taught that all plants gathered for medicines or charms must be taken from the ground at a moment when the planet governing them was in a favorable position in the heavens, and they insisted that some plants were more potent if dug up at night. For example: "Roots of hemlock digg'd i' the dusk, slips of yew, slivered in the Moon's eclipse." These were two things needed to make a witch's curse really effective.

Physicians at one time used to collect their own medicinal plants and were thought to require a real knowledge of astronomy to assist them. The following is from a book written by a sixteenth-century doctor: "Above all things next to grammar a physician must have surely his Astronomye, to know how, when and at what time, every medicine ought to be administered."

To this day the African herbalist and witch-doctor for the most part collects his own herbal and medicinal requirements. Early herbalism was much affected by superstitious beliefs, as for that matter religious rites often were. Pagan rites and superstitions have often persisted despite all the efforts of the churches.

CHRISTIANITY

With Christianity came a rewriting of the old herbals by the monks, who left out heathen rites and ceremonies connected with the picking of plants and substituted prayers and psalms. They also taught that herbs were God's gift, given to man for his health and well-being. Thus, while gathering them, it would be holy to say a few paternosters, a psalm, or the following salutation: "Hail thou holy herb growing in the ground."

Before the Reformation, monks were already great herbalists. All monastery gardens were well stocked with physic herbs, which were given to the poor who came knocking at the gates when ill or in trouble. One of the special little prayers recited when picking herbs went as follows: "Thou art good for manie a sore and healest manie a wound. In the name of sweet Jesus, I take thee from the ground."

The plants belonging to Venus, the Goddess of Love, were usually the flowers specially dedicated to lovers: for example, dark damask roses, violets, primroses, and all vineyards with their "rich purple fruit."

Rosemary is an emblem of remembrance. Ophelia, the daughter of Polonius in Hamlet, says: "There's rosemary, that's for remembrance." According to ancient tradition, this herb strengthens the memory. As Hungary Water, it was once taken very extensively to quiet the nerves. It was much used at weddings, and to wear rosemary in ancient times was as significant of a wedding as the wearing of a white favor nowadays. In the language of flowers, it means "fidelity in love." A rosemary bush is said to have sheltered the Holy Family. It was the Blessed Virgin's mantle which, thrown upon it, turned the flowers into such an exquisite blue. Because of the legend, mothers used to put a sprig of rosemary into the baby's cradle to give it safety and peaceful dreams. A sprig of rosemary has been worn as a charm against black magic.

The Romans are said to have brought the nettle to England, and the Roman soldiers used to rub their bodies vigorously with them to keep themselves warm in a climate which they considered cold and uncongenial. The nettle, by the way, belonged to Mars, the God of War, and so it may be that these Roman soldiers thought that a dose of one of the herbs of Mars would make valiant soldiers of them.

WITCHCRAFT

Witchcraft is also strongly associated with plants. It is the exercise of supposed supernatural powers by persons in communion with the devil or with evil spirits. Kluckhohn (4) gives what I imagine is a most satisfactory definition of witchcraft: "the influencing of events by supernatural techniques that are socially disapproved." The practitioners of witchcraft are known as witches if female and wizards if male, but the word *witch* is also used in the sense of embracing both male and female. In Africa, the name *witch-doctor* is much used in this connection. But it means a magician or sorcerer whose business is to detect witches and to counteract the effects of their magic. Witch-doctors have also been described as those who profess to cure disease and to counteract witchcraft by the use of magic arts.

In Africa all living things, human, animal, and plant, are believed to have a "spirit" which, on the whole, may remain dormant and therefore benign. Such spirits may, however, change to malign activities by their misuse by a magician or sorcerer. Even inanimate objects, such as mountains, streams, rocks, and stumps of dead trees, may be the home of spirits. Evidence of this is found in the common practice of addressing inanimate objects.

In this connection, it is instructive to examine the views of Margetts (5). He says that, for practical purposes, primitive African magical practitioners or "ritual experts" may be classified as follows:

A. Practitioners of good intent ("white" magicians)
   1. Medicine men, healers or folk doctors (an archaic term for healer still used sometimes is "leech").
      (a) Surgeons (who use cutting procedures and include midwives, obstetricians, genital operators—circumcisers, excisers and infibulators, tribal-mark scarifiers, trepanners).
      (b) Physicians (who do not use cutting procedures and include herbalists, exorcisers, hypnotists, religious healers).
   2. Other magicians: diviners, seers, soothsayers, mystics, mediums, priests, prophets, oracles, exorcists, dance masters, finders or "smellers-out" of evil magicians, necromancers, rainmakers and

other weather-controllers, blacksmiths, garden and hunt magicians, amulet-makers.

B. Practitioners of evil intent ("black" magicians)
   1. Witches and wizards or sorcerers
   2. Vampires
   3. Ghouls

There is considerable overlap between "good" and "evil" intent ("white" and "black" magic), as every field Africanist knows. However, this division has much to commend it because it indicates *motive.*

SACRED GROVES

Friede (3) tells us " that the Greeks animated their forests with tree-nymphs; that the old Germans venerated their gods in sacred groves; that among the Celts of Gaul, the Druids chose groves of oak for their solemn service." He points out that "reverence for trees and tree-spirits is often of great importance in primitive religion." A tree, to begin with, may be viewed as the body of the tree-spirit, but ultimately it becomes its abode which the spirit can quit at will.

Burial groves are always held to be sacred and thus taboo, but, in some parts of Africa, there may be a particular tree in a grove which is held in reverence to such an extent that an annual sacrifice is made at its foot. These sacred groves and revered trees have been important factors in the survival of the indigenous arboreal flora of Africa. Sacred groves, the burial places of important people like chiefs, have often flourished because of the taboos imposed on the local populace. Even the gathering of firewood is interdicted.

According to Harley (6), the Mano of Northern Liberia seldom have wayside shrines (such as I have seen at Khujiar, high up in the Himalayas on the road to Chamba), but make contact with ancestors in secret either through *ma* or at an annual sacrifice made at the foot of a great tree in the sacred grove. *Ma* are wooden masks, part fetish and part symbol, of *Poro,* a sort of secret society connected with initiation. The sacred tree may be *Bombax, Afzelia,* or *Khaya.*

In West Africa, the cola tree, *Cola acuminata* Schott. & Endl., and possibly other species of *Cola,* is venerated as a spirit tree when the nut has been planted on a grave or at a crossroad as a sacrifice. In Tanganyika, *Jatropha curcas* L. is planted by some tribes to mark grave sites (7).

Every compound in Ashanti has an altar to the Sky God in the shape of a forked branch cut from *Alstonia boonei* De Wild. (*A. congensis* Engl.) The Ashanti name for the tree *nyame dua* = God's tree (8).

Another sacred area for some tribes is the hut where the great tribal talisman is housed. The Tonga of Portuguese East Africa regard the shrub *Annona senegalensis* Pers. (*A. chrysophilla* Boj.) as a good "war" medicine, and they administer the root to a suckling child to wean it from its mother's breast, as it is supposed to cause forgetfulness. The Shangana have a similar use for the *Annona* root. What is more, it is the duty of the first wife of the chief of a Tonga clan to keep a perpetual fire burning in the sacred hut where the great tribal talisman is housed. This fire is so sacred that even its embers are taboo. If by mischance it should die out, it has to be relit with the aid of two friction sticks of the *Annona* plant. These are not used for lighting ordinary fires. Furthermore, of the four sticks planted around a low circular shrine, in which the sacred stones are embedded, two are from *Annona senegalensis* (9).

The medicine-man of the Shangana, tribal neighbors of the Tonga, uses the wood of the same shrub for a sacred fire, but ordinary folk must not use the wood for any purpose. Further north in Africa, we find that the Bemba tie amulets of the wood around the neck of their hunting dogs as a charm to protect them from being gored by warthogs.

In Ghana, *Costus afer* Ker. is placed on a cultivated field or a path or at the entrance to a house for protection against evil, and it is also planted in sacred groves (10). In Ashanti, it is used in religious ceremonies, and the odor is regarded as inimical to ghosts and evil influences (8).

SACRED TREES

Trees often, and other types of plants less frequently, are regarded with reverence or as sacred for a variety of reasons. In some cases, the reason is lost in the antiquity of the people. Sometimes, however, the reason is clear, as in the case of the Herero myth which ascribes the origin of man to *Combretum primigenum* Marl. & Engl. (11). Friede (3) suggests that the worship of trees by primitive man may have originated in arid areas like South West Africa, because of the preciousness of the occasional tree. For many years, only the stump of such a venerated tree survived near the South of Ovamboland. All Hereros and Ovambos, passing close to the stump, would take a bunch of leaves or some grass, spit on it, rub the forehead with it and thrust it into a hole in the stump, saying, "I greet thee, O my father! Grant me a prosperous journey!" The reverence is nowadays extended to any specimen of this tree by Herero, Ovambo, and Damara (3). Baines (12) writes about the "Mother-tree of the Damara," which he thought was the same species.

Veneration for trees or the association of them with religious and other ceremonies, particularly magical, is known in various parts of the world. For example, jambolan (*Syzygium cumini* Skeels) is "venerated by Buddhists" and "is commonly planted near Hindu temples, because it is considered sacred to Krishna" (McCann quoted in Morton [13]). "The leaves and fruits are employed in worshiping the elephant-headed god, Ganesha or Vinaijaka, the personification of 'Pravana' or 'Am,' the apex of Hindu religion and philosophy" (Benthall quoted in Morton [13]).

Sacred trees are well known to the African. On the Ivory Coast, *Afzelia africana* Smith is regarded as sacred and is seldom used alone in a medi-

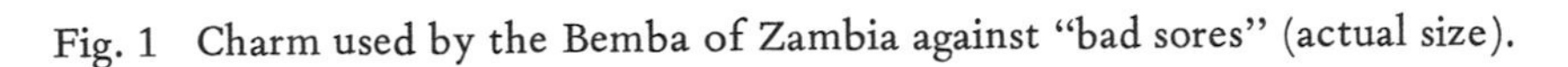
Fig. 1 Charm used by the Bemba of Zambia against "bad sores" (actual size).

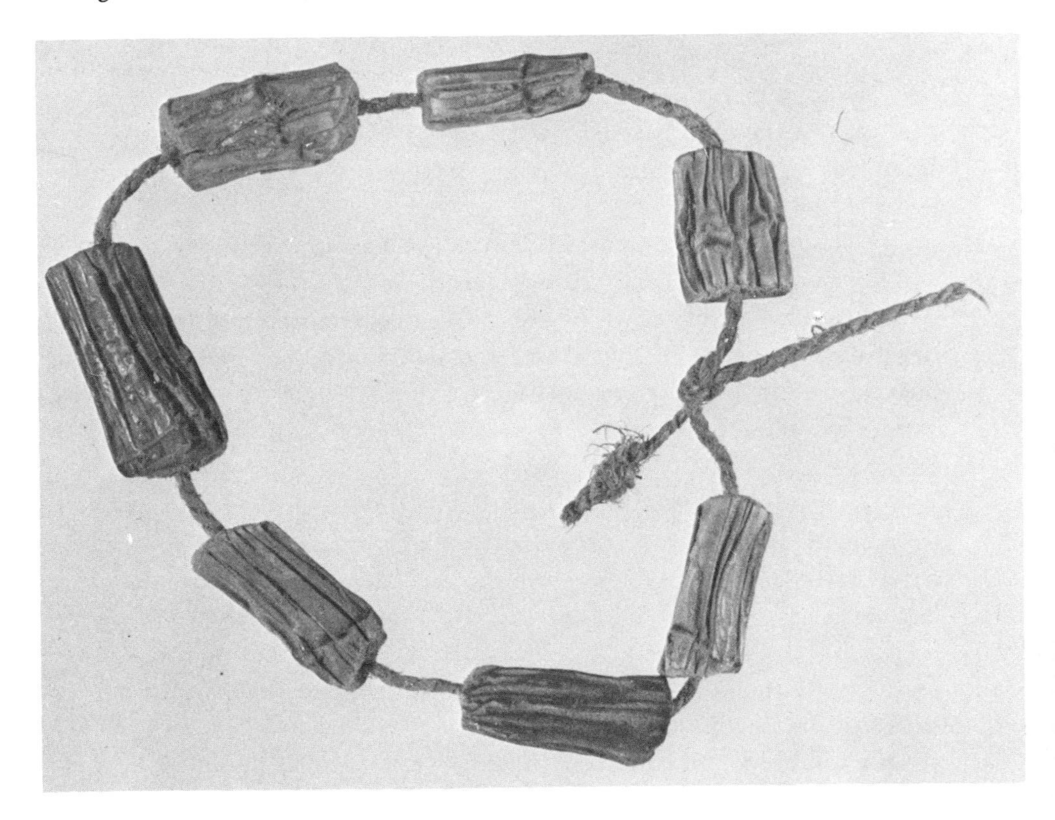

cine (14). It is not surprising, then, to find that necklaces of the seed are worn by Hausa children as a charm against bad luck (10). The Hausa know the seed as *fasa daga* or *fasa maza* (meaning scattering of men in war) and use it in various ways as a charm in battle, to dispel evil and to avert injury by weapons (10). The selection of this seed may be due to its curious appearance.

The Thembu of South Africa burn a twig of *Acokanthera venenata* G. Don in the hut to keep away *impundulu,* a visiting spirit which chokes people in their sleep and causes bad dreams, possibly a projection of its known toxic nature.

Baphia nitida Lodd. has certain sacred and superstitious uses in West Africa: for example, the dyeing of sacred objects (14). Thomas (quoted by Irvine [14]) has recorded that the red dye is put on the forehead of a man during a particular ceremonial dance, indicating that he has killed a man or a leopard.

Blighia sapida Konig, or akee, is another sacred tree in parts of the Ivory Coast (14). The unopened unripe fruit and the cotyledons are known in Cuba and Jamaica to be poisonous, producing vomiting and circulatory collapse (15).

In the Transvaal, on the Leydsdorp road, there was for a long time a solitary palm tree, *Borassus flabellifer* L. var. *aethiopum* Warb., in which the spirit of the chief Magoeba was thought to reside. The tree, about sixty feet high, has a natural swelling on the trunk in which the spirit was thought to dwell.

The Thlaping and Tswana of Botswana (Bechuanaland) have a considerable regard for the magical properties of *Boscia albitrunca* Gilg & Benedict. The cutting down of the tree is discouraged, and the wood is never burnt because "this has the effect of causing the cows to produce bull calves." If the fruit withers before the ripening of the kaffircorn (a variety of millet, *Sorghum dochna* Snowden), the crop fails due to blight [9]).

In Senegal, portions of *Calotropis procera* Ait.f., an irritant and cardiotoxic plant, are placed over the doorway of a hut as protection against witchcraft (10). The pod and gum of *Cassia sieberiana* DC. are ingredients in the Hausa sorcerer's rod (14).

In Ashanti, *Chlorophora excelsa* Benth. & Hook.f. is regarded as a sacred tree by the Ibo and as one of the trees credited with furnishing souls for the newborn. The household god of the Ibo is always carved from a solid block of *iroko,* the trade name for the timber. The bark is an ingredient in the brass bowl used by the Ibo in the ceremony of making a new shrine (10).

The Lobedu of the Transvaal place a pole of the wood of *Combretum suluense* Engl. & Diels. in the roof so as to protrude perpendicularly from the top of the hut (16). This is intended to keep out witches sent in by lightning.

According to Walker (17), *Copaifera religiosa* J. Leonard has a reputation throughout Gabon as a magic tree of the "Bouiti" sect. The seed, wood, and bark emit the odor of coumarin, but I do not know if this has any connection with the religious regard for the tree. Walker (17) records that *Senecio gabonensis* Oliv. & Hiern is a magical plant much in vogue among the adepts of "Bouiti." He also states that *Tabernanthe iboga* Baill. is a plant of initiation of the sect "Bouiti" in West Africa. In the Congo it is used as an aphrodisiac and stimulant. An interesting use is the taking of "iboga" extracts by African hunters while stalking game. This is done to enable them to remain motionless for as long as two days while retaining mental alertness. This is not an impossible idea, for a suitable dose of nicotine produces a similar static effect on animals. There are other plant products which act in like manner.

In somc parts of West Africa, *Elaeis guineensis* Jacq. var. *idolatrica* A. Chev. is regarded as sacred (14). *Ceiba pentandra* (L.) Gaertn. is another of the sacred trees of West Africa (14), and, on the Ivory Coast, its magic property is used in remedies for rickets, pregnancy, and languor (18).

A decoction of the root of *Elephantorrhiza elephantina* Skeels is sometimes sprinkled about an African kraal to hinder evil doers. The Lobedu impale the root stock of *Hypoxis villosa* L.f. on a stick of this plant placed outside a kraal in order to repel witches (16).

The cycad *Encephalartos barteri* Carruth, the ghost or hosanna palm, is known in Ghana by the Krobo as *kpadei-atah* which means "palm of the spirits," indicating that it is only of use to the shades of the departed (10, 14).

According to Ferreira (9), the Kikuyu have used *Ficus* trees from generation to generation as shrines or places of sacrifice. Irvine (14) says that fig trees have various uses in superstitious medicinal practice in West Africa, *Ficus capensis* being used as a sacrifice plant.

In Central Africa, *Kigelia pinnata* DC. is regarded with reverence, and religious meetings are held in its shade. Charms are often cut from this tree (16). The Zulu use an infusion of the fruit in the pre-battle preparation of warriors (19).

Another tree, *Loesnera kalantha* Harms, is held in superstitious regard in West Africa, and the cutting of it is forbidden under severe penalty. "In olden times potent amulets were made from it and it is a reputed fetish reviver." Nowadays it is regarded "as a counsellor and the people pay large sums to witch-doctors to sit by it and tell it their troubles, plucking a few leaves" (10).

Pentaclethra macrophylla Benth. is often a sacred tree to the Ibo and is one of the large trees "credited with furnishing souls for the new-born" (Talbot quoted in Dalziel [10]). Incidentally, in the Kasai Province of the Congo, a decoction of the pod is taken as a remedy for trypansomiasis (20).

SHINGIRA

Cory (21), in a study of the medicines of the Sukuma in Tanganyika (now part of Tanzania), threw a flood of light on the reasons underlying the selection of the ingredients in magical medicines. The secret societies of Sukumaland include ancestor worshippers, soothsayers, snake-charmers, porcupine-hunters and elephant-hunters in addition to the usual practitioners of indigenous medicine and divination.

There are two classes of ingredients: the first represents the *person* who orders the medicine; the second, known as *shingira* (sing. *chingira*), determines the specific objective of the combination, *protective, assertive, creative,* or *aggressive.* It is this second ingredient which endows the medicine with its magic quality, the principles underlying their selection being the "law of similarity," but the "aggressives" are selected by the "law of contact." For the sake of brevity, I shall give only a few examples:

A. Medicines representing the person.
Acacia fischeri Harms has large roots and so represents a mother. *Acacia orfota* Schweinf. is a thorn tree and so represents the abstract idea of protection.

B. Shingira or magical ingredients.
(a) *Protective* shingira, which fall into two categories, *lukago* and *makili.*

(i) The function of *lukago* is to protect the client, his family and property from evil *influences* and from *lions.* For example: any root dug up at a spot where a snake or a lion passed a person without harming him. As these aggressive animals avoided a person, so will evil pass without doing harm.

(ii) The function of *makili* is to protect a child from the evil influence which the adultery of its parents has had on its health. An example is made from a small piece of each of the following: the root of *Combretum parvifolium* Engl. and of *Harrisonia abyssinica* Oliv. treated with semen of the father, secretion from the vagina of the mother, and ghee (clarified buffalo milk butter).

(b) *Assertive* shingira are known as *samba* of which there are three groups (i) general, (ii) love-philtres and (iii) medicines ensuring professional success. Examples are:

(i) Foam of beer brewed in the house, unbeknownst to the neighbors, ensures that profits from a journey will not arouse the envy of neighbors.

(ii) Splinters of the wood of *Pterocarpus chrysotrix* Taub., which is reddish resembling blood which comes from the heart, represents the feeling of love. The inflorescence and pod of this tree have greenish sticky hairs which trap insects and so there is a possible symbolic use to trap the loved one.

(iii) Hunting medicines are mostly of animal origin but the grass *kiroto* represents the fact that the hunter, on sighting game, takes an arrow from his quiver as easily as he pulls up such a grass. A medicine for a *thief* is the root from a plant growing on the grave of a blind man, which means that the thief remains invisible to his victim.

(c) *Creative* shingira are used to produce fertility in man, animal, and crops.

(i) an example of a human aphrodisiac is a splinter of a root which is found growing across a path. It is a hard object which emerges from the grass (pubic hair) at one side and disappears into the grass (the female) at the other.

(ii) an example of a medicine for fertility in domestic animals is a splinter from a tree in which monkeys sleep, the significance being that the monkey is very fertile.

(d) *Aggressive* shingira of black magic are evil in their intent, for example, homicide, the production of sterility, and the spoiling of fields. In addition to many bizarre ingredients, the following are used: grass from the thatch of the victim's house on the "law of contact"; a splinter from a tree felled by a storm produces a rapid, potent action of the medicine; the empty shell of the seed of *Entada abyssinica* Steud. produces sterility in a woman.

In addition to the physical side of magic medicines, Cory (21) records that the Sukuma use both *actions* and *incantations* as shingira. One example of each illustrates their application: a person, on receiving a protective amulet from the practitioner, is asked by him to hold it against his neck and then allow it to fall, saying, "You, my amulet, repel all hostile powers. May any of my enemies who come near me fall down"; a woman seeking a medicine to make her sterile is given a medicine containing as shingira a tree fungus and a splinter from a male pawpaw tree (*Carica papaya* L.). During the process of preparing the medicine, the woman repeats, "I want to be like the male pawpaw tree and like the fungus which does not multiply. May I bear no offspring."

FERTILITY

The importance of fertility is evidenced by the large number of plants and other materials used as aphrodisiacs, as magical stimulants to reproduction, and as charms to aid crop production or to prevent harmful bewitchment of crops.

Roberts (22) gives a neat description of the observance of fertility rites. He points out that "witchcraft forms the normal basis on which Native thought is built up, in much the same way as science is the foundation of life and thought in our 'Western' civilization." The Bantu in South Africa spare no effort "in their endeavour to weave a protective web of witchcraft about the fields and flocks and homes of family and tribe." They believe in a close interrelation between all the reproductive powers of nature.

Thus, there are many fertility rites, often taking place in the open fields where crops are to be grown. In some tribes, as for example in the Northern Transvaal, there is a fertility hut of phallic design called *ntola oa koma.* It contains the *dikowana* of magic drums consecrated by human sacrifice and the store of sacred seed sited beyond the drums. A few grains of this seed are given to each householder at sowing time.

According to Friede (3), *Adansonia digitata* L. is worshipped in West Africa and in the Sudan as a "fertility tree," probably on account of the size of its trunk. Many hundreds of miles further south in the Soutpansberg in the Limpopo Valley, a rock-painting at Mabala Tutwa, the cave of the spotted giraffe, depicts female figures with life-sized "seedpods" of the baobab instead of breasts (Van Riet Lowe quoted in [16]). Evidently in this area the tree has magical significance (3).

In Southern Botswana forked sticks of *Celtis kraussiana* Bernh. are sometimes used to stir meat while it is being cooked. This is said to encourage rapid increase in their domestic herds (8).

Cenia hispida B. & H. and *Rhus discolor* E. Mey. are plants which are burnt in Lesotho to increase crop yield (23). Among the Lenge and Chopi of Portuguese East Africa, the female initiates carry a twig of *Dodonaea viscosa* Jacq. in the "Txuruvula" dance, which seems to be symbolic of spring and possibly of fertility. The shrub is peculiar in that when the new season's flowers appear the fruits of the previous season are still on the bush (24). A Zulu stimulant to production is to sprinkle a watery preparation of *Elephantorrhiza elephantina* Skeels over kaffir-corn fields when they are in blossom (25).

Ficus capensis Thunb. and probably other species of *Ficus* are widely used among the tribes of Africa as a fertility charm to promote conception and the yield of crops (9). The idea probably originated from the abundant clustered fruits. Another mimetic use is dependent upon the presence of a latex in all parts of the tree. The Fulani feed both fruit and leaf to produce an increase in the herd and to increase the yield of milk, and the Zulu administer an infusion of these organs to cows with an insufficient flow of milk (9).

According to Kerharo and Bouquet (18), the Senonfo people of the Ivory Coast sometimes hang the sausage-like fruit of *Kigelia africana* Benth. in the hut as a fertility charm, probably with phallic significance from the length and size of the fruit.

The Sukuma of Tanzania administer the root of *Markhamia hildebrandtii* Sprague to relieve female barrenness (Bally quoted in [16]). In the absence of any known rational cause, the effect must be regarded as of magical significance. In Tanganyika, the bark is chewed for the relief of toothache (7). The Venda of Northern Transvaal administer the powdered bark of *Sclerocarya caffra* Sond. to regulate the sex of children about to be born, that from a male tree for boys and that from a female tree for girls (9).

The Ramah Navaho use the berry of *Solanum triflorum* Nutt. to increase the productivity of watermelon seed. The dried berry is stored until spring, soaked in water and planted with the watermelon seed, because it has lots of berries (26). The leaf and berry of *Solanum villosum* Mill. is soaked in water and put on watermelon seed by the Ramah Navaho to ensure a good crop (26).

The cardiotoxic plant *Strophanthus gerrardi* Stapf. is employed by the Zulu of Natal to ensure good crops (27). The Kgatla of the Transvaal value *Terminalia sericea* Burch. ex DC. as a fertility charm and do not cut down the tree when the crops are growing, as this is believed to bring hail (9).

Bangueria infausta Burch. enters into the cattle fertility rites of the Ndebele, Tswana and Kgakla (19). The wood is not, as a rule, used as a fuel, owing to the belief that, if this were done, cows would bear male calves only. The Culwicks (28) record that Bena women drink an infusion of the leaf of *Ximenia americana* L. as an aid to fertility. The active principles are tannin and resin (29), so there can be no true pharmacological effect. The Kgatla use a decoction of the plant in cattle fertility rites (19).

ANCIENT PERU

According to Vargas (30) plant motifs were freely used in pre-Columbian times in the decoration of various objects, such as vessels of clay and wood, textiles and stones. As a result, the ancient Peruvians placed on record their knowledge of the relationship and uses of the plants in their surroundings as well as their emotional appeal. It is thought that they may even have been influenced in some respects by their pantheistic religion in their use of plants for ornamentation. Garcilaso Inca de la Vega (quoted in [30]) even went the length of writing: "And thus they worship herbs, plants, flowers, trees of all kinds."

I shall restrict myself to mentioning only those plants which have some relationship to witchcraft and folk medicine. Stone objects imitating cultivated products such as the potato (*Solanum tuberosum* L. subsp. *andigenum* Yuz. & Buk.) and maize or sara (*Zea mays* L') are well known in Peru. Such stone objects, known as *conopas* (pl.), as well apparently as clay representations of domestic edible plants (Table 1), were placed in the fields on farms where these were cultivated, so that the crops would be abundant. These *conopas* have been found chiefly in the sierras and are thought to be essentially Andean in origin.

Table 1. Food Plants in Inca and other Pre-Columbian Andean Cultures

Ananas comosus (L.) Merr.	*Lycopersicon esculentum* Mill.
Annona cherimolia Mill.	*Lycopersicon pervianum* Dun.
Annona muricata L.	*Manihot esculenta* Crantz.
Arachis hypogaea L.	*Oxalis tuberosa* Mol.
Canna edulis Ker.	*Pachyrrhizus tuberosus* Spreng.
Capsicum sp.	*Persea americana* Mill.
Carica çandicans Gray	*Phaseolus lunatus* L.
Carica papaya L.	*Phaseolus vulgaris* L.
Chenopodium quinoa willd.	*Solanum muricatum* Ait.

Cucurbita moschata Jacq.	*Solanum tuberosum* L. subsp. *andigenum* Yuz. & Buk.
Inga feuillei DC.	*Tropaeolum tuberosum* R. & P.
Ipomoea batatas Lam.	*Ullucus tuberosus* Lozano
Lucuma bifera Mol.	*Zea mays* L.
Lupinus mutabilis Sweet	

KAVA

Kava is the most commonly used name for a Polynesian beverage made from the root of *Piper methysticum* Forst. There is no fermentation and so no alcohol is present in the beverage. It is, nonetheless, intoxicating on account of its content of water-insoluble pyrones, methysticin, dihydromethysticin and dihydrokawain, which all produce a sedative action (31). Water-soluble fractions from the root have been found also to depress spontaneous motor activity in the mouse (31).

The Fijian name for kava is *yagona* (gs"ng" sound) (32, 33), the Raratonga *kava*, the Samoan *'awa* (34) or *'ava* (32), the Tahitian *ava*, the Manu'a *'ava* (34), the Tongan *kava* (33), the Hawaiian *awa* (32) and Cook Islands *kava* and *'awa* (33). In the New Hebrides, which includes the Sandwich group, *kava* is the name used (35), while it is known by the Marind–Anim people of southwest New Guinea as *wati* (Dam-Bakker et al., quoted in [35]). *Kawakawa* is the Maori name for *Piper excelsum* Forst., not used in making a beverage (32).

Traditionally, kava was a ceremonial beverage reserved for great occasions. Nowadays, it has become not only a social drink but also a convivial one. In many parts, the historical restrictions on the making of kava and on the drinking of it have disappeared. Indeed, kava saloons are found in many urban centers (34), and kava is reported as being prepared in some government offices during the forenoon for the regalement of its employees (33, 34). Holmes (34) gives reports on the extent of kava-drinking in the Pacific area. Samisoni (33) ascribes the degradation from ceremonial use to the modern free availability to the influx of Indian settlers, apparently for the most part Hindu, who did not have the Polynesian regard for the ceremonial nature of the beverage and its preparation.

The home island of Samisoni is Rotuma, which is politically linked with Fiji but is not truly Fijian. The language of Rotuma is not Fijian but an offshoot of Polynesian. On Rotuma, which has not been "invaded" by the Indian, kava has remained very definitely a ceremonial drink and is never drunk as a routine or convivial beverage. Its use is reserved for a great occasion, such as the welcoming of an important visitor or visiting chief, meetings of the Council of Chiefs, the marriage of a daughter or son, and at burial services according to rank, or at the spectacular ceremony of dancing with kava, which is known as *meke yagona* in Fiji only, not Rotuma. The dance is made by one male, lasts fifteen minutes, and is performed only at the welcoming of exceedingly distinguished chiefs.

Kava is made in a special bowl of large size with a minimum of three legs. The minimum size is twelve to eighteen inches, the maximum up to several feet in diameter (Fig. 2). The very large bowls may have many more legs than the smaller sizes (up to ten or twelve). On the other hand authentic much-used kava bowls of large size in the Queensland Museum have only three legs.

The bowl is called *tanoa* and is carved from a solid block of the timber of either *vaivai* (a rain tree) or *dilo* (a very hard wood), but plastic ones are nowadays commercially available. The drinking bowl may be half a coconut shell, especially prepared and carved and finished with a black polish. It may also be carved from a suitable piece of wood. Tanoa are usually reserved for ceremonial occasions but in villages and "commercial joints" any old container suffices.

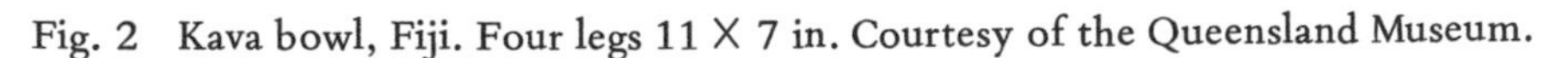

Fig. 2 Kava bowl, Fiji. Four legs 11 × 7 in. Courtesy of the Queensland Museum.

At the marriage ceremony on Rotuma there is a lengthy ceremonial feast, lasting, in olden times, several days and starting when the bride and bridegroom return from the "churching." It commences with a ceremonial announcement of all the riches of the bride's father, who gives the marriage feast. The list begins with *fon kava* (the number of kava bushes growing within the property of the bride's father). This is followed by an enumeration of such items as the number of cattle, the number of pigs, and the number of poultry, and so on. Nowadays, the presentation and cost of the feast is shared by the bridegroom's people who, in that event, also have an announcer, usually an older woman who has to chant.

In Fiji there are witches, always male, who use kava in their ceremonial known as *sova yagona* (sova = to spill). They may be consulted for the purpose of causing ill to some designated person or to assist in attracting a desired damsel. The kava made during the ceremony is not drunk but is thrown on to the ground at its conclusion, which means that it is being offered to the selected spirit of the witch who is one of the latter's familiars. He may have quite a few to carry out his behests. The suppliant gives an offering of sufficient merit to the witch before the latter will undertake the service. In Tikopia, a Polynesian outlyer in Melanesia, kava is "not drunk but is poured on the ground as a libation to the gods" (32).

Holmes (34) has explored the function of kava in modern Samoan culture. In the Manu'a island group of American Samoa, kava is served at all formal and informal meetings of chiefs. The Manu'a name is *'ava,* and it is prepared in the modern Polynesian way by steeping the pulverized root in a prescribed amount of water until a cloudy khaki-colored liquid is produced.

The ancient traditional method of preparation involved the chewing of the *fresh* root by maidens. The masticated material, together with saliva, is ejected into a bowl of a size suitable for the occasion. It contains some water so that the masticated material is diluted somewhat. It seems that lack of local availability of fresh root in some parts and the increasing commercialization of kava production have stimulated the use of the dried root. All reports emphasize that kava made from this is less potent than that made from the fresh root.

I have stated that kava is used in witchcraft ceremonial on Fiji, which makes one suspect a linkage with religious practices. In a detailed study, Holmes (34) confirms this and states that kava-drinking and its attendant ceremonies have a long history and are "intimately related to indigenous religious practices and village social and political organizations. Mythology relates how kava-drinking was given to mortals by the first high chief, *Tagaloa Ui,* and prescribes the form for modern kava ceremonies."

One of the most significant developments of recent times has been the linking of kava with the ritual of the "cargo cult." Lawrence (36), in his book "Road Belong Cargo," points out that the "cargo cult" is of religious significance and that there are recurrent resurgences. In latter times the cult has become linked with antipathy to the missionary and to the European and associated not infrequently with the free use of kava.

DEATH BEWITCHMENT IN NEW GUINEA

According to Stopp (37), in New Guinea all sickness results from a magic spell which requires to be broken to bring about a cure. Only skin infections, tooth troubles, and febrile conditions are recognized and treated as physical diseases. Many plants, however, are used for "the prenatal determination of sex, for chasing away unwanted concubines and for increasing the beauty of the newly born." The "spell of death" means that a person may be killed by supernatural forces which concentrate in certain materials, examples of which are human hair cut into short lengths and a certain type of argillaceous earth. Deaths from old age, miscarriage, external wounds, and accidents are outside the scope of the "spell of death."

A concept similar to this "spell of death" has been recorded among the Bantu of Africa. Another similarity is seen in the New Guinea lack of sharp differentiation between truly medicinal and magical uses of plants.

In New Guinea, it seems that the cures used to avert the "spell of death" are usually drastic purgatives which relieve constipation and colics of the bewitched victim (38). No specific plants are mentioned in this context, but Straatsmans (38) states that seven plants are used as purgatives. Stopp (37) records the following remedies for the "spell of death": *Saccharum sp.* (*bo mongkepa*), which apparently is mildly purgative; *Ficus sp.* (*mbon*), leaf eaten; *Lycopodium serratum* Thunb. (*weipo*), produces vomiting and diarrhea; *Rhododendron macgregoriae* F. Muell. has similar effects and is used with the same objective as *Lycopodium.* Interestingly enough, the leaf of *Alphitonia incana* (Roxb.) T. & B. (*poketa*) is used to wrap a corpse and is considered to be an excellent bearer of the "spell of death." An unidentified plant, known as *nde kumpe ogl bogla borem* (the bark that cuts off the belly) is used to impart the "spell of death."

TOBACCO

Bishop (39) is of the opinion that it is highly probable that tobacco smoking originated in America as a religious rite. In his own words: "The burning of incense and spices has had a place in worship from time immemorial and there is little doubt that the custom of smoking developed from the inhalation of the fumes from tobacco burnt as an offering to the Great Spirit. Tobacco was regarded by the aboriginal peoples of America as a sacred plant and as the special gift of the gods to man and, when the plant reached the Old World, this attitude was adopted by Europeans along with the Indians' ideas concerning the efficacy of tobacco as a remedy against almost all bodily ills." Space does not permit me to delve into the fascinating history of this subject, which has been fully reviewed previously (6, 16, 18, 39-42).

By the end of the sixteenth century, the custom of smoking was known throughout France and the Iberian peninsula, whence it is supposed to have spread into Eastern Europe, Egypt and India. The introduction to the East may have been more direct, for the Arabs are thought perhaps to have been on the American continent before Columbus (43).

The use of tobacco is widespread throughout the continent of Africa, both as a social habit and for magical and ceremonial use. Laidler (42) maintains that the smoking of *dagga* (marijuana) and of tobacco was unknown to the aboriginals of Southern Africa until the early settlers from Holland introduced the pipe and tobacco in the seventeenth century (Fig. 3). Be that as it may, the Paris Mission at Morija in Lesotho records that tobacco was one of the few plants cultivated by the Southern Sotho when their missionaries arrived in Lesotho. Their opinion is that tobacco was introduced probably from the Portuguese of Mozambique. Hemp was known, apparently, before tobacco. Both *Nicotiana tabacum* L. and *N. rustica* L. are cultivated and smoked in Africa.

In New Guinea, there is only one definitely identified plant employed for smoking and this is *Nicotiana tabacum* L., but it is known that other unidentified plants are also used (38). In India *Nicotiana* is freely used.

Early explorers of South and Central America have reported the chewing of the leaf of tobacco and of other plants: *coca* in Venezuela and Colombia and *yaat* in Nicaragua (39). *Lobelia inflata* L. was the tobacco

Fig. 3 A Mpondo pipe, Pondoland, South Africa. After an etching by an unknown artist.

smoked by the North American Indian in earlier times. *Nicotiana attenuata* Torr., mountain tobacco, is smoked by the Ramah Navaho in times of shortage of commercial tobaccos (26).

Australian "native tobacco" is derived from *Nicotiana suaveolens* sens. lat. and has a pharmacological action similar to that of tobacco, *N. tabacum* L., and *pituri, Duboisia hopwoodii* F. Muell. (44). Both *N. suaveolens* and *Duboisia hopwoodii* were chewed by the aborigine for their narcotic effects. There is no evidence that they smoked tobacco before the advent of the European, and they have apparently not taken to snuffing or dipping. The Queensland Museum has an aborigine tobacco pipe from Cape York, North Queensland, which I am assured shows New Guinea influence (Fig. 4).

Fig. 4 Australian aborigine tobacco pipe, Cape York, North Queensland. Courtesy of the Queensland Museum.

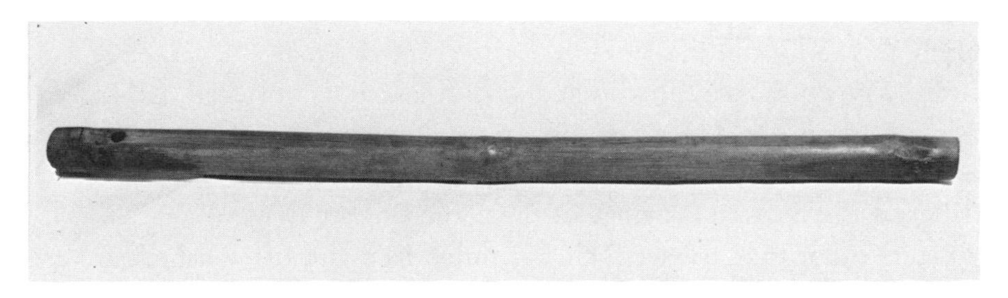

In Ghana, the dried older leaf of *Carica papaya* L. is smoked like tobacco (14). In New Caledonia this plant is used as a tobacco substitute, and in Mauritius it is smoked for the relief of asthma (45). In Eastern Sudan, the Arab has the curious custom of chewing the seed of *Adenopus breviflorus* Benth. while smoking tobacco in order to produce a kind of intoxication (10).

SNUFF

When one recollects the history of tobacco and its ceremonial uses, one is not surprised to find that tobacco snuff has significant magical applications in the practice of divination and the countering of witchcraft. Schultes (46) states that the snuffing of plant materials, especially for hallucinogenic effects, seems to be peculiarly New World.

The primary ingredient in many African snuffs is tobacco, which, in older times, was usually home-grown but nowadays may be obtained through trade channels. Two kinds are known, *Nicotiana tabacum* L. and *N. rustica* L., both commercial sources, but home-grown leaf partic-

Fig. 5 Xhosa snuff box made from ox horn with a leather bung. Height 4 in. Eastern Province, Cape, South Africa. In the author's collection.

ularly of the former in common in the country kraals. *N. glauca* R. Grah., known as *tree tobacco* and *wild tobacco,* is not used. All three originate from the Americas (39). In my boyhood, snuffing and dipping, as well as smoking, were universal among the Xhosa and the Mfengu tribes of the Eastern Province of the Cape Colony (Fig. 5).

Plants other than tobacco, always pulverized and often ashed, are used as an admixture to the finely powdered tobacco or as a substitute for it. Some of these additives are highly prized. In my experience, by far the commonest is the leaf of some species of aloe. *Agave americana* L., *Aloe aristata* Haw., *A. ferox* Mill., *A. humilis* (L.) Mill., *A. kraussii* Bak., *A. latifolia* Haw. [invalid, prob. *A. saponaria* (Ait.) Haw.], *A. marlothii* Berger, and *A. saponaria* Haw. var. *ficksburgensis* Reynolds are all used in Lesotho, except *A. marlothii* Berger, which is employed in Natal Swaziland and Zululand. *Agave americana* L. is also valued in Natal. In the absence of aloe, the incinerated dried cob of *Zea mays* L. or of the dried stalk of *Sorghum* may be used in Swaziland and Lesotho respectively.

Campbell and Cooper (47) record that some of the snuffs of the African consist of calcined aloe leaf only. This is particularly the case in the Transvaal (except for the Venda, who add tobacco) and among the Zulu. I have found the Zulu, especially in the hinterland of the Natal South Coast as far South as the Umzimvuna River, mix tobacco and *Aloe marlothii* in making the snuff.

Amarantus caudatus L. is in general use in Southern Africa as a snuff plant, while *Amarantus spinosus* L. is used in Lesotho, Swaziland, and the Transvaal (Venda) in the proportion of 50 percent with tobacco. *Artemisia afra* Jacq. is used in Lesotho.

In West Tropical Africa, the pulverized shell and peduncle of the fruit of *Adansonia digitata* L. and the pulverized fruit of *Xylopia aethiopica* A. Rich. are valued as an additive to snuff to increase its pungency (10).

In the Transvaal, the Lobedu use *Cussonia spicata* Thunb. as a snuff plant and in Mokeetsi (Transvaal) *Senecio longiflorus* Sch. Bip. is used for a like purpose. In Lesotho, the following are all employed: *Rhus erosa* Thunb., *Salvia* sp., *Senecio coronatus* Harv. and *Tagetes minuts* L. *Asclepias decipiens* N.E. Br. is used alone by the Southern Sotho of Lesotho as a snuff (23).

The pod of *Pentaclethra macrophylla* Benth. is burned to produce ash used in native soap-making and indigo-dyeing. The residue, after extraction of the potash from the ash, is pulverized and mixed with tobacco snuff (10). On the Ivory Coast, the powdered leaf of *Cola caricaefolia* (G. Don) K. Schum. is mixed with powdered tobacco to make a snuff (18).

There are so many sternutatory snuffs in use in Africa that it is impossible to mention more than one or two of the more usual ones. Such snuffs are employed primarily for the relief of headache, but I suspect that often there may be a touch of magic behind their use. I have in mind the old saying "God bless you" when a person sneezes. Here are two interesting examples. The powdered bark of *Erythrophloeum guineense* G. Don is a Zulu sternutatory, and quantities are sold in Durban for this purpose. The tree is well known for its use for *trial by ordeal*, and in Northern Nigeria the Jukun value the bark as a charm against witchcraft (9). The Xhosa relieve headache by sneezing as a result of snuffing the powdered wood of *Ptaeroxylon obliquum* Radlk (9). It is a violent sternutatory, and its common name is sneezewood.

A further group of snuffs of a very different caliber are those which are taken in order to produce narcosis, a hallucinatory effect, or a psychomimetic action. These seem to be peculiar to the New World, especially South America. Divination and oracular seances seem to be devoid of such aids in Africa, where the witch-doctor may certainly work himself into a hysterical frenzy but seems not to rely on drug assistance.

The following data, except where otherwise indicated, are taken from Schultes (46) and Schultes and Holmstedt (48), who point out that the most important snuff in pre-Conquest America was made from tobacco. *Nicotiana tabacum* L., "from which comes most of the tobacco that is smoked, snuffed and chewed at the present time, was likewise the source of most of the narcotic in pre-Conquest South America, Middle America and the West Indies." *N. rustica* L., a native of North America, "was smoked and probably snuffed by the Indians of Mexico and North America before the arrival of the European." Curiously enough *N. tabacum* L. was introduced from the Old World to North America long after the Conquest.

The bean of *Anadenanthera peregrina* (L.) is the source of the snuff *yopo* or *niopo* of the Orinoco basin of Venezuela and Colombia (46, 49, 50). The West Indian *cohoba* snuff is identical with it (Safford cited in [46]). Schultes (46) draws attention to the report by Lévi-Strauss that, amongst the tribes of the Guaporé River in Amazonian Brazil, tobacco snuff was mixed with crushed *angico* leaves from leguminous trees, especially *Anadenanthera.* He also reports that he has observed the addition of other plant material (other than ashes) on two occasions only, namely, that the Witotos of the Rio Igaraparaná and the Yukunas of the Rio Miritiparaná of Colombia added powdered coca (*Erythroxylon coca* Lam.). Other additions have not the potent action of coca, but the ash of *Theobroma* bark and *Cecropia* leaf are added, possibly, to prevent the absorption of moisture by the snuff in the very wet atmosphere of those parts.

Another hallucinogenic snuff, known as *yá-kee, yá-to,* or *pa-ree-ká* in the far northwest Amazon in Colombia, is made from the resin which oozes from the bark of *Virola calophylla* Warb., *V. calophylloidea* Markgr. and possibly *V. elongata* (Bth.) Warb (48). *Virola calophylloidea* is the source of the *epená* snuff of the Waika Indians of the Rio Negro basin of Brazil. The hallucinogenic effect is due to tryptamine derivatives. Virola snuff is consumed only by the medicine-men.

DIVINATION

I am not sure that I am able to give a completely satisfactory short definition of *divination* which would embrace all aspects or types of it. The *Oxford English Dictionary* defines it as "the foretelling of future events or the discovery of what is hidden or obscure by supernatural or magical means." There is more to it than this, however, in the continent of Africa and, I suspect, elsewhere in the world (Fig. 6). In Africa, certainly, it is closely linked with the magical aspect of the indigenous prac-

Fig. 6 A Kgatla divination session. *Left:* Kempe Sekale of Mathubudukwane. *Right:* Ramokobelo Molefi of Mahalane. Photograph by Dr. G. Teichler, at Mathubudukwane, Botswana, near Derdepoort, Transvaal.

tice of medicine, so that many herbalists are diviners as well. In addition, the diviner also occupies an important position in the social structure of the tribe. Anderson (51), speaking of Kordofan, says that "the borderline between purely medicinal and general superstition is absolutely indefinite and both are so intimately blended with religious rite that it is impossible to touch on one without encroaching on the other." These people lived in dread of "ever present evil, seen and unseen, emanating from man, ghost and devil."

The objectives of divination are: the foretelling of future events; the discovery of what is hidden or obscure; the tracing of an errant wife or her lover; and the finding of lost cattle—all by supernatural or magical means. The most serious aspect of this divination, and the one pregnant with serious consequences, is the "smelling out" of the person supposedly responsible for a death; it is also used with other objectives.

The techniques of divination vary considerably as can be gauged by the following examples: (a) by gazing into a dish containing water and gaining thereby inspiration, practiced by some Southern African diviners; (b) the rubbing board oracle (Iwa) used among the Azande and Nyam-Nyam of the Soudan; (c) by revelation in a dream or a trance; (d) by intoxication or under the influence of a drug; (e) by the throwing of divinatory dice, often referred to as "the bones." The bone used is invariably the talus but substitution is common. There is always at least a pair of each "bone," a right and a left, and these represent the totem animal of the tribe or clan to which the person belongs; (f) by divinatory bowls.

To the diviner who uses dice, the preservation of the integrity and perspicacity of these divinatory dice is of great moment. On opening them out for a seance, he invariably flicks some snuff over them. In addition, this treatment is accorded the dice at intervals between consultations. The undermentioned may be so used by diviners, particularly to protect their dice from evil influences projected by witchcraft (Fig. 7).

In Lesotho, *Haplocarpha scaposa* Harv. (23) and *Ilex mitis* Radek. (16) are associated with the use of the divinatory dice, the latter being especially used to counter witchcraft. The Kgatla dice-thrower uses the root of *Mundulea sericea* A. Chev. as a charm to ensure good divination. The plant is highly toxic.

The Mano of Liberia blow tobacco snuff up the nose for convulsions

Fig. 7 Pungashe, a Shangana diviner of Robollo Kap, North Transvaal. "Making medicine" preparatory to "throwing the bones" for the author.

or coma. If the sufferer responds by sneezing, there is still hope of recovery (6).

Divination in one form or another is one of the oldest characteristics of early religion, but its association with bowls in particular is of great interest (52). These are of three kinds: (i) the bowl used in the oracle of a god—for example, among the Yoruba of Nigeria; (ii) the divining bowl used for the detection of a witch or wizard who has caused a death. These are quite complicated and are found among the Venda of the Northern Transvaal and the Karanga of Rhodesia. The Yao of Malawi apparently have divining "bowls," but any suitable container is used. (iii) The bowl used for purely magical purposes: for example in Calabar, West Africa.

According to Davis (52), the same principle underlies the use of all three; namely, that the movements on the surface of the fluid are determined by some spirit which controls the instrument.

The divining bowl of the Venda, according to Ferreira (9), is made of the wood of *Sclerocarya caffra* Sond., and so are their drums. The bowl, about 12 to 18 inches in diameter, is carved from a solid piece of wood to a depth of one and a half to two inches with a broad flat border. The border and the bed of the bowl are decorated with designs symbolizing phases of the religious and magical ideas of the Venda as well as the sibs of the tribal group. In the center of the bowl is a small protuberance with a large cowrie shell embedded in it. This is the *mukhombo* (umbilicus), representing the mother's spirits.

To divine the murderer, "the bowl is filled with water in which from four to six pips are floated. By watching the movements of these pips and noticing the symbols at which they touch as they float around the bowl, the diviner is able to reconstruct the motives and history of the crime and to discover the sib to which the guilty man belongs."

This use of the wood of *Sclerocarya caffra* is interesting, as *morula* beer, made from the fruit, plays an important part in the "first fruits" and other ceremonies and rites of the Thonga tribes. The powdered bark is also mixed with water and administered to regulate the sex of children to be born, bark of the male tree for boys and the bark of the female tree for girls (9).

The leaf of *Salvia divinorum* Epling & Jativa is used by the Mazatec Indians as a divinatory or psychotomimetic agent. The native name is *hojas de la Pastora,* and it may be identical with *pipilzintzintli,* a divine plant used as an hallucinogen in pre-Conquest Mexico. Apparently its use as an hallucinogen is limited to times when psilocybin-containing mushrooms are unavailable (53).

The botanical identity of Soma-Homa, the divine plant of the ancient Aryans in the Indus Valley, remains a mystery. It may be *Amanita, muscaria* (Linn. ex Fr.) S. F. Gray. The juice of this plant was used to prepare an intoxicating beverage, used as a stimulant and aphrodisiac. It even made the gods dance with joy and was used with reverence on all religious and ceremonial occasions (53).

The indigenous inhabitants of French Equatorial Africa chew the root of *Tabernanthe iboga* Baill. to produce an epileptic-like madness during which prophetic words are said to be uttered (53). Ibogaine, one of the twelve alkaloids known to occur in the plant, has a central CNS stimulant action with weak but definite anticonvulsive properties.

According to Tyler (53), in the highlands of Mexico the name *sinicuichi* is applied to *Heimia salicifolia* Link & Otto and to an intoxicating beverage made from its leaf. This drink produces a mild intoxication characterized by slight giddiness, apparent darkening of the surroundings, deafness and withdrawal from reality—a state of pleasant drowsiness. Peculiarly enough, habitués of the beverage are able to remember with great exactitude happenings and experiences of their remote past and even claim to recall events experienced by their ancestors before the birth of the individual.

CONCLUSION

I am aware that I have not even touched upon several aspects of my topic. This is not due to lack of interest but to lack of space. Fortunately, one of them was the topic of a symposium held at the University of California in 1967, the proceedings of which were published under the title *Ethnopharmacologic Search for Psychoactive Drugs* (54). The subjects covered include nutmeg (*Myristica fragrans* Houtt); tryptamine derivatives; *Banisteriopsis; Peganum harmala* L.; and *Amanita muscaria* (Linn. ex Fr.) S. F. Gray.

In the case of *Datura,* I refer you to Safford's studies published in the Smithsonian Report for 1920 (55), Gustav Schenk's "The Book of Poisons and the Use of Datura in School Initiations," reported by Watt and Breyer-Brandwijk (16). Then there are such procedures as rain-making and witch-finding, in which plants do not necessarily play a part. Magic and charms are also involved in "doctoring" the fighting man, the production of invisibility, neutralizing the memory, and in initiation ceremonies.

Appendix

Plants Mentioned in the Text

BOTANICAL NAME	COMMON NAME
Acacia fischeri Harms	Mgushi (Sukuma)
Acacia orfota Schweinf.	Msese (Sukuma)
Acokanthera venenata G. Don	Boesmangif
Adansonia digitata L.	Baobab
Adenopus breviflorus Benth.	—
Afzelia africana Smith	Afzelia Faza dega Faza masa
Agave americana L.	Agave (in S. Africa)
Aloe aristata Haw.	Serelei (So. Sotho)
Aloe ferox Mill.	Bitter aloe
Aloe humilis (L.) Mill.	—
Aloe kraussii Bak.	Lekhalana (So. Sotho)
Aloe latifolia Haw. [invalid, prob. *A. saponaria* (Ait.) Haw.]	—
Aloe marlothii Berger	Umhlaba (Zulu)
Aloe saponaria (Ait.) Haw.	Soap aloe
Aloe saponaria Haw. var. *ficksburgensis* Reynolds	—
Alphitonia incana (Roxb.) T. & B.	Poketa (N. Guinea)
Alstonia boonei De Wild.	Nyame dua (Ashanti)
Alstonia congensis Engl. = *Alstonia boonei* De Wild.	—
Amanita muscaria (Linn. ex Fr.) S. F. Gray	Fly agaric
Amarantus caudatus L.	Love lies bleeding (red variety)
Amarantus spinosus L.	Prickly amaranth
Anadenanthera colubrina (Vell.) Brenan	—
Anadenanthera colubrina (Vell.) Brenan var. *cebil* (Griseb.) Altschul	—
Anadenanthera peregrina (L.) Spreg.	—
Ananas comosus (L.) Merr.	Pineapple
Annona cherimolia Mill.	—
Annona chrysophylla Boj.	Wild custard apple
Annona muricata L.	Soursop
Annona senegalensis Pers.	Custard apple
Arachis hypogaea L.	Ground-nut

BOTANICAL NAME	COMMON NAME
Artemisia afra Jacq.	Wilde als
Asclepias decipiens N.E.Br.	–
Banisteriopsis sp.	–
Baphia nitida Lodd.	–
Blighia sapida Konig	Akee
Bombax sp.	Bombax
Borassus flabellifer L. var. *aethiopum* Warb.	Palmyra palm
Boscia albitrunca Gilg & Benedict	Caperbush
Calendula officinalis L.	Marigold
Calotropis procera Ait. f.	–
Canna edulis Ker.	–
Cannabis sativa L.	Marijuana
Capsicum sp.	Capsicum
Carica candicans Gray	–
Carica papaya L.	Pawpaw
Cassia sieberiana DC.	Mbaraka (Swahili)
Cecropia sp.	–
Ceiba pentandra (L.) Gaertn.	Kapok tree
Celtis africana Burm. f.	Camdeboo
Celtis kraussiana Bernh. = *Celtis africana* Burm. f.	–
Cenia hispida S. & H.	Mokubetso (S. Sotho)
Chenopodium quinoa Willd.	–
Chlorophora excelsa Benth. & Hook. f.	Iroko
Citrus aurantium L.	Orange blossom
Cola acuminata R.Br.	Kola
Cola acuminata Schott. & Endl.	Kola
Cola caricaefolia (G. Don) K. Schum.	Kola (not true Kola)
Cola vera K. Schum.	Kola
Combretum parvifolium Engl.	–
Combretum primigenum Marl. & Engl.	Mother-tree of the Damara
Combretum suluense Engl. & Diels	Zulu bush-willow
Conium maculatum L.	Hemlock
Copaifera religiosa J. Léonard	–
Costus afer Ker.	Bush Cane (Sierra Leone)
Cucurbita moschata Jacq.	–
Cussonia spicata Thunb.	Cabbage tree
Datura sp.	Datura
Dodonaea viscosa Jacq.	Tsekatseki (Chopi and Lenge)
Duboisia hopwoodii F. Muell.	Pituri

BOTANICAL NAME	COMMON NAME
Elaeis guineensis Jacq. var. *idolatrica* A. Chev.	African oil palm
Elephantorrhiza elephantina Skeels	Elandsboontjie
Encephalartos barteri Carruth	Ghost or Hosanna palm
Entada abyssinica Steud.	–
Erythrophloeum guineense G. Don	Ordeal tree, Red water tree Sassy bark
Erythroxylum coca Lam.	Coca
Fraxinus excelsior L.	Ash tree
Ficus sp.	Fig Mbon (N. Guinea)
Ficus capensis Thunb.	Cape wild fig
Haplocarpha scaposa Harv.	–
Harrisonia abyssinica Oliv.	Mkussu (Sukuma)
Heimia salicifolia Link & Otto	–
Hypoxis villosa L.f.	Inkbol
Ilex mitis Radlk.	Cape Holly
Inga feuillei DC.	–
Ipomoea batatas Lam.	Sweet potato Yam
Jatropha curcas L.	Purging or Physic nut tree
Khaya senegalensis A. Juss.	African Mahogany
Kigelia africana Benth.	Sausage tree
Kigelia pinnata DC.	Sausage tree
Laurus nobilis L.	Bay
Lobelia inflata L.	Indian tobacco (N. Amer.)
Loesnera kalantha Harms	–
Lucuma bifera Mol.	–
Lupinus mutabilis Sweet	–
Lycopersicon esculentum Mill.	Tomato
Lycopersicon pervianum Dun.	–
Lycopodium serratum Thunb.	Weipo (N. Guinea)
Manihot esculenta Crantz. = *Manihot utilissima* Pohl	–
Markhamia hildebrandtii Sprague	–
Mundulea sericea A. Chev.	Fish poison bush
Myristica fragrans Houtt.	Nutmeg
Nicotiana attenuata Torr.	Mountain tobacco

BOTANICAL NAME	COMMON NAME
Nicotiana rustica L.	Tobacco
Nicotiana suaveolens sens. lat.	Australian "native" tobacco
Nicotiana tabacum L.	Tobacco
Oxalis tuberosa Mol.	–
Pachyrrhizus tuberosus Spreng.	–
Peganum harmala L.	Harmala
Pentaclethra macrophylla Benth.	Oil bean tree Atta bean
Persea americana Mill.	Avocado pear
Phaseolus lunatus L.	Lima bean
Phaseolus vulgaris L.	Kidney bean
Piper methysticum Forst.	Kava
Piper excelsum Forst.	Kawakawa
Primula vulgaris Huds.	Primrose
Ptaeroxylon obliquum Radlk.	Sneezewood
Pterocarpus chrysotrix Taub.	–
Rhododendron macgregoriae F. Muell.	–
Rhus discolor E. Mey.	Mohlohlwane (Sotho)
Rhus erosa Thunb.	Besembos
Rosa sp.	Rose
Rosmarinus officinalis L.	Rosemary
Saccharum sp.	Bo mongkepa (N. Guinea)
Salvia divinorum Epling & Jativa	–
Salvia sp.	Salvia
Sclerocarya caffra Sond.	Marula
Senecio coronatus Harv.	–
Senecio gabonensis Oliv. & Hiern	–
Senecio longiflorus Sch.-Bip.	–
Solanum muricatum Ait.	–
Solanum nigrum L.	Black nightshade
Solanum triflorum Nutt.	Cut-leafed nightshade
Solanum tuberosum L. subsp. *andigenum* Yuz. & Buk.	Potato
Solanum villosum Mill. = *Solanum nigrum* L.	–
Sorghum dochna Snowden	Kaffircorn
Sorghum sp.	Millet
Strophanthus gerrardi Staph.	–
Syzygium cumini Skeels	Jambolan
Tabernanthe iboga Baill.	Iboga
Tagetes minuta L.	Blackjack (S. Africa) Cobbler's pegs (Australia)

BOTANICAL NAME	COMMON NAME
Taxus baccata L.	Yew
Terminalia sericea Burch. ex DC.	Assegai wood
Theobroma sp.	Cacao
Triticum aestivum L.	Wheat
Tropaeolum tuberosum R. & P.	–
Ullucus tuberosus Lozano	Nasturtium
Urtica dioica L., poss. *U. pilulifera* L.	Nettle Roman nettle
Vangueria infausta Burch.	Wild medlar
Viola odorata L.	Violet
Virola calophylla Warb.	–
Virola calophylloidea Markgr.	–
Virola elongata (Spr. ex Bth.) Warb.	–
Vitis vinifera L.	Grape
Ximenia americana L.	Wild olive Tallow nut
Xylopia aethiopica A. Rich.	–
Zea mays L.	Maize Sara

References

1. Katherine L. Oldmeadow, *The Folklore of Herbs* (Birmingham, Cornish Bros., 1946).

2. Ailsa Garland, "The Bride and her Bouquet" (Editor of *Vogue* speaking in a B.B.C. broadcast). Reported in *The Outpost: Magazine of the British South African Police 41* (1963), 41.

3. H. M. Friede, "Sacred trees of the Bantu," *Trees S. Afr.* (1953), 4, 6, 12.

4. Clyde Kluckhohn, *Navaho Witchcraft* (Boston, Mass., Beacon Press, 1967).

5. Edward L. Margetts, "Traditional Yoruba healers in Nigeria," *Man 102* (1965), 115-118.

6. George Way Harley, *Native African Medicine with Special Reference to Its Practice in the Mano Tribe of Liberia* (Cambridge, Mass., Harvard University Press, 1941).

7. J. P. M. Brenan and Peter J. Greenway, *Check Lists of the Forest Trees and Shrubs of the British Empire,* No. 5, Tanganyika Territory, pt. II (Oxford, Imperial Forestry Institute, 1949).

8. R. S. Rattray, *Ashanti* (Oxford, Clarendon Press, 1923).

9. F. H. Ferreira, *The Trees and Shrubs of South Africa,* pts. 1 and 2 (Roneo, 1952).

10. J. M. Dalziel, *The Useful Plants of West Tropical Africa* (London, Crown Agents for the Colonies, 1937).

11. Rudolph Marloth, *The Flora of South Africa* (London, Wesley, 1913-1932).

12. Thomas Baines, *Explorations through South West Africa* (Ridgewood, N.J., Gregg Press Inc., 1864).

13. Julia F. Morton, "The Jambolan (*Syzygium cumini* Skeels)" *Proc. Florida State Hortic. Soc. Miami 76* (1963), 328-338.

14. Frederick Robert Irvine, *Woody Plants of China with Special Reference to Their Uses* (London, Oxford University Press, 1961).

15. Robert H. Dreisbach, *Handbook of Poisoning: Diagnosis and Treatment,* 5th ed. (Los Altos, Calif., Lange Medical Publications, 1966).

16. John Mitchell Watt and Maria Gerdina Breyer-Brandwijk, *The Medicinal and Poisonous Plants of Southern and Eastern Africa,* 2nd ed. (Edinburgh, E. & S. Livingstone, 1962).

17. André R. Walker, "Usages pharmaceutiques des plantes spontanées du Gabon," *Bull. Inst. Étud. Centrafr.* (n.s.) *4* (1952), 181-186; *5* (1953), 19-40; *6* (1953), 275-329.

18. J. Kerharo and A. Bouquet, *Plantes médicinales et toxiques de la Côte-d'Ivoire-Haute Volta* (Paris, Vigot Freres, 1950).

19. John Mitchell Watt, "Magic and trees," *Trees S. Afr. 8* (1956), 3-15.

20. P. Staner and R. Boutique, "Materiaux pour l'étude des plantes médicinales indigènes du Congo Belge," *Mem. Inst. Colon. Belge; Sect. Sci. Natur et Méd. 5*(6). (1937), 1-228.

21. H. Cory, "The ingredients of magic medicines," *Africa 19* (1949), 13-32.

22. Noel Roberts, "Native education from an economic point of view," *S. Afr. J. Sci. 14*(2) (1917), 88-100.

23. E. P. Phillips, "The botany of the Leribe plateau," *Ann. S. Afr. Mus. 16* (1917).

24. E. Dora Earthy, "Valenge medicines," *Inf. Bull. Afr.* no. 2, 8 (1932).

25. Fr. Mayr, "The Zulu kafirs of Natal," *Anthropology 2* (1907), 392-399.

26. Paul A. Vestal, "Ethnobotany of the Ramah Navaho," *Peabody Mus. Amer. Archeol. Ethnol., Harvard University 40* (1952), no. 4.

27. Jakob Gerstner, "The arrow poison *Strophanthus* in Southern Africa," Letter in *S. Afr. Med. J.* (1949), 390.

28. A. T. Culwick and G. M. Culwick, *Ubena of the Rivers* (London, Allen & Unwin, 1935).

29. Thomas S. Githens, *Drug Plants of Africa* (Philadelphia, University of Pennsylvania Press, 1949), African Handbook, 8.

30. César Vargas, "Phytomorphic representations of the ancient Peruvians," *Econ. Bot. 16* (1962), 106-115.

31. Joseph P. Buckley, Anglo R. Furgiuele, and Maureen J. O'Hara, "Pharmacology of Kava," in D. Efron, Bo Holmstedt, and N. S. Kline, eds. *Ethnopharmacologic Search for Psychoactive Drugs* (Washington, D.C., Public Health Serv. Publ. No. 1645, 1967).

32. Clellan S. Ford, "Ethnographical aspects of Kava," in D. Efron, ed., *Ethnopharmacologic Search for Psychoactive Drugs* (Washington, D.C., Public Health Serv. Publ. No. 1645, 1967).

33. Jim Samisoni, Personal communication to the author, 1968.

34. Lowell D. Holmes, "The function of Kava in modern Samoan culture," in D. Efron, ed., *Ethnopharmacologic Search for Psychoactive Drugs* (Washington, D.C., Public Health Serv. Publ. No. 1645, 1967).

35. Carleton D. Gajdusek, "Recent observations on the use of Kava in the New Hebrides," in D. Efron, ed., *Ethnopharmacologic Search for Psychoactive Drugs* (Washington, D.C., Public Health Serv. Publ. No. 1645, 1967).

36. Peter Lawrence, *Road Belong Cargo* (Manchester, Eng., Manchester University Press, 1964).

37. Klaus Stopp, "Medicinal plants of the Mt. Hagen people (Mbowamb) in New Guinea," *Econ. Bot. 17* (1963), 16-22.

38. W. Straatmans, "Ethnobotany of New Guinea in its ecological perspective," *J. Agric. Trop. Bot. Appl. 14* (1-2) (1967), 1-20.

39. W. J. Bishop, "Some early literature on addiction, with special reference to tobacco," *Brit. J. Addiction 46* (1949), 49-65.

40. Hewitt Grenville Fletcher Jr., "The history of nicotine," *J. Chem. Educ. 18* (1941), 303-308.

41. John Mitchell Watt, "The tobacco habit," *S. Afr. J. Clin. Sci. 4*(2) (1953), 94-105.

42. Percy Ward Laidler, "Pipes and smoking in South Africa," *Trans. Roy. Soc. S. Afr. 26* (1938), 1-23.

43. M. D. W. Jeffreys, Personal communication to the author.

44. Leonard James Webb, "Guide to the medicinal and poisonous plants of Queensland," *Bull. C.S.I.R. Australia* (1948), 232.

45. Peter J. Greenway, "*Carica papaya* L.," *E. Afr. Agric. J. 13* (1947), 8, 98, 228.

46. Richard Evans Schultes, "The botanical origins of South American snuffs," in D. Efron, ed., *Ethnopharmacologic Search for Psychoactive Drugs* (Washington, D.C., Public Health Serv. Publ. No. 1645, 1967).

47. J. M. Campbell and R. L. Cooper, "The presence of 3:4-benzpyrene in snuff associated with a high incidence of cancer," *Chemy Ind.* (1955), 64-65.

48. Richard Evans Schultes and Bo Holmstedt, "The vegetal ingredients of the myristicaceous snuffs of the Northwestern Amazon," *Rhodora 70* (1968), 113-160.

49. Siri von Reis Altschul, "A taxonomic study of the genus *Anadenanthera*," *Contr. Gray Herb. 193* (1964), 3-65, and unpubl. thesis, Radcliffe College, Cambridge, Mass., 1961.

50. *Ibid.*

51. R. G. Anderson, "Medical practices and superstitions amongst the people of Kordofan," *Rep. Wellcome Trop. Res. Labs. 3* (1908), 280-322.

52. S. Davis, "Divining bowls: their use and origin," *Man 55* (1955), 132-135.

53. Varro E. Tyler, Jr., "The physiological properties and chemical constituents of some habit-forming plants," *Lloydia 29* (1966), 275-292.

54. D. Efron, ed., *Ethnopharmacological Search for Psychoactive Drugs* (Washington, D.C., Public Health Serv. Publ. No. 1645, 1967).

55. William E. Safford, "Daturas of the Old World and New: an account of their narcotic properties and their use in oracular and initiatory ceremonies," *Smithsonian Report* (1920), 537-567; Publ. 2644 (1922).

The Future of Plants as Sources of New Biodynamic Compounds

RICHARD EVANS SCHULTES

Professor of Biology; Director and Curator of
Economic Botany, Botanical Museum of Harvard University

Nunc vos potentes omnes herbas deprecor. exoro maiestatem vestram, quas parens tellus generavit et cunctis dono dedit: medicinam sanitatis in vos contulit maiestatemque, ut omni generi identidem humano sitis auxilium utilissimum.

Precatio Omnium Herbarum

INTRODUCTION

The relationship between man and plants has been very close throughout the development of human cultures. Through most of man's history, *botany* and *medicine* were, for all practical purposes, synonymous fields of knowledge, and the shaman, or witch-doctor—usually an accomplished botanist—represents probably the oldest professional man in the evolution of human culture.

At no time in the development of mankind, however, has there been more rapid and more deeply meaningful progress in our understanding of plants and their chemical constituents than during the past quarter century. And this is curious, especially in view of the somewhat earlier deprecation in pharmaceutical chemistry of any emphasis on plants. The gradual sophistication of phytochemistry in the last half of the nineteenth century and the exaggeration of hope for specific remedies from vegetal sources for any and all ills set up a counter-current, a tendency to disparage any data concerning the potential value of physiologically active plants. The importance and exclusiveness of synthetic chemistry was exalted, and its potentialities were held to be so great that the Plant Kingdom could be sloughed off without ceremony. The "Coal Tar Age" was assured in therapeutics, wherefore there would be no need of harking back to remnants or even hints from earlier ages that counted on natural sources for their medicinal and other products. DeRopp has expressed it well, in discussing the delay until 1947 for the discovery by western science of such an ancient remedy as *Rauwolfia:*

> This situation results, in part at least, from the rather contemptuous attitude which certain chemists and pharmacologists in the west have developed toward both folk remedies and drugs of plant origin . . . They further fell into the error of supposing that because they had learned the trick of synthesizing certain substances, they were better chemists than Mother Nature who, besides creating compounds too numerous to mention, also synthesized the aforesaid chemists and pharmacologists. Needless to say, the more enlightened members of these professions have avoided so crude an error, realizing that the humblest bacterium can synthesize, in the course of its brief existence, more organic compounds than can all the world's chemists combined.

Then, as we can all recall, the discovery, almost within a decade, of a series of so-called "Wonder Drugs," nearly all from vegetal sources, sparked a revolution. It crystallized the realization that the Plant Kingdom represents a virtually untapped reservoir of new chemical compounds, many extraordinarily biodynamic, some providing novel bases on which the synthetic chemist may build even more interesting structures. The startlingly effective drugs that have come from this decade or two of discovery are scattered throughout the Plant Kingdom. They range from muscle relaxants from South American arrow poisons, antibiotics from moulds, actinomycetes, bacteria, lichens and other plants; rutin from a number of species; cortisone precursors from sapogenins of several plants, especially from *Strophanthus* and *Dioscorea;* hypertensive agents from *Veratrum;* cytotoxic principles from *Podophyllum, Vinca* and other sources; khellin from *Ammi Visnaga;* reserpine from *Rauwolfia;* hesperidin from the citrus group; bishydroxycoumarin from *Melilotus;* and sundry others—not to mention the numerous psychoactive structures of potential value in experimental psychiatry, some new, some old, from many cryptogamic and phanerogamic sources.

Not only have new drugs from vegetal sources been discovered, but new methods of testing and refined techniques have led to the finding of novel uses for older drugs.

As a result of these advances, nearly one half of the 300,000,000 new prescriptions written currently in the United States contain at least one ingredient of natural plant origin. Even if we exclude the antibiotics and steroids, well over 17 percent of all American prescriptions filled in 1960 used one or more kinds of plant products—either produced directly from plants or discovered from plant sources and later synthesized. A more up-to-date analysis of American prescriptions covering over one billion written in 1967 gives the following breakdown: 25 percent contained principles from the higher plants; 12 percent were microbial-derived products; 6 percent were animal-derived substances; 7 percent had minerals as the active ingredient; 50 percent of the active principles were synthetic.

If so many revolutionary discoveries have been made in little more than a quarter century, what logical reason is there to presume that the end to such good fortune has come? Since even the flurry of phytochemical research engendered by the "Wonder Drugs" has barely scratched the surface of the Plant Kingdom, let us consider briefly the potentialities offered by this vast assemblage of species.

THE EXTENT OF THE PLANT KINGDOM

Linnaeus wrote that the "number of plants in the whole world is much less than is commonly believed," calculating that their number "hardly reaches 10,000." Another early estimate was made by Lindley who, in 1847, credited the Plant Kingdom with a total of nearly 100,000 species in 8,935 genera. He assigned 1,194 species to the algae; 4,000 to the fungi; 8,394 to the lichens; 1,822 to the bryophytes; 2,040 to the ferns and fern allies; 210 to the gymnosperms; and 80,230 to the angiosperms.

Intensification of exploration during the last century obliged taxonomists gradually to increase their horizons. But the estimates have not kept pace with botanical collecting and taxonomic research. Most of the currently accepted calculations have not been substantially altered since the early years of this century. They allow the Plant Kingdom between 250,000 and 350,000 species. This aggregation is usually thought to have the following distribution: Algae—18,000; Fungi (including Bacteria)—90,000; Lichens—15,000; Bryophytes—from 14,000 to 20,000; Pteridophytes—6,000 to 9,000; Gymnosperms—about 675 species in 63 genera; Angiosperms—about 200,000 species in some 300 families, of which 30,000 to 40,000 are Monocotyledons.

As a taxonomist and plant explorer, I have long felt this estimate to be totally unrealistic, especially in view of the continued description of 5,000 new species and varieties each year. Perhaps it is significant that botanists with long field experience in the tropics are unhappy with contemporary calculations. May not even the highest currently accepted census for the angiosperms—200,000 species in 10,000 genera and some 300 families—be deficient?

Richard Spruce, the British explorer of the Andes and Amazon for over 15 years during the last century, estimated that the vascular plants of the Amazon Valley numbered some 60,000 species, and, considering the sparsity and superficiality of plant-collecting up to his day, he wrote that there might "still remain some 50,000 or even 80,000 species undiscovered."

In the early years of the present century, Jacques Huber, the Swiss specialist on the Brazilian Amazon, set the arborescent flora of the eastern part of this area at some 2,500 species—even though he felt that Spruce's estimates were overly optimistic. It is interesting to note that Huber's student, the great modern authority on the Amazon flora, Adolpho Ducke, described more than 762 species and 45 new genera of higher plants from the Brazilian sector of the hylea.

Several contemporary plant explorers who have worked in northwestern South America express agreement that "there remains today little better or more exact judgment than that expressed by Spruce" more than a century ago.

José Cuatrecasas has, for example, described upwards of 800 new species. Bassett Maguire calculates that the "vascular flora of the New World tropics would be considerably in excess of 150,000 and may indeed reach the 200,000 figure" and that "certainly, there are yet 25% to be uncovered." Thus, Maguire expects from 37,000 to 50,000 species to be discovered and described from tropical America. Julian Steyermark estimates the flora of Venezuela to comprise between 20,000 and 30,000 species of phanerogams. I have calculated that Colombia, undoubtedly the richest country in number of plants in the New World, may have in the neighborhood of 50,000 species of flowering plants and ferns.

Even outside of the truly tropical areas, certain floras are extremely rich. The Brazilian state of Rio Grande do Sul, for example, has 4,300 species, calculated to comprise about one-tenth of the flora of the whole country. This means that a census of the Brazilian flora would be placed somewhere near 43,000 species. Martius' monumental *Flora Brasiliensis,* written by many specialists over a period of 66 years and finished in 1906, enumerated 22,766 species in 2,253 genera. Our concept of Brazil's flora has, consequently, doubled in only half a century.

In view of the probability that the Plant Kingdom is more extensive than generally thought, a review of the possible size of its several groups might be significant.

Contemporary specialists now believe that some 1,500 species of Bacteria are known. Approximately 12 percent of them are ubiquitous marine types. Despite their great importance as causative agents of human ills, the classification of the Bacteria has been less thoroughly studied than that of other groups, and they are often more readily discerned from their biological effects than from their morphology.

The Fungi are now variously estimated at from 30,000 to 100,000 species. Ubiquitous, a highly successful group of plants, albeit one of the most ancient, they are dominant organisms in today's flora, notwithstanding their usual microscopic size. A contemporary specialist states that even 100,000 might be a conservative estimate, and that the grand total could be well over 200,000. A conservative estimate recently considered the Fungi to be constituted as follows: Phycomycetes, 1,000 species; Ascomycetes, 12,000 (with some calculations as high as 40,000); Basidiomycetes, 13,500; Fungi Imperfecti, 10,500 (with some calculations up to 30,000).

Although the Fungi may be one of the most significant groups of plants in human affairs, they have been relatively insignificant as sources of medicinally valuable compounds. Many are highly toxic to man; some are narcotic. It has long been thought that alkaloids are nearly absent from the Fungi, but modern work refutes this concept. Recent research has shown that 2 percent of a random sampling of fungi are alkaloidal, that quaternary compounds (choline, muscarine, etc.) occur widely in mushrooms, and that betaines, simple amines, and peptides are not uncommon.

Three recent developments have brought the Fungi into sharp relief: 1) the large number of antibiotic substances found in this group; 2) the importance and ubiquitous occurrence of aflotoxins and other food toxins; 3) the study of hallucinogenic mushrooms with their unusual and potently biodynamic constituents—the curious narcotic hydroxindole alkylamine, psilocybine, from a number of "sacred" Mexican mushrooms and derivatives of ibotenic acid from the hallucinogenic *Amanita muscaria.* The study of Fungi as allergens and as indirect poisons is still embryonic. Since the discovery in 1960 of the toxic and carcinogenic aflotoxin compounds due to fungal infection of foodstuffs, more than 700 papers on the difurano-coumarin constituents in the Fungi have engendered interest in aflotoxins and mycotoxins in general, opening new avenues of research.

The Algae, all aquatic but about 56 percent marine, comprise a most varied group of organisms estimated at from 19,000 to approximately 32,500 species. One group alone—the Diatoms—accounts for some 6,000 to 10,000 species. Recent research has indicated that these plants are promising indeed for a variety of biodynamic effects: antibiotic activity; central nervous and neuroactive effects; antiviral, wound healing, anticoagulant, antiyeast, antifungal and antihelminthic properties. Many of the Algae are toxic, especially among the Blue-green Algae, of which 150 species are known to be poisonous. The importance of the algal groups as sources of biologically active compounds is destined to increase as studies of marine organisms is intensified, since this represents the most significant group of plants living in the sea.

Attention has recently been focused on the Lichens as a neglected source of biodynamic constituents. They number from 16,000 to 20,000 species in about 450 genera. Bacteria-inhibiting properties have been found in many of them. About one-half of the temperate zone Lichens possess lichen acids capable of inhibiting gram-positive Bacteria and even tuberculosis Bacilli and some Fungi.

The Bryophytes—mosses and liverworts—have, it seems, been somewhat neglected from the phytochemical point of view. They may number from 14,000 to 25,000 species concentrated in the wet tropics, where taxonomically they are very poorly known. I am convinced that many surprises await future chemical studies of the Bryophytes.

The Pteridophytes, occasionally valued in folk medicine because of their mucilage content, have not contributed many biodynamic constituents. It is true, however, that phytochemical research has been far from intense, partly because of the lack of toxic and other biological effects. Modern techniques of analysis may reveal compounds of significance in many of the nearly 10,000 species in 250 genera of ferns and fern allies. A hint of the unexpected results that could lie in store is provided by the recent isolation from the comparatively small family *Lycopodiaceae* of nearly 100 extraordinarily interesting alkaloids.

The Spermatophytes, dominant on land today, comprise two divisions that are highly unequal from the point of view of biodynamic constituents. The Gymnosperms, with 700 species in 65 genera, have been of medicinal interest primarily as sources of essential oils and resins, the most notable exception being *Ephedra* with its unusual alkaloidal constitution. The Angiosperms, on the other hand, have been man's prime source of medicinal and, indeed, of all other economic plants, partly because, since they are the obvious part of land vegetation, man is most familiar with them. Even though most of man's medicines have historically an angiospermous origin, the potentialities of this group of plants have really been but superficially examined.

It may be of interest here to indicate, in order of decreasing importance, the principal drug-yielding families of Angiosperms according to a survey of more than one billion American prescriptions written in 1967: Dioscoriaceae, Papaveraceae, Solanaceae, Scrophulariaceae, Rutaceae, Rubiaceae, Liliaceae, Bromeliaceae, Rhamnaceae, Caricaceae, Plantaginaceae, Sterculiaceae, Gramineae, Leguminosae, Umbelliferae, Ericaceae.

Estimates of the extent of the Angiosperms vary greatly. Most botanists have accepted 200,000 to 250,000 species in 300 families and 10,500 genera. The Monocotyledons are credited usually as making up a quarter as many species as the Dicotyledons. I believe that we must allow somewhere near half a million species for the Angiosperms; but, even if we accept the smaller estimates, the opportunities for phytochemical research are almost virgin and unlimited. The field for expanded investigations for the near future nowhere holds greater potentialities than amongst the Angiosperms, unless perhaps the Fungi, botanically so much less thoroughly understood, may prove to be equally promising. Wherefore, I want to consider a few random aspects of the Angiosperms as isolated indications of the future that the Plant Kingdom as a whole offers in any search for new biodynamic compounds.

ALKALOIDS AND GLYCOSIDES IN PLANTS

Perhaps the most conspicuous phytochemical research during the past decade has been directed toward alkaloids in the Angiosperms. The activity of alkaloid research cannot better be shown than by the growth in the number of structures isolated. In 1800, alkaloids were not known; in 1949, about 1,000 were recognized; ten years later, this figure was increased to 2,175; by 1969, the total stands at 4,350. Of this total, 256 occur in non-angiospermous plant groups: Agaricaceae, 34; Hypocreaceae, 51; other Fungi, 40; Algae, 5; Cycadaceae, 6; Equisetaceae, 5; Lycopodiaceae, 95; Pinaceae, 2; Gnetaceae, 9; Taxaceae, 9. This concrete figure of 3,094 alkaloids in the Angiosperms means that approximately 94 percent of the alkaloids are known from this segment of the Plant Kingdom, against only 6 percent from all other plant groups.

Within the Angiosperms, there is an apparent disparity in occurrence of alkaloids between the Monocotyledons and the Dicotyledons. The latest figures indicate that only 488 of the total of 3,094 alkaloids in the Angiosperms, or slightly under 16 percent, are found in the Monocotyledons, distributed amongst 12 of the 45 families.

Since the Monocotyledons are generally considered to represent about one quarter of the total number of species in the Angiosperms, this apparent disparity may be appreciated as of doubtful significance, especially in view of the fact that, in general, many of the large monocotyledonary families, such as the Orchidaceae, have not been so thoroughly investigated by phytochemists as the major and certain alkaloid-rich dicotyledonary families.

In the Monocotyledons, the richest alkaloid families have seemed to be the related Amaryllidaceae and the Liliaceae. In the neighborhood of 200 alkaloids are known from each family. In the former, with 1,050 species in 90 genera, alkaloids have been isolated from 33 genera; in the latter, with 3,700 species in 250 genera, alkaloids occur in 31 genera. These two families have long been thought to represent the only outstandingly alkaloid-rich Monocotyledons. This assumption may well have been fallacious. That there is still much to do in alkaloid studies in the Monocotyledons cannot better be illustrated than by the largest monocotyledonary family—in fact, the largest angiospermous family—the Orchidaceae, with 25,000 to 35,000 species in about 600 genera. A recent survey of plant alkaloids cited only 24 alkaloidal species in 11 genera, but actually only two alkaloids—dendrobine and nobiline from *Dendrobium nobile*—had ever been isolated. If the Orchidaceae have, as botanists believe, stemmed from a common remote ancestor with the Amaryllidaceae and Liliaceae, there might be justification in expecting the orchids to have a higher percentage of alkaloid-bearing species. A spot test survey for alkaloids which I carried out on 1,454 species on herbarium specimens in the Orchid Herbarium of Oakes Ames indicated 66 definitely positive, 48 doubtful or with a trace—for a total of 114, or 8 percent. Actual phytochemical work carried out recently on fresh orchid material has shown the presence of alkaloids in 324 species of 1,073 examined—a much larger proportion, 32 percent. An extrapolation of this figure would indicate that the Orchidaceae are far richer in alkaloids than even the Amaryllidaceae or Liliaceae.

Amongst the Dicotyledons, alkaloid distribution has been shown to be seemingly much more complex and capricious. At least one family is

100 percent alkaloidal—Papaveraceae. The Himantandraceae—a family of one genus with one or perhaps two species—has yielded 21 alkaloids. Alkaloids are almost lacking in some families, such as the Rosaceae and Labiatae. One family, the Apocynaceae (admittedly the most intensively studied of all plant families since the discovery of reserpine) accounts for 18 percent of all known alkaloids, with 765 having been isolated from 26 percent (46 of 180) of its genera.

Alkaloid distribution in the Dicotyledons probably may look less erratic when more is known about some of the minor or rare families. At the present time, however, even in the larger families, good information is available, in general, for those which, because they have yielded valuable medicinal compounds, have been intensely studied: for example, Apocynaceae, Solanaceae. But even in most of the larger families (Compositae, 13,000 species; Leguminosae, 13,000; Rubiaceae, 5,500; Euphorbiaceae, 4,000), only a small proportion of the species has as yet been analyzed.

Recent and current research is fast increasing our knowledge of the number of families in which alkaloids are present. Field spot tests indicate that the rare monocotyledonary family Velloziaceae should be studied. The Salvadoraceae, Elaeocarpaceae, Lythraceae, Convolvulaceae, and Myristicaceae are known to be decidedly interesting to the alkaloid chemist, with the potentiality, in several of these groups, of the discovery of new types of alkaloids. Other families, long known to have possessed alkaloids, are now shown to be definitely richer than once thought or to yield interesting new structures—e.g., Euphorbiaceae, Rhamnaceae.

There is no need here to elaborate further on the alkaloids nor to delve deeply into the many other secondary organic plant constituents. In almost all, if not all, cases, the same potentialities are inherent and are made manifest by modern investigations. Next in importance as biodynamic constituents, especially as toxic compounds in plants, are the several classes of glycosides. They may be more widespread in the Plant Kingdom than the alkaloids. Among the most valuable are the cardiac glycosides and genins, of which some 300 are known from the Angiosperms, concentrated especially in the Apocynaceae, Asclepiadaceae, Liliaceae, Moraceae, Ranunculaceae, Scrophulariaceae, and other families. Many of the plants containing these cardenolides have been used in primitive societies as arrow- and ordeal-poisons. Steroidal sapogenines have, in recent years, attracted much attention in medicine, although pharmacological knowledge of this group of glycosides long antedates their chemical elucidation. Some of the species containing these compounds (e.g., *Strophanthus*) have, likewise, been valued for ages in the preparation of arrow-poisons. An extensive search through the Plant Kingdom for biodynamic sapogenins has, over a period of twelve years, screened 6,000 species representing 208 families in 1,397 genera—almost all Angiosperms—from nearly all parts of the world. This screening program, which represented two-thirds of the families recognized in Engler and Prantl, indicated that the most conspicuous genera for steroidal sapogenine were *Agave, Yucca, Dioscorea.* These 6,000 samples, incidentally, were screened for other organic constituents as well, and 10 percent were found to contain alkaloids, 64 of which were new structures. Approximately 150 natural flavonoids are known, occurring as heterosides, and more than 33 biological activities for 30 of these plant pigments have been reported. While rutin has been the outstandingly important flavonoid, other activities, including antiviral and cytotoxic effects, have also been noted for this group of compounds.

Many other categories of biodynamic plant constituents commonly occur in the Angiosperms, but the examples cited suffice to indicate what a future lies ahead when the Plant Kingdom is fully and systematically investigated. It should be borne in mind that there are well over 2,000 organic plant principles of known structure that lie outside of the alkaloid-glycoside classification and that many of them have or have had some application in or bearing on medical problems. The census of terpenoids, coumarins, anthraquinones, phenolic compounds such as tannins, essential oils, and many other organic constituents of Angiosperms are most certain to be increased greatly in number and novelty when, with sophisticated modern techniques of chemistry, the Plant Kingdom is screened as intensively for them as it has been recently for alkaloids.

ETHNOBOTANICAL SEARCH FOR NEW DRUGS

Were one not forewarned, one might assume that this grand advance in our understanding of the distribution of organic vegetal compounds had resulted from an organized search on the part of the botanist and chemist alone. Such an assumption would fall far short of giving a true picture of this modern awakening, which has been possible primarily because of the implementation of the interdisciplinary outlook in science. Great strides toward a breakdown of narrow "compartmentalization" and an upbuilding of the interplay of sundry, often seemingly unrelated, disciplines have led directly to the success that has been evident in the field.

Perhaps in no other effort has the interdisciplinary approach been so essential and so effective. The basic disciplines involved, naturally, have been botany, chemistry, and pharmacology; but anthropology, archaeology, linguistics, history, sociology, comparative religion, and numerous other specialities have likewise contributed appreciably to the search for new biodynamic plants. This intertwining of data and points of view from sundry fields has often been called *ethnobotany* or, in respect to drug plants, *ethnopharmacology*.

A number of avenues are open in a search for new biodynamic plant constituents. The most obvious, perhaps, is a random or semi-random screening of plants. This method is expensive in time and money, but it has been employed in several recent surveys, some of which have concentrated on a search for specific constituents—alkaloids, sapogenins, flavonoids, and so on. Others have been geographically limited but have sometimes been chemically more inclusive—covering alkaloids, saponines, flavonoids, including leucoanthocyanins, tannins, cyanogenic and cardiac glycosides, unsaturated sterols, triterpenoids, organic acids and phenols, essential oils and their components, etc. More or less extensive surveys have been carried out on plants from Australia and New Zealand, Borneo and Papua, Malaya, Hawaii, Taiwan, parts of Brazil, Colombia, Russia, and other countries, not to mention specific sections of the United States and Europe.

Other random surveys have been centered upon a search for plant constituents active for specific diseases or for very definite biological properties. Thus, there have been surveys for antiviral and antibacterial activity, for cytotoxic effects, and for a broad spectrum of other properties. Perhaps one of the most ambitious random searches has been the screening of plants for possible antineoplastic activity by the Cancer Chemotherapy National Service Center, which has tested more than 26,000 plant extracts, representing some 6,500 species. Farnsworth has truly said that "in the light of present knowledge and experience . . . a random selection and testing of plants selected from a broad cross section of families and genera will prove of greatest value in attempts to discover new entities for the treatment of clinical malignancies." Faced with the amazing size of the Plant Kingdom—even of only the higher plants—this random sampling requires the investment of great sums of money and the availability of many kinds of specialists, and demands as well relatively long periods of time: even if such a program could be carried out at the present time, it could be realized only by a large company or by a government agency.

Another approach concentrates intensive investigation on plants with folk uses reported in both ancient and modern literature. This literature, diffuse and often—in fact, usually—uncritical and of hearsay nature, is found in many fields: anthropology, ethnology, history, a variety of chronicles, exploration and travel, missionary reports, and many other human activities where direct contact has been sustained with primitive societies.

Ethnobotanical literature goes back 3700 years to the Code of Hammurabi. Many specialists might think it folly to examine such old literature. Yet, had we critically evaluated the writings of the Egyptian papyri, we might not have had to wait until the 1940's for an acquaintance with the antibiotic properties of certain fungi.

There is still probably much to be learned from a careful sifting of the medieval European herbals, which are usually passed over as having been thoroughly exhausted. Two other literature sources can be of great help. One—and its wealth is still underestimated—the accumulated writings of the early explorer-naturalist-physician-herbalist researchers of the sixteenth and seventeenth centuries: those intrepid, enquiring, insatiable chroniclers of things and events in new lands. They set down everything. In many cases it is still possible to winnow the grain from the chaff—yet this task still begs for a thorough effort. The Dutch botanist of the seventeenth century, Rumphius, for example, whose work is basic to the natural history of the East Indies, accumulated notes on folk uses of more than 700 species of plants. Insisting on verifying personally whenever possible what natives reported, he nonetheless wrote down what he considered fact as well as hearsay, or, as he said, "fables, superstitions and old women's babblings . . . certainly not, as it were, that I put a firm trust in them," but because "in those faery tales always some grain of Truth, some unseen natural virtue lies hidden, and to excite amateurs to diligent search, I assure them that in these lands many secrets of nature are revealed daily erstwhile unknown to Europeans and seemingly unworthy of belief."

The writings of naturalists and others of the New World—some still in manuscript form—are replete with yet unverified reports of therapeutic and biodynamic values of plants. In some cases, the specific identification of the plant is relatively easy. A concerted attack on such repositories of first-hand information, old though they be, might profitably be intensified. One shining example is that incredible treasury of folk knowledge of ancient Mexico: *Nova plantarum, animalium et mineralium mexicanorum historia,* written from data gathered personally in the field between 1570 and 1575 by Dr. Francisco Hernandez, physician to the King of Spain.

This approach, though often frustrating because of the casualness of diagnosis of specific ills or, more commonly, the lack of proper botanical identification of source plants, has been outstandingly, even spectacularly, successful in a number of instances. While one or two examples suffice to prove the point, many are the instances where folk uses of plants, had they been seriously followed up, might have led much earlier to valuable discoveries. The rediscovery of the use in Mexico and the exact botanical identification of several potent hallucinogens—especially the "sacred" mushrooms and morning glories—all fully outlined in detail in early historical chronicles and missionary reports, is perhaps the most recent example of ancient folk reports' having led to significant phytochemical advances.

An example from the Old World that has sparked a revolutionizing discovery in medical chemistry is provided by the uses in ancient and modern India of the snake root, *Rauwolfia,* clearly set forth in the Vedas dating from 1500 B.C., which should have led experimenters at a much earlier date to study the isolation of reserpine and the hypotensive properties of the constituents of this plant.

Another, and perhaps even more important, source of literature is that made up by modern ethnobotanical writings. This source is far more extensive and reliable than is usually suspected. It ranges from complete studies of the uses of and beliefs about plants in primitive societies to incidental but oftentimes highly significant remarks on one or two species of plants by travellers, missionaries, or explorers. A first step in evaluating and utilizing this scattered information on living cultures would be to gather the sources together in one large bibliography—a seemingly colossal task but actually not incommensurate with the data on folk medicine which could conservatively be expected to accrue therefrom. Botanists, anthropologists, and other investigators have been increasingly aware of ethnobotany. The past century has been unusually prolific in this type of literature. A realistic consolidation of this wealth of modern literature might indicate that, geographically and culturally, there exists a better coverage, spotty though it undoubtedly is, than has hitherto been expected.

Even from the use of modern literature, however, ethnobotany has its limitations. It is sometimes botanically uncritical or anthropologically unsophisticated—deficient even from both points of view. Seldom are voucher specimens cited to permit verification of identifications, and the diagnosis of ills and diseases is usually open to serious question. Frequently, when ethnobotanical data are found in floras, no specific tribe or group of people or no accurate geographic location is cited for the uses.

All in all, it would appear to be unwise and optimistic to base—as pharmaceutical houses have done on occasion—an entire natural products program on literature reports alone. In this day of computerization, however, a thorough utilization of the far-flung literature annotations of an ethnopharmacological nature ought not to be beyond the realm of easy possibility and could, in conjunction with other interdisciplinary approaches, help immeasurably in an extension of research into part of the Plant Kingdom's still hidden chemical wealth.

Perhaps the outstanding example, at least in modern times, of the use of the literature, complemented by data from other sources, is the Cancer Chemotherapy National Service Center screening of plant extracts for antineoplastic activity or cytotoxic properties. The world literature, systematically searched "for references since the beginning of writing," indicated "how completely the history of the herbal treatment of cancer has been identified with the history of medicine, indeed of civilization." The earliest of the literature studied was dated 2838 B.C. Many unpublished sources were likewise tapped in this survey, including archival material, solicited and unsolicited letters from around the world, and field notes of cancer therapeutic interest from herbarium records. Against the background provided by this exhaustive search, the random screening, already mentioned, of more than 6,500 species of plants for antineoplastic activity provided a truly interdisciplinary attack on the problem—and one giving the greatest hope for success, since it was not based wholly on one approach.

Another approach lies in tapping the wealth of ethnopharmacological data on labels in the world's herbaria. Information hidden away on these labels represents first-hand observations made in the field by the plant collectors; thus it is usually of much greater reliability than literature sources. And, since the information is physically attached to a specimen, no uncertainty about the plant's identity can arise. The most ambitious search, shortly to be published, has been carried out by Altschul and consists of a sheet by sheet search of the 2,500,000 specimens in the several herbaria at Harvard University—resulting in more than 7,500 reports of medicinal or toxic uses of plants in primitive societies the world over.

By far the best avenue of approach, however, is ethnobotanical field work among as yet intact aboriginal societies. Here a great challenge arises: there are not enough trained ethnobotanists to carry out the necessary research against the rapidity of disintegration of primitive cultures. Botanists usually are too occupied with the vast effort of collecting essential to their own phytogeographic or monographic studies to spend time in the slow detective work necessary for assembling the pieces of an ethnobotanical puzzle. While they often do make observations of far-reaching value in pointing out avenues of approach for later intensive work, too many botanists have manifested definite hostility to the thought of paying heed to native lore. The anthropologist, likewise, normally so deeply committed to unraveling obscure or complicated sociological enigmas, is occasionally able to signalize an important point of departure for the ethnobotanist. The anthropologist is, unfortunately, often discouraged from the pursuit of proper ethnobotanical research because of the collecting of voucher specimens, a chore which has sometimes been erroneously portrayed by professional botanists as so complex and burdensome as to be distasteful to or even impractical for an anthropologist. Since there does not appear to be much hope for the immediate and prompt training of a sufficient number of men specifically in ethnobotany, the botanist and, to a lesser extent perhaps, the anthropologist must take the initiative, in view of the fast ebbing away of time, to study many aspects of man and his knowledge and use of plants with biodynamic properties.

What really explains the recent advances in ethnopharmacological studies and the discovery, during the present century, of so many more interesting and promising plant constituents than ever before in the history of pharmacy and medicine? The primary cause is undoubtedly the development and application of the interdisciplinary approach. No longer are those often very divergent fields of science that impinge upon man and his relationship to his ambient vegetation so highly compartmentalized that the specialist in one field never has an opportunity of speaking with a colleague from another. Many examples illustrate the efficacy of this approach. I want to use but one from my own work: the intricately interdisciplinary story of the hallucinogenic morning glories of Mexico.

The early chroniclers in Mexico, shortly after the Conquest, wrote about one of the strangest of the sacred Aztec psychotomimetics: the vision-inducing seed of a vine called *ololiuqui*. Several early sources crudely illustrated it as a convolvulaceous plant, and Hernandez offered a sophisticated drawing which left no doubt that *ololiuqui* was a morning glory.

Religious persecution drove the use of ololiuqui into hiding. For four centuries ethnologists failed to find a morning glory employed in Mexico in magico-religious rites. Furthermore, botanists and phytochemists were confounded, since no intoxicating substances were known to exist in the Convolvulaceae. Then, in 1916, the American ethnobotanist Safford suggested that ololiuqui must have been a species of *Datura;* that the Indians, in an attempt to protect their sacred ololiuqui from prying and irreverent Spanish eyes, had deliberately misled and duped the early chroniclers. He reasoned that 1) no morning glory was known to be narcotic; 2) the flowers of the morning glories were tubular and might be substituted by the natives for those of a *Datura;* 3) the narcosis described for ololiuqui agreed well with that caused by *Datura;* and 4) *Datura* was and, indeed, still is widely employed in aboriginal Mexico as a divinatory hallucinogen. Safford's identification was readily accepted, and, even to this day, is relatively well established in the scientific literature.

Despite occasional botanical and anthropological articles in Mexico insisting that ololiuqui was, in effect, a morning glory, Safford's "identification" gained a foothold because no voucher specimens of a convolvulaceous plant so used were ever cited, even though preliminary pharmacological study of seeds of a wild Mexican morning glory called *piule* indicated that, in frogs, they produced a kind of "half-narcosis" and that an active principle—possibly an alkaloid linked with a glycoside—might be present.

In 1939, ethnobotanical field work which I carried out in Oaxaca finally provided an identifiable convolvulaceous specimen from a plant cultivated in a curandero's dooryard as his source of narcotic seeds for use in divination. It was *Rivea corymbosa.* This discovery and a review of ololiuqui's history was published. Several pharmaceutical houses soon investigated the seed but found no indication of intoxicating principles.

Against the background of this impasse, a psychiatrist—Osmund—decided to experiment and found that seeds of *Rivea corymbosa* are, in reality, highly hallucinogenic. In 1955 he reported that the seeds induce apathy and anergia, together with heightened visual perception and increased hypnagogic phenomena; he found no mental confusion but instead an acute awareness combined with alteration of time perception, followed a few hours later by a period of calm, alert euphoria. Here, it is appropriate, perhaps, to harken back to Hernandez' statement that, through ololiuqui, Aztec "priests communed with their gods . . . to receive a message from them, eating the seeds to induce a delirium when a thousand visions and satanic hallucinations appeared to them." He wrote, further, that the narcotic was so powerful that "it will not be wrong to refrain from telling where it grows, for it matters little that this plant be here described or that Spaniards be made acquainted with it."

Then, the phytochemical investigation of *Rivea corymbosa* seeds fortunately fell into the hands of the chemist Hofmann, discoverer of LSD, who had studied deeply the related ergot alkaloids of *Claviceps purpurea.* Amazingly, he discovered, in the seed of this morning glory, member of one of the most advanced groups of the Angiosperms, several ergoline alkaloids allied to the synthetic LSD, some of them identical with the hallucinogenic alkaloids isolated from ergot, a comparatively primitive fungus. Thus, modern, highly sophisticated psychiatry and chemistry came into play to vindicate ancient folklore reports and contemporary ethnobotanical studies.

Further field work, also in Oaxaca, has established the use as a sacred hallucinogen in divination rites of yet another morning glory, *Ipomoea violacea,* probably the *tlitliltzen* of the ancient Aztecs; and the same ergoline alkaloids have been isolated from *Ipomoea violacea* as from *Rivea corymbosa.*

This whole research, incidentally, has led to a phytochemical examination of the Convolvulaceae. It is now known that these chemical structures are not at all uncommon in the family, occurring in both Old and New World convolvulaceous genera. Thus, the early ethnobotanical writings of Spanish missionaries and chroniclers have had a hand, four centuries later, in sophisticated phytochemical investigations of a whole family of plants!

THE FUTURE

Now: what about the future of the Plant Kingdom as a source of biodynamic constituents?

When it is realized that there are more than 50 categories of secondary organic constituents known from the world of higher plants alone and that only a small fraction of the higher plants have been phytochemically investigated—and then usually only for one or two categories of constituents—the wide-open field for research must be obvious. Sophisticated modern microtechniques, furthermore, can greatly amplify the horizon attainable by today's phytochemists. Even though it represents nothing more than an "educated guess," I would venture to say that less than 10 percent of the organic constituency of the Angiosperms are known, that fully 90 percent remains for discovery and investigation.

Justification for doubt might exist. When one combines, however, the spectacular advances in the chemistry of secondary organic vegetal constituents with the many pointers provided by modern ethnobotany—then, and only then, does this "guess" appear perhaps not to be extravagant.

A few examples of phytochemical advances have been mentioned, and many more are, under any circumstances, well known to all of us who work directly or tangentially in this field of research.

What is certainly not so obvious are the very numerous and diverse indications of as yet uninvestigated biodynamic activity from ethnobotanical observations. Long lists of such indications might be cited, but, in conclusion, I shall enumerate only a few that have resulted from my own ethnobotanical field observations and from those of some of my colleagues and students made in the New World tropics, mainly in the Amazon Valley. When one realizes how closely these observations are circumscribed, in number of researchers, in time and in geographical area, an understanding of how many thousands might be available from a large number of investigators working over a long time and across the five continents may be appreciated.

New and chemically unexplained arrow poison plants, recently collected in the Amazon, belong to species of the myristicaceous *Virola,* the thymelaeaceous *Schoenobiblos,* and the lauraceous *Ocotea.* The Flacourtiaceae, especially *Ryania,* are numbered amongst the most toxic plants of the Amazon. Several species of Caryocaraceae, Humiriaceae, Marcgraviaceae, and Quiinaceae, from which alkaloids and other highly active constituents are not yet known, are elements of the toxic flora of the northwest Amazon. An as yet undescribed and chemically unstudied species of the bombacaceous *Patinoa* possesses ichthyotoxic principles and is employed by the Tikuna Indians to stupefy fish. Members of the Bignoneaceae and Monimiaceae are frequently indicated as poisonous elements in the neotropical flora. A recently collected species of *Piper* is chewed by the Motilone Indians of Colombia for its tongue-numbing properties. The Connaraceae, allied to the richly alkaloidal family Leguminosae, has species recognized or utilized as poisons, but phytochemical studies have largely neglected this promising small family. The Ericaceae, especially in the Andean highlands, deserves deeper investigation. *Aristolochia* has recently attracted attention because of interesting new compounds in some of its species and several Amazonian species are important folk medicines. The Acanthaceae call for more intense chemical study as a result of several unexpected reports of the use of several genera as fish poisons and as an ingredient of an hallucinogenic snuff. The recent discovery in a South American *Psychotria* of *N,N*-dimethyltryptamine argues for a new examination of great sections of the Rubiaceae. Furthermore, a number of rubiaceous plants have been reported as toxic, even though nothing is known of their active principles: *Duroia, Psychotria, Palicourea, Retiniphyllum,* among other genera. In spite of its well-known alkaloidal content, the Solanaceae bears much deeper study, especially some of the minor and rarer genera of the Andes, some of which have significant folk uses in medicine or magic: *Cestrum, Latua, Iochroma, Brunfelsia.* The Araceae, poorly known from the taxonomic point of view and even less understood phytochemically, is put to many interesting uses by Indians in tropical America, and many species are well worth careful study for their reputation as poisons and medicines, such as oral contraceptives. The Lythraceae should be placed high on the priority list for phytochemical examination, partly because of the interesting phytochemical studies on *Heimia,* an auditory hallucinogen. Even such improbable families as the Cucurbitaceae, Amaranthaceae, and Olacaceae might prove surprisingly rewarding in various biodynamic principles. A thorough phytochemical study of the Malpighiaceae is long overdue. Folk beliefs concerning the properties of many tropical American species of apocynaceous genera that have never been investigated, in spite of the preeminence of this family as an alkaloid-rich group, indicate promise in a search for still more active compounds. Nothing is known of the chemistry of the interesting Andean families Gomortegaceae and Desfontainaceae, members of which are employed as intoxicants in Chile.

I would fain extend this list. To do so, however, would only emphasize and re-emphasize the near virginity of the field that, even in this one part of the world, awaits the ethnopharmacologist and phytochemist willing to follow interdisciplinary lines of approach. When the opportunities around the world are considered, the almost limitless potentialities may easily be appreciated.

Even though the Father of Systematic Botany, Linnaeus, thought that there were no more than some 10,000 species of plants in the world, he did comprehend the great potentialities ahead when, in 1754, he wrote what has aptly been described as his creed:

> Man, ever desirous of Knowledge, has already explored many things; but more and greater still remain concealed; perhaps reserved for far distant generations, who shall prosecute the examination of their Creator's work in remote countries, and make many discoveries for the pleasure and convenience of life. Posterity shall see its increasing Museums, and the knowledge of the Divine Wisdom, flourish together; and at the same time all the practical sciences . . . shall be enriched; for we cannot avoid thinking, that what we know of the Divine works are much fewer than those of which we are ignorant.

Selected References

In view of the many fields touched upon in the foregoing pages, a complete bibliography to cover all aspects discussed and their ramifications would be excessively long and involved. Consequently, a very brief list of chosen references is offered with the purpose only of suggesting avenues for continued consideration of some of the points raised in the discussion.

Altschul, Siri von Reis, "Psychopharmacological notes in the Harvard University herbaria," *Lloydia 30* (1967), 192-196.

Archer, W. Andrew, "Adolpho Ducke, botanist of the Brazilian Amazon," *Taxon 11* (1963), 233-242.

Bate-Smith, E. C., and T. Swain. "Recent developments in the chemotaxonomy of flavenoid compounds," *Lloydia 28* (1965), 313-331.

Bohonos, Nestor, "Microbial biodynamic agents" in J. E. Gunckel, ed., *Current Topics in Plant Sciences* (New York, Academic Press, 1969), 289-302.

Brown, Stewart A., "Chemotaxonomic aspects of lignins," *Lloydia 28* (1965), 332-341.

Correll, Donovan S. et al., "The search for plant precursors of cortisone," *Econ. Bot. 9* (1955), 307-375.

Dawson, E. Yale, *Marine Botany* (New York, Holt, Rinehart and Winston, 1966).

del Pozo, Efrén C., "Empiricism and magic in Aztec pharmacology," in D. Efron, ed. *Ethnopharmacologic Search for Psychoactive Drugs.* Public Health Service Publ. no. 1645 (Washington, D.C., Government Printing Office, 1967), 59-76.

Der Marderosian, Ara H., "Current status of marine biomedicinals," *Lloydia 32* (1969), 438-465.

——and Heber W. Youngken, Jr., "The distribution of indole alkaloids among certain species and varieties of *Ipomoea, Rivea* and *Convolvulus* (Convolvulaceae)," *Lloydia 29* (1966), 35-42.

De Ropp, Robert S., *Drugs and the Mind* (New York, St. Martin's Press, 1954).

Fairbairn, J. W., "The anthracene derivatives of medicinal plants," *Lloydia 27* (1964), 79-87.

Farnsworth, Norman R., L. K. Henry, G. H. Svoboda, R. H. Blomster, M. J. Yates, and K. L. Euler, "Biological and phytochemical evaluation of plants. I.: Biological test procedures and results for two hundred accessions," *Lloydia 29* (1966), 101-122.

——"Biological and phytochemical screening of plants," *J. Pharmacol. Sci. 55* (1966), 225-276.

——"Drugs from higher plants," *Tile and Till 55* (1969), 33.

Fernández Pérez, Alvaro, "The past and future of medicinal plants of Colombia," in *Phytochimie et plantes médicinales des terres du Pacifique.* Colloques Intern. Centre Nat. Recherche Scient., no. 144 (1966), 37-48.

Gibbs, R. Darnley, "A classical taxonomist's view of chemistry in taxonomy of higher plants," *Lloydia 28* (1965), 279-299.

Gray, William D., *The Relation of Fungi to Human Affairs* (New York, Henry Holt, 1959).

Harshberger, J. W., "The purposes of ethno-botany," *Bot. Gaz. 31* (1896), 146-154.

Hartwell, Jonathon L., "Plants used against cancer," *Lloydia 30* (1967), 379-436; *31* (1968), 71-170; *32* (1969), 79-107, 153-205.

Hegnauer, R., *Chemotaxonomie der Pflanzen,* 1 (1962); 2 (1963); 3 (1964); 4 (1966); 5 (1969). (Basel, Switzerland, Birkhauser Verlag.)

——"Chemotaxonomy, past and present," *Lloydia 28* (1965), 267-278.

Hofmann, Albert, "The active principles of the seeds of *Rivea corymbosa* and *Ipomoea violacea,*" *Bot. Mus. Leaflets* (Harvard University) *20* (1963), 194-212.

——"Psychotomimetic agents," in A. Burger, ed., *Chemical Constitution and Pharmacodynamic Action 2* (1968), 169-235.

Holmstedt, Bo, "Historical survey" in D. Efron, ed. *Ethnopharmacologic Search for Psychoactive Drugs.* Public Health Service Publication no. 1645 (1967), 3-32.

Jackson, Daniel F., ed., *Algae and Man* (New York, Plenum Press, 1964).

Jiu, James, "A survey of some medicinal plants of Mexico for selected biological activities" *Lloydia 29* (1966), 250-259.

Jones, Kenneth L., "The antibiotics from a botanical viewpoint," Smithsonian Institution, Ann. Rept. no. 1963 (1964), 369-380.

Liener, Irvin E., "Seed hemaglutinins," *Econ. Bot. 18* (1964), 27-33.

Lindley, John, *The Vegetable Kingdom,* 2nd ed. (1847).

Linnaeus, Carolus, *Species Plantarum.* Ray Society Facsimile Edition. *Introduction by W. T. Stearn* (London, Quaritch Ltd., 1957), p. 155.

Lüning, Björn, "Studies on Orchidaceae alkaloids. I," *Acta chem. scand. 18* (1964), 1507-1516.

——"Studies on Orchidaceae alkaloids. IV" *Phytochemistry 6* (1967), 857-861.

Maguire, Bassett. "Organization for Flora Neotropica," *Brittonia 18* (1966), 225-228.

Murça Pires, J., Th. Dobzhansky, and G. A. Black, "An estimate of the number of species of trees in an Amazonian forest community," *Bot. Gaz. 114* (1953), 467-477.

Raffauf, Robert F., "Mass screening of plants for alkaloids," *Lloydia 25* (1962), 255-256.

——"Some notes on the distribution of alkaloids in the Plant Kingdom," *Econ. Bot. 24* (1970), 34-38.

——*Handbook of Alkaloids and Alkaloid-containing Plants* (New York, John Wiley, 1970.
Scagel, Robert F. et al., *An Evolutionary Survey of the Plant Kingdom* (Belmont, Calif., Wadsworth Publishing Co., Inc., 1965).
Scheindlin, Stanley, "New developments in plant drugs," *Amer. J. Pharm. 136* (1964), 216-226.
Schultes, Richard Evans, *A Contribution to our Knowledge of Rivea corymbosa, the Narcotic Ololiuqui of the Aztecs* (Cambridge, Mass., Botanical Museum of Harvard University, 1941).
——"Hacía un censo de la flora de Colombia" Univ. Nac. Colombia, no. 23 (1958), 77-102.
——"Tapping our heritage of ethnobotanical lore," *Econ. Bot. 14* (1960), 257-262.
——"The role of the ethnobotanist in the search for new medicinal plants," *Lloydia 25* (1962), 257-266.
——"The widening panorama in medical botany," *Rhodora 65* (1963), 97-120.
——"The search for new natural hallucinogens," *Lloydia 29* (1966), 293-308.
——"The place of ethnobotany in the ethnopharmacologic search for psychotomimetic drugs" in D. Efron, ed., *Ethnopharmacologic Search for Psychoactive Drugs,* U.S. Public Health Service Publ. no. 1645 (1967), 33-57.
——"The Plant Kingdom and modern medicine," *Herbarist,* no. 34 (1968), 18-26.
Spruce, Richard, *Notes of a Botanist on the Amazon and Andes,* E. R. Wallace, ed. (London, Macmillan, 1908).
Usdin, Earl, and Daniel H. Efron, *Psychotropic Drugs and Related Compounds,* U.S. Public Health Service Publication no. 1589 (1967).
Vogelenzang, E. H., "Arzneipflanzen und moderne Arzneibücher," *Planta Medica 15* (1967), 347-356.
Von Reis, Siri, "Herbaria: sources of medicinal folklore," *Econ. Bot. 26* (1962), 283-287.
Wasson, R. Gordon, "Notes on the present status of ololiuhqui and the other hallucinogens of Mexico," *Bot. Mus. Leaflets* (Harvard University) *20* (1963), 161-193.
——"Soma: divine mushroom of immortality" (New York, Harcourt, Brace & World, 1968.
Willaman, J. J., and H. L. Li, "General relationships among plants and their alkaloids," *Econ. Bot. 17* (1963), 180-185.
——and Bernice G. Schubert, "Alkaloid hunting," *Econ. Bot. 9* (1955), 141-150.
——*Alkaloid-bearing Plants and their Contained Alkaloids,* USDA Technical Bulletin no. 1234 (1961).
Willis, J. C., *A Dictionary of the Flowering Plants and Ferns,* 7th ed. [Revised by H. K. Airy Shaw] (Cambridge, England, Cambridge University Press, 1966).
Youngken, Heber W., Jr., "The biological potential of the oceans to provide biomedical materials," *Lloydia 32* (1969), 407-416.

The Significance of Comparative Phytochemistry in Medical Botany

TONY SWAIN
Cabinet Office, Whitehall, London, and
Royal Botanic Gardens
Kew, Richmond, Surrey, England

INTRODUCTION

Comparative phytochemistry, chemotaxonomy, chemical plant taxonomy, or biochemical systematics is the study of the distribution in plants of chemical compounds and the biochemical operations involved in their biosynthesis and metabolism. In spite of its high-sounding title, it is an ancillary science. But, depending on one's viewpoint, it can help to throw light onto the classification and evolution of plants, the biosynthetic relationships between naturally occurring organic compounds of diverse structure, or help to pinpoint those plant families, genera or species which are likely to yield compounds of medicinal interest. It is important, therefore, to trace its growth from its early beginnings to see how its present diverse roles have come about, and how it has developed into a distinct discipline, albeit an embryonic one, which gives promise of becoming increasingly important among the biological sciences.

The impetus behind the study of plants lies in their importance to man as supplier of food and, more especially, drugs. Certainly the number of plants that were used in folk medicine far exceeded those cultivated for food. If men had been autotrophic organisms, and plants had had no function in their lives other than the contribution which they make to the beauty of Nature, then the need to identify and classify separate plant species would probably have been postponed until the Renaissance or later. But the uses of plants in combating physical and mental diseases was early recognized, as apparently was their ability to survive natural disasters, for it is not recorded that Noah took any plant species aboard the Ark! Their use in folk medicine in both Old and New World early cultures led to their being given specific vernacular names and often to their being grouped on the basis of their specific curative actions (1).

The need to classify plants thus proceeded in a reasonably well organized but rudimentary way until the seventeenth century. It was not until two centuries later that comparative phytochemistry came into the picture, when it was realized that one might discover relationships in plants, due to their possessing common elementary processes of metabolism, by a study of the distribution of their stable accumulated end-products. The occurrences of specific chemical compounds in plants and the fact that certain related chemical structures were duplicated in members of the same plant taxa, then gave the promise of a new aid to plant classification and of the possibility that such knowledge could be used in a predictive way to search for new active medicinal principles. In what follows, then, we will examine the historical beginnings of medical botany and try to discover how this relationship between medicine and botany, once so strong, was weakened in the eighteenth century and has now been resuscitated by the recent increasing interest in comparative phytochemistry.

HISTORICAL PERSPECTIVES

Drugs of plant origin have been used since time immemorial. The Sumerians, as early as 4,000 B.C., referred to the poppy, *Papaver somniferum,* as the "joy-plant." The ancient Egyptians had a well-developed pharmacopoeia; the Ebers Papyrus of 1500 B.C. lists over 800 prescriptions for use in various disorders. Most of these contained ingredients, such as snake's blood, which exemplified the imagination of the physicians of the day, but many called for plant preparations which we now know contain active principles useful against the diseases for which they were prescribed (1).

In the early Greek literature are also many references to the use of drugs of plant origin; Homer's Odyssey, for example, refers to the use of euphoric principles and to arrow poisons. And we are all aware of the use by the Greeks of hemlock as a means of execution. Herodotus (484-424 B.C.) mentions how the Scythians threw hemp (hashish) seed on red hot stones and inhaled the vapour under felt coverings (2). He said that the method was akin to that used in the Grecian vapor baths and mentioned that the "Scythians were delighted and shouted for joy. The vapor serves them instead of a water bath for they never, by chance, wash their bodies with water"; a practice which seems to be followed by some present-day devotees of the drug. Hippocrates (460-355 B.C.) was the first to list plants by their use, but the most complete and orderly Greek biological work was that of Theophrastus (370-287 B.C.), a pupil of Aristotle, who was also a keen observer of plants but whose writings on the subject are unfortunately lost to us (1, 3).

A list of drug plants was prepared by Celsius (25 B.C.-A.D. 35), but the first classification of drugs according to their therapeutic use was composed by Dioscorides (A.D. 64-120?), the Greek physician of Nero. Later Galenos (131-201) published a detailed list of polypharmaceutical drugs, mainly from plants, most of which are still grouped together as galenicals. His treatise, like that of Dioscorides, also listed essential oils, perfumes, condiments, and so on, and they both classified plants according to these properties. Avicenna (980-1077), in his medical "Canon," listed 760 drugs which served as the main guide to medical practice in western Europe from the twelfth to the seventeenth centuries (4).

The first faltering steps to the publication of a full Materia Medica were taken by Paracelsus (1493-1541), who attacked galenical polypharmacy and suggested the need for a critical examination of plants used in medical practice. Further progress was made by Leonhart Fuchs, L'Obel, and Valerius Cordus whose "Dispensorium," published in 1515, can be regarded as the first true pharmacopoeia. The first *official* pharmacopoeia appeared in the city of Augsburg in 1564, and others quickly followed. Nevertheless, it was not until 1772 that a State Pharmacopoeia was published, in Denmark, and the earliest in English was not produced until 1820, when the first edition of the Pharmacopoeia of the United States of America was published in Boston (5).

THE EARLY CLASSIFICATION OF PLANTS

The early users of plant drugs naturally had to be able to recognize and describe the plants they needed and to distinguish them clearly from others which were either ineffective or poisonous. In doing this, they assembled a great deal of information about the form and habit of plants in general. Indeed, Theophrastus distinguished several different types of flower morphology, although his analytical division of plants was based mainly on their habit (3). With the increasing knowledge of different types of plants, separately identifiable, there was also a need for some classification. As already mentioned, Dioscorides and Galenos both classified plants on the basis of their active principles, and, except for a few observations by Albertus Magnus (1193-1280), it was not until the sixteenth century that plant classification for its own sake was begun (3). Even then, it was the more sophisticated needs of mediaeval medicine which led the way. The apothecaries and physicians of that day needed to identify plants more accurately than could possibly be done from the traditional descriptions and crude woodcuts given in the older herbals, and, by providing new descriptions and detailed illustrations made by observations on living specimens, they greatly increased our knowledge of plants. However, bedeviled as they were by the "Doctrine of Signatures," which led them to ascribe specific medicinal properties to plants which bore some resemblance to parts of the human body, their classification was still to a large extent based on the use to which plants were put. Several early pharmacists, however, notably Nehemiah Grew (1641-1712) and James Petiver (1658-1718), drew attention to the relationship between the medicinal properties ("vertues") that the plants contained and their morphology (6).

A century earlier, however, Caesalpinus (1519-1603), professor at the University of Pisa, had first molded an elementary classification of plants based on morphological characters; he was followed by the great English botanist John Ray (1628-1705), who formulated the general principle that all parts of the plant should be used for classification, a dictum which is now recognized as the cornerstone of the natural taxonomic system (3). However, until the middle of the eighteenth century, the influence of medicine on botany, and vice versa, continued as strongly as ever. For example, the Edinburgh medical school was founded by a group of doctors in 1670 who, besides starting a college for the training of physicians, decided to lay out an herb garden so as to ensure that the drugs they used were pure (7). Six years later, a Professor of Botany was appointed to teach the subject to medical students. Even Carl von Linné (Linnaeus, 1707-1778) himself was trained as a physician and held the Chair of Botany and Medicine at Uppsala.

THE RISE OF MODERN BOTANY

If, with the publication of Linnaeus' "Species Plantarum" in 1753, it appeared that the worlds of botany and medicine were starting to diverge, they were soon to be united again through the developments in natural product chemistry. The search for the active principles of drugs led the way. Morphine (I) was isolated in a pure form from the opium poppy in 1806 by F. W. Sertürner (1783-1841), and this was quickly followed by the isolation of a whole host of other drugs and other compounds from a wide variety of plants.

Later on, in the nineteenth century, the rapid development of organic chemistry led to the understanding and determination of chemical structure and subsequently to the replacement of plant drugs by those of synthetic origin. This change is easily seen if one examines the early pharmacopoeias; the earliest French Official Pharmacopoeia (Codex Medicamentarius sive Pharmacopoeia Gallica, 1818), for example, lists 923 drugs, 820 (89 percent) of which were of plant origin. Today, however, the majority of the drugs listed are synthetic; the latest edition of the French Pharmacopoeia (VIIth Edition, 1949-1954) having only 26 percent of drugs from plants, while in the latest British (Xth Edition 1963) and American (XVIIth Edition, 1960-1962) Pharmacopoeias, only 6.5 and 5.3 percent respectively of the drugs listed are of plant origin.

The change to synthetic drugs together with the major advances in our knowledge of human physiology (1) inaugurated by Magendie (1783-1855) and his distinguished pupil Claude Bernard (1813-1878), and recognition of the function of microorganisms in the transmission of disease following the work of Pasteur (1773-1856) and Robert Koch (1843-1910), led finally to the medical sciences, including biochemistry, outstripping botany and occupying the major position in biology that they do today (8).

Nevertheless, the discoveries over the last twenty years or so of the antibiotics, the increasing use of steroids, especially in oral contraception (9), and the recent focus on hallucinogenic drugs (10) and other compounds either directly obtained from plants or synthesized from plant products, has brought medicine and botany closer than ever. Furthermore, the more extensive exploration of tropical flora, especially those species used in folk medicine, has shown that a whole new world of chemical compounds of medicinal interest waits to be examined (11).

Of course, important advances were made in botany during the nineteenth century. De Candolle (1778-1841) monographed all the then known species in a magnum opus which finally ran to 17 volumes. He can also be regarded as the true instigator of the principles of chemotaxonomy, since in his "Essai sur les plantes médicinales comparées avec leur formes extérieures et leur classification naturelle" he built on the work of Petiver and Grew and pointed to the possibility of using chemical characters in plant classification (3, 6).

The influence on botanical thought of the publication of *The Origin of Species* in 1859 by Charles Darwin (1809-1882) who, although he was the son of a country doctor, was trained in botany and geology, was undoubtedly less than its effect on zoology, even though it was the famous Harvard botanist Asa Gray (1810-1886) who helped to establish Darwin's claim to priority. The recognition of evolutionary change did not lead to major alteration in taxonomic groups, but to preoccupation with phylogenetic speculation with little evidence to support it. Darwin himself also grasped the importance of variable populations instead of types to represent species, but this concept became clear only after the discovery in 1900 of the publication of the work that Gregor Mendel (1822-1884) had carried out on the genetics of *Pisum sativum* 35 years earlier (12). The population concept probably had a greater impact on plant taxonomy than any other during the last hundred years. It led eventually to the recognition of geographical and ecological subspecies, and to the development of the modern biological concept of species (13).

THE BEGINNINGS OF COMPARATIVE PHYTOCHEMISTRY

Until organic compounds could be isolated in a pure state and their structure determined, there could be no real science of comparative phytochemistry. It is not, therefore, surprising to find that few, if any, contributions of value in the field were published before the late nineteenth century (6). The work of the early pioneers of chemotaxonomy has been well documented, but we should remind ourselves of their foresight. Helen Abbot, surely one of the most remarkable American women scientists of any era, writing in the 1880's, drew attention to the great diversity of the structures of chemical compounds in plants and pointed to their possible use in classification. The work carried out in the pharmacology laboratory established in 1888 in the famous tropical botanic garden in Buitenzorg (now Bogor), Java, also shows a truly modern approach; Greshoff, for example, stressed the importance of studying biosynthetic pathways as a tool in taxonomy (14). He suggested that using such knowledge one might discern whether a compound isolated from widely different groups of plants had the same or different biological origin, or whether different compounds that were produced by closely related groups of plants were the result of further biosynthetic elaboration from a common precursor. Greshoff also pointed out the need to know more about the evolutionary tendencies of metabolic pathways and groups of chemically related plant constituents. These speculations resulted from the work that he and his predecessor Eykman carried out on the distribution of alkaloids in the Lauraceae and Hernandiaceae, and later his own work at Kew on the comparative phytochemistry of cyanogenetic compounds, saponins, and tannins in a wide variety of plants.

Another important landmark in chemotaxonomy was the extensive study started in 1897 by Baker and Smith on the distribution of essential oils of the *Eucalyptus* and its relatives which eventually covered 176 species (6). Baker and Smith showed several correlations between the occurrence of certain monoterpenoids, such as phellandrene (II) and pinene (III), and the taxonomy of the plants which yielded them. They also suggested, on phytochemical grounds, that *Eucalyptus* had arisen from *Angophora,* another genus of the same subtribe, Leptospermae, of the family Myrtaceae.

Such early studies were of course descriptive in character; that is, they were concerned merely with recording the isolation, structural clucidation, and distribution of chemical constituents in the Plant Kingdom. Hallier, in 1913, clearly differentiated this approach from a more dynamic one in which comparative studies are made of biosynthetic pathways, thus echoing Greshoff's suggestions. Of course, the time was not then ripe for such studies. Indeed, the tools available to the chemist were still relatively primitive, and many years of laborious work were spent in determining the structure of compounds which any modern graduate student using the battery of methods now at his disposal and with the hard-won foundation that the earlier chemists laid down could solve within a week.

Indeed, the biosynthetic relationships between the majority of chemical constituents which were isolated in the nineteenth century remained unrecognized for several decades. In many cases it was not even possible for later workers to isolate compounds in sufficient quantity for the elucidation of their structure, either because the original source had been wrongly identified or wrongly named by the original investigator, or because of the fact that they were accumulated in the plants in too small an amount to be obtained in a pure state using the laborious methods of purification available. Often, too, they or their derivatives or degradation products were too labile or had extremely complex assemblages

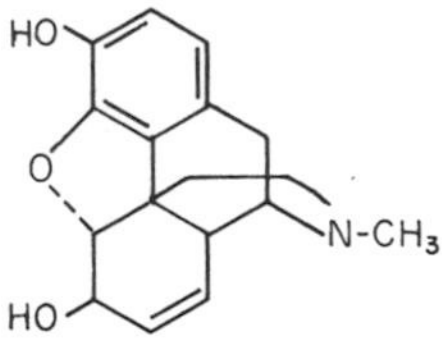

I. Morphine

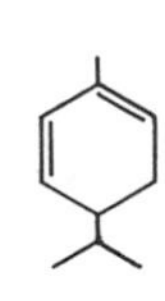

II. Phellandrene

III. α-Pinene

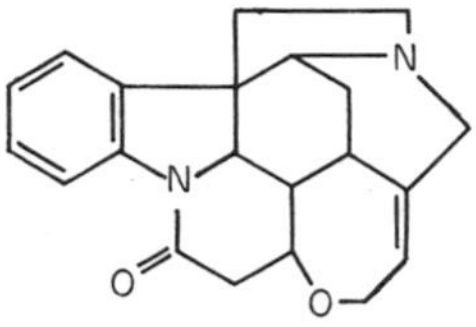

IV. Strychnine

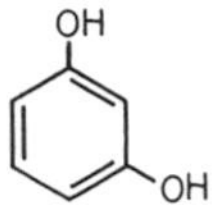

V. Resorcinol

CH_2FCOOH

VI. Fluoracetic acid

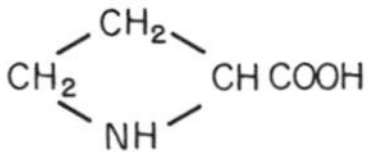

CH_2—CH_2—$CH\,COOH$ (ring closed through NH)

VII. Azetidine-2-carboxylic acid

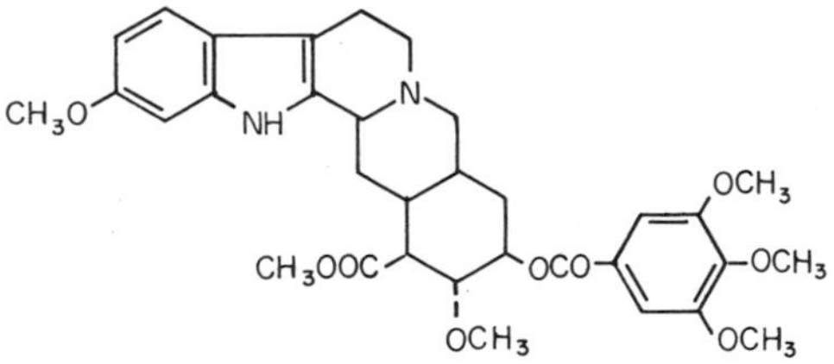

VIII. Reserpine

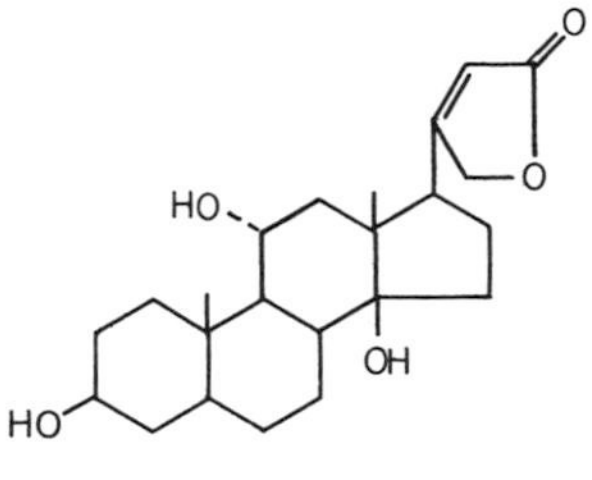

IX. Sarmentogenin

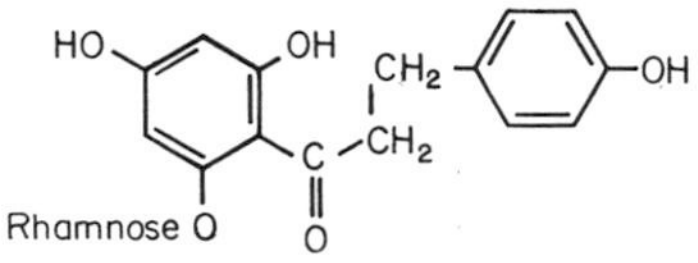

X. Glycophyllin

of ring structure. For example, strychnine (IV) was first isolated in 1818, but its structure was not finally elucidated until 1945 (15).

Nevertheless, similarity in the empirical formulae of the terpenes then isolated (e.g. II) led Bertholet in 1860 to comment on their possible relationship, which was later clarified by Wallach, who put forward his isoprene rule in 1914. As the structures of more and more compounds were established, organic chemists postulated a number of other biogenetic generalizations, and many of those suggested by Robinson, Ruzicka, and Winterstein were later shown to be wholly correct or to give useful leads in the design of later biosynthetic experiments (16).

In the 1920's, however, the main interest in comparative biochemistry as an aid to taxonomy shifted away from individual simple chemical compounds to a study of serological techniques which involved observation of the variation of the reaction of test animals in producing antibodies when foreign substances, antigens, contained in plant extracts were injected into their bloodstream. The development of these methods arose from the observations of Nuttall at the turn of the last century. It was believed that these techniques were of greater fundamental importance than the mere study of the occurrence of simple molecules, or even diverse morphological characters, because the causative antigens were regarded as proteins. At first, much credence was placed in this method by a number of plant taxonomists, but the bitter controversy which arose between the two main schools of the science in Germany in the late 1920's, threw the method into disrepute from which it is only now slowly recovering (3).

MODERN TAXONOMY

In recent years, the unifying concepts of cell-biology have done much to draw attention to the common biochemical heritage of all living organisms. At the same time, they have enabled us to focus our attention more closely on the probable nature of the underlying biochemical causes of the differences which divide them. The plant sciences have had a new resurgence, and the rapid increase in our knowledge of the physiology, biochemistry, and chemistry of plants has had an equally profound influence on taxonomy (17).

Modern taxonomy, in spite of the views of some of its critics, should not in any way be regarded as a mere moribund activity carried out in herbaria in the observation of minute morphological differences of mounted specimens of dead plants with the help of a hand lens. Of course, the majority of taxonomists quite properly still work in herbaria, but the experimental approach to the subject has given new impetus and help in solving many otherwise intractable problems. Experimental taxonomy is an expanding and, above all, an exciting discipline drawing on electron microscopy for its examination of anatomical and cytological features, and on biochemistry and chemistry in its restless search for new characters as aids in identification and classification (18). Its practitioners are engaged in extensive observations of populations of living plants in order to elucidate the underlying causes of the variation, hybridization, and distinctness of species and in the investigation of modern methods of information storage and retrieval needed to utilize the enormous amount of taxonomic data which is available. Thus, if he is to make progress, the taxonomist has to familiarize himself with a wide variety of disciplines other than botany and needs to approximate to the Renaissance ideal of the universal man.

It is as well to remind ourselves of the aims of taxonomy which have been briefly touched on and then to consider how comparative studies on chemical compounds might meet these aims. It should be noted that taxonomy really implies the study of classification rather than the end product, the actual system of classification produced. Strictly, one should use the wider term systematics, if the study of the kinds and diversity of organisms and the relationship between them is meant. But despite the discrete meanings put on the terms taxonomy and systematics, they are usually treated as synonymous (3, 19). Taxonomy then involves the recognition of a number of basic categories of plants (taxa), which form discrete breeding populations isolated by inherent reproductive barriers, and which can be clearly distinguished from all other such basic categories on the basis of the number of constant characters they possess in common. The grouping of these basic categories (species) into a hierarchical structure of genera, families, orders, and so on, again is based on the possession of common characters, the number of such characters obviously diminishing as the groups get larger. Such groupings are preferably made on the basis of presumed natural affinities or relationships which are determined either on the basis of overall similarity (phenetic relationships) or by implying nearness of descent (phylogenetic relationships), rather than by some purely artificial system such as those used in popular handbooks which, however, may have advantages in ease of use.

This definition of taxonomy is, of course, much simplified and ignores many of the most interesting and knotty problems of modern plant systematics. The taxonomist, in forming his groups, must obviously be able to identify and name plants—that is, determine into which category they should be placed and also to communicate his results to other workers. The main problems which face taxonomists are, first, deciding which characters shall be used; second, whether one character should be given a greater or lesser weighting than others; and third, determining natural affinities (3). Taxa which have been grouped on the basis of overall similarity may, nevertheless, because of convergence, be unrelated in the phylogenetic sense. The aim of most taxonomists is to produce a classification which has an evolutionary meaning, even though one must recognize that this could only be speculative on present knowledge. Indeed, it is not really very useful to speculate, except among the larger categories and then only in a general way, whether one group of plants is more primitive than another on the sole basis of varying specialization of the usually employed morphological characters when we do not know what biochemical transformations bring about such variations or what change in genetic information they represent.

COMPARATIVE BIOCHEMISTRY

Recent advances in our knowledge of the basic ways in which the genetic information of an organism is stored and transferred through the arrangement and replication of the bases of deoxyribonucleic acid, DNA, and of the ways in which such information may be changed through mutation should ultimately give us a lead to many of the underlying problems of systematic biology. But such knowledge is still a long way off. Many elegant experiments have shown that the genes specify the complete structure of proteins and thus the biochemistry of the entire cell (20). It has also been shown that the genes are composed of DNA, the sequence of bases in which determines the sequence of amino acids in proteins. The role of ribonucleic acid, RNA, in transferring the genetic information to the site of protein synthesis has also been elucidated. More importantly, experiments have shown that the code of base sequences in DNA is universal throughout living matter, three bases in the same sequence determining the positions of the same amino acid in all proteins. The number of distinct triplets (codons) is 64, and each of the 21 protein amino acids are coded by at least two triplet sequences; some have four or more. It has been suggested that this variation in codon number might be of evolutionary significance. Be that as it may, it is obvious that a greater knowledge of the comparative chemistry of these macromolecules must ultimately lead to the most accurate image that we can obtain of phylogeny using present-day living organisms.

But the determination of nucleotide sequences is only just beginning and, with the methods available, is technically exacting. It will undoubtedly prove even more difficult to determine such sequences in the nucleic acids of higher plants due to the presence of larger amounts of extraneous interfering materials than occur in microorganisms on which the bulk of the studies on the nucleic acids have so far been carried out. The more recent use of hybridization techniques for determining similarities between the DNAs and RNAs from different organisms has already given much useful information in the case of bacteria and higher animals (21). But before these observations can be of use in taxonomy we need to know more about the underlying basis of hybridization and the restrictions imposed by variations in the kinds and distribution of nucleotides along the polymeric chains. Furthermore, there is the problem of cell differentiation; a flower petal cell and a root hair cell have, presumably, the same genetic composition. We know that a good part of the genetic information in eucaryotic cells is not expressed (22). But, except at the enzyme level, the causes and the controlling mechanisms of such suppression are not fully understood. The application of DNA base structure sequences to morphological problems must wait until such mechanisms are elucidated.

The sequence of amino acids in proteins has been extensively studied over the past 10 years, especially since the development of automatic amino-acid analyzers. These studies have shown that proteins having a similar enzymic function, for example, as oxygen carriers, isolated from different species show a heterogeneity which is roughly proportional to the difference one might expect on phylogenetic grounds. For example, the two cytochromes *c* from man and monkey show only one amino acid difference, whereas the cytochromes of man and horse show 12, and those of vertebrates and yeasts show nearly 50. Similar studies have been made on a large number of other proteins (23). Such exciting discoveries will undoubtedly be applied, one day, to problems of plant taxonomy.

In the meantime, the development of electrophoretic methods for the separation of proteins on a micro scale has been used to study the heterogeneity of sngle enzymes and the variation in the mobility of proteins having similar enzymic activities from a wide variety of plants. Again, these studies are only in the elementary stages, and we need to know more about the variation of enzyme structure (isozymes) in populations before these characters can be used in taxonomy (21).

One last biochemical technique, which has been receiving increasing attention, is the survey of the occurrence of different metabolic mechanisms in related taxa. One excellent example is the variation in aromatic oxidation by *Pseudomonas* and related species (24): another is the variation in the detoxification of foreign organic substances when they are added to the intact plant through the cut stem or the petioles of the leaf. For example, the addition of simple phenols, like resorcinol (V), to a wide variety of plants showed that, whereas ferns and flowering plants converted the compound to a β-glucoside, lower plants had no such mechanisms (14). Obviously, surveys of this kind take time, but they may be valuable in delineating large groups and suggesting their phylogenetic affinities.

COMPARATIVE PHYTOCHEMISTRY

Where does comparative phytochemistry stand in all this? First, let us remind ourselves of what we are after. In classical taxonomy, using the term in its restricted sense, many different types of character are used: morphological, including studies on pollen morphology (palynology), cytological, anatomical, and embryological (3). Depending on the taxa under investigation, characters are chosen on different grounds with reference to the degree of differentiation which is sought. For example, the characters which distinguish between large groups will obviously not be of use to delineate within the group. Exactly the same is true of chemical characters. The occurrence of simple common metabolites such as glucose or aspartic acid are of value only if we want to distinguish cellular from noncellular forms of life.

Descriptive plant chemotaxonomy has centered mainly on the examination of so-called secondary metabolites: that is, compounds which are, so far as we know, not essential to the functioning of the living plant cell, and certainly not part of the biochemical transformations which are common to all cells. We can thus ignore all the simple sugars and organic acids which form part of the glycolytic and tricarboxylic acid cycles and associated pathways of metabolism, and likewise the basic nucleotides involved in DNA and RNA and the 20-odd amino acids which are the constituents of all proteins (20). We can also ignore the common lipids. We can eliminate from consideration the biosynthetic precursors of all these compounds and the enzymes and co-factors associated in their synthesis and degradation. Of course, the accumulation of any of these compounds in vast excess may be important chemotaxonomically, as this indicates that some abnormal synthesis or utilization must be occurring in the plant in question.

Up to the beginning of the last war, the chemistry of many naturally occurring secondary substances had been determined, but it was not until the introduction of column chromatographical methods for the separation of carotenoids by Zechmeister in the early 1930's, and, more importantly, the development of paper chromatography for the separation of polar compounds by Consden, Gordon, and Martin in 1944, that the full flowering of the descriptive comparative phytochemistry of secondary compounds could properly begin. Over the last 15 years, many other powerful techniques have been added to the armory of the natural product chemist, such as gas- and thin-layer chromatography, electrophoresis, ultra-violet, infra-red, nuclear magnetic resonance, mass and X-ray spectroscopy, and the application of radioactive tracer techniques for the elucidation of biosynthetic pathways (25).

In the last 10 years, with such techniques, comparative data have been accumulated on a wide variety of secondary chemical compounds from simple alkanes to complex polysaccarides and including acetylenes, fatty acids, flavonoids, quinones, nonprotein amino acids, many different types of alkaloids, porphyrins, and the sulphur-containing glycosides (26–29).

All the secondary compounds which have been examined so far are obviously accumulated in plants in sufficient concentration to enable them to be unambiguously identified. We have little knowledge why they accumulate or whether they contribute, or once did contribute, to the fitness of the taxa; or if they are, or were, involved in the adaptive changes of populations. We also need to know more about their *rate* of accumulation or degradation, and whether they are used as reserve materials or are intermediates in the synthesis of other secondary compounds (14). We thus know little of their function. This does not mean, of course, that these compounds have no function, and indeed the biological meaning of many of these plant constituents needs to be more extensively studied. Some plants accumulate compounds which are toxic to general cell metabolism. For example, members of the Dichapetalaceae accumulate fluoroacetic acid (VI), which interferes with acetate metabolism, while certain species of the Liliaceae accumulate azetidine-2-carboxylic acid (VII), which interferes with proline metabolism. Such plants must have some means of neutralizing the effects of such metabolites, presumably either by possessing effective detoxication mechanisms or by storing them in organelles separate from the main stream of cellular metabolism. It is evident that all these are important questions which biochemists and physiologists must try to answer (14).

VARIATION IN SECONDARY PLANT PRODUCTS

The value of any given chemical character in taxonomy, and, indeed, the value of such observations to medicine, lies in the constancy in which such a compound occurs in a given population growing under a wide range of natural conditions. The actual production of any compound by a plant under a given set of conditions is, of course, normally under strict control, and the ability to regulate steps in biosynthesis is held to confer an evolutionary advantage. But we are all aware of the wide variation in the quantity of drugs accumulated in a given species grown in different environments. *Rauvolfia vomitoria* grown in the Congo contains 10 times more reserpine (VIII) than that grown in neighboring Uganda; *Ammi visnaga,* when grown in Arizona, was found to contain none of the typical chromones and coumarins of the Egyptian-grown plants (30).

We really know too little about the effect of edaphic and climatic factors on the metabolism of secondary compounds or how they vary in populations due to genetic factors. Nor do we have enough information on their distribution within the various organs of the plant, or how their distribution changes with ontogeny, or whether they are all subject to diurnal or seasonal variation (31). All these factors are important before we can consider whether a given constituent is valuable as a character in plant classification of whether a given plant would be a useful source of substances of medicinal value.

The numerous studies on the effect of adaptive and climatic factors on the variation in secondary products in plants indicates that, under average conditions, the plant may not be able to carry out all the biosynthetic reactions of which it is capable. In other words, all the potential genetic information they possess is not being used. Thus, most secondary products are increased in plants grown in full sunlight over those grown under shade, and the addition of extra nutrients in the form of fertilizers, besides increasing the general vigor of the plant, often has an even more marked effect on secondary substances. Thus, the content of alkaloids can often be vastly increased by the application of nitrogen. Similar observations have been made on other groups of medicinally important compounds, but such studies have been sporadic, and more needs to be done to investigate the underlying causes of quantitative variation.

There are many examples where examination of several samples of the same species gathered from different geographical locations has shown that a given compound is present in some samples and totally absent from others. Other cases occur where populations in one locality show a constant difference from other populations elsewhere in the quantitative accumulation of a given group of secondary plant products.

Perhaps the most striking example of chemical variation from the medicinal point of view is the search for the steroidal sapogenin sarmentogenin (*IX*) in *Strophanthus sarmentosus,* which was of interest as a possible precursor in the synthesis of cortisone. Twenty *Strophanthus* species and 23 samples of *S. sarmentosus* were collected after a 16,000-mile trek through 12 countries in central Africa, but none contained the desired compound. Reichstein examined 50 different African samples of the species and found four distinct chemical races, or chemovars, only one of which made sarmentogenin in reasonably detectable amounts (30).

Another interesting example is the dihydrochalkone glycoside, glycophyllin (X), which was first isolated in 1866 from *Smilax glyciphylla* (Liliaceae), a native Australian plant growing over a wide range of locations from north to south. A reexamination of a sample of the plant obtained from Queensland in the 1950's showed no trace of the substance, and it was concluded that the original reporter might have mis-identified his specimen. A search for glycophyllin in related Australian species proved fruitless, and the problem was finally solved only when a further specimen of *S. glyciphylla,* this time collected near the site of the original sample in New South Wales, proved to contain the compound in relatively large amounts (32).

The existence of chemovars which vary relatively in the chemical constituents that they contain is well known, especially in relation to the distribution of the terpenes. The early work of Baker and Smith on essential oils of *Eucalyptus* took no note of intra-specific variation (6). Later, Penfold and Morrison showed that chemical races of eucalypts plainly did exist, distinguishing between four types of *E. dives* on their relative content of piperitine (XI), phellandrene (II), piperitol (XII), and cineole (XIII). The occurrence of chemovars in *Eucalyptus* has been confirmed by the extensive work of Hillis on their flavonoid constituents (33).

Mirov, in his outstanding contribution to the terpene chemistry of the

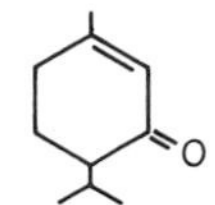

XI. Piperitone

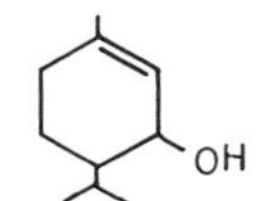

XII. Piperitol

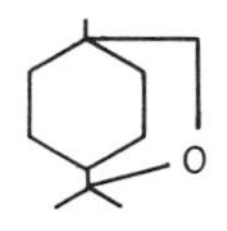

XIII. Cineole

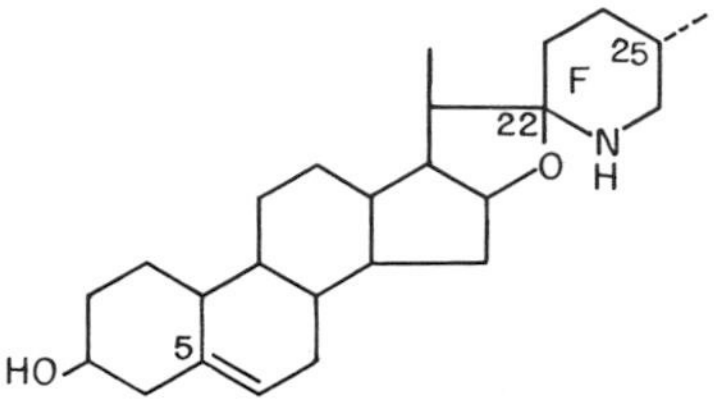

HO 5 22 F 25 O N H

XIV. Solasodine

CH_3COOH (a) Acetic acid

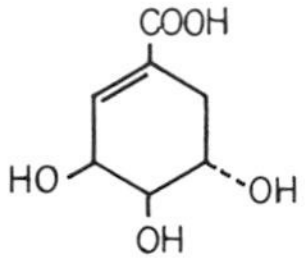

COOH HO OH OH (b) Shikimic acid

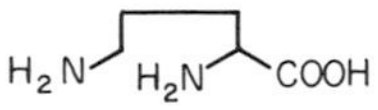

H_2N H_2N COOH (c) Ornithine

H_2N H_2N COOH (d) Lysine

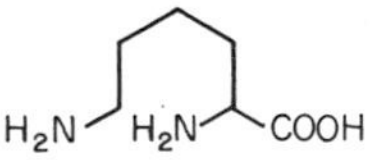

XV. Elementary precursors

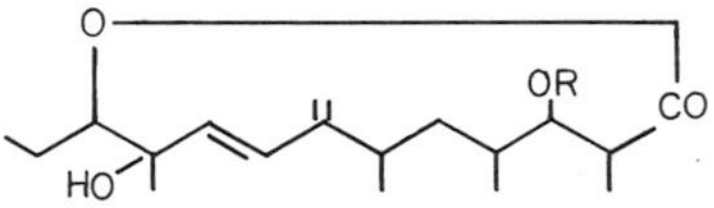

O OR CO HO

XVI. Methymycin

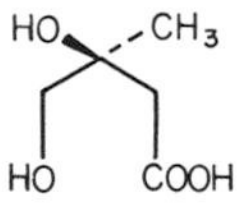

HO CH_3 HO COOH

XVII. Mevolonic acid

R—CH_2·CH·COOH NH_2

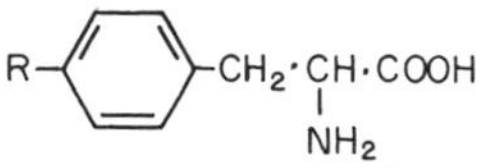

XVIII. R=H, Phenylananine
XIX. R=OH, Tyrosine

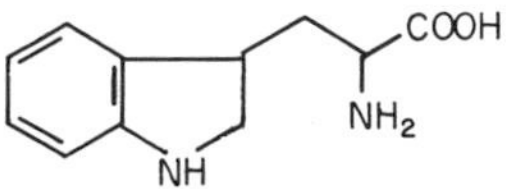

COOH NH_2 NH

XX. Tryptophan

pines, has also shown the existence of several chemovars, and has discussed the possible origin of certain subspecies of different genera on the basis of their relative content of monoterpenes (34).

Another interesting example of medicinal interest is the occurrence of three chemovars of *Solanum dulcamara.* In the young fruits, all three contain the steroidal alkaloid solasodine (XIV); one type contains, in addition, 5-dehydrotomatidine and yamogenin; the second, soladulcidine and tigogenin; and the third, diosgenin. (These other compounds differ from XIV by replacement of O for N in ring F, the stereochemistry at C_{22} and C_{25}, and the possession of the double bond Δ^5.) As the fruits develop to the full green stage, solasodine predominates in all three types, but, as the fruits ripen, this alkaloid disappears, and the other steroids again prevail (35).

The inheritance of certain groups of compounds in hybrids is also of interest. Genetic evidence, mainly from the crossing of cultivars of individual species of flowering plants, shows that the genes for flower pigments formation usually exhibit dominant-recessive relationships, although there are exceptions, especially in polyploids. In *Baptisia leucanthe* = *B. sphacrocarpa* hybrids, however, the flavonoid pattern was additive for each parent (36), that is, without epistasis, and the same was found to be true of the phloroglucinol derivatives in hybrids of the ferns *Dryopteris clintoniana* = *D. goldiana* (37). Such observations are not necessarily the general rule, but are of great interest in the study of hybridization in the field.

THE BIOSYNTHETIC PATHWAYS OF SECONDARY METABOLISM

The main groups of simple secondary compounds arise from only four elementary precursors (38): acetate, shikimate, and the two diamino acids, lysine and ornithine (XV). Acetate, as is well known, gives rise to the commonly occurring 16 or 18 carbon straight chain fatty acids and their congeners, by direct head-to-tail linkage with concomitant reduction. The same route leads also to the other polyketides: long chain alkanes, acetylenes, and the alcohol constituents of plant waxes. Acetate is also involved in the formation of part of a number of other polyketide structures, such as the macrolide antibiotics of the mycobacteria, the main backbone of which is formed from propionate (e.g. XVI). Polyketides from acetate also form the backbone of several aromatic compounds, including simple benzoic acids, and related ketones, naphthalene derivatives, anthraquinones, xanthones, chromones, and one of the aromatic rings of both flavonoids and stilbenes. Acetate is also the progenitor of the branched chain six carbon dihydroxy acid, mevalonic acid (XVII), which is the immediate precursor of the isopentenyl and 3,3-dimethylallyl isoprenoid residues which condense to form the terpenoids, sterols, sapogenins, steroidal alkaloids, and the aglycones of the cardiac glycosides.

Shikimate is the progenitor of the aromatic ring in the amino acids phenylalanine (XVIII), tyrosine (XIX), and tryptophan (XX), which are themselves important intermediates in the formation of several important classes of alkaloids. Phenylalanine and tyrosine give rise directly to hordenine (XXI) and mescaline (XXII), and by further condensation to alkaloids containing the isoquinoline nucleus such as papaverine (XXIII), berberine (XXIV), morphine (I), and colchicine (XXV). Tryptophan gives rise to the indole alkaloids, from the relatively simple psilocybine (XXVI) to the more complex structures such as yohimbine (XXVII), vomicine (XXVIII), and lysergic acid (XXIX). Phenylalanine, or perhaps its intermediates, are also important progenitors of cinnamic acids, coumarins and lignin, and of the aromatic B ring in flavonoids. Shikimic acid can also yield, directly, simple hydroxy benzoic acids such as gallic acid which, with its congeners, is the precursor of the hydrolysable tannins.

The two aliphatic diamino acids, lysine and ornithine (XVc&d), are both important precursors of a number of simple alkaloids; ornithine yielding nicotine (XXX), cocaine (XXXI), and the necine alkaloids (e.g. XXXII) and lysine giving rise to anabasine (XXXIII), pseudopelletierine (XXXIV) and the lupinine (XXXV) and sparteine (XXXVI) bases.

In many cases, the same basic steps are used in biosynthesis (38). Thus, in the formation of alkaloids, decarboxylation of the amino acid gives rise to the corresponding amine, which is then further transformed; mescaline (XXII) is formed in this way from phenylalanine (XVII). Or, two molecules of the amino acid are first either decarboxylated or oxidatively decarboxylated, and then the resulting amine and corresponding aldehyde coupled to give more complex structures, as in the formation of papaverine (XXIII).

Almost all terpenoids are formed by simple head-to-tail condensation of two to four isoprenoid units with either subsequent cyclization of the resulting polymer to give mono- (e.g. III), sesqui- or diterpenes, or the three- and four-unit polymers are dimerized by tail-to-tail condensation to give tri- and tetra-terpenoids, the sterols (e.g. XIV) and carotenoids (XXXVII) respectively.

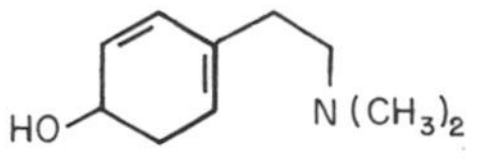

XXI. Hordenine

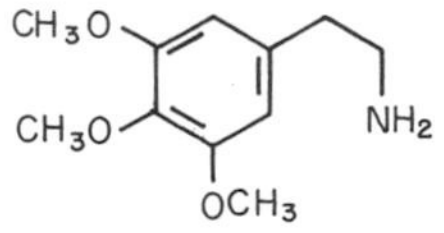

XXII. Mescaline

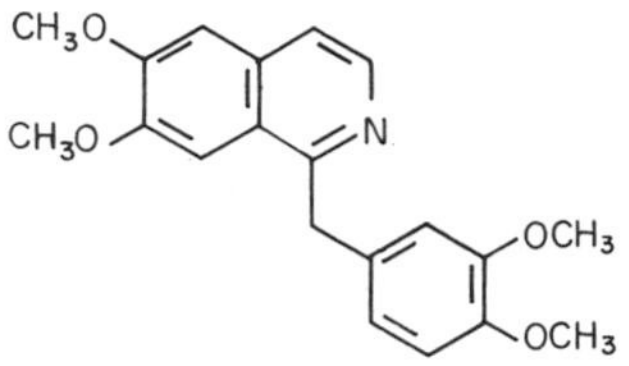

XXIII. Papaverine

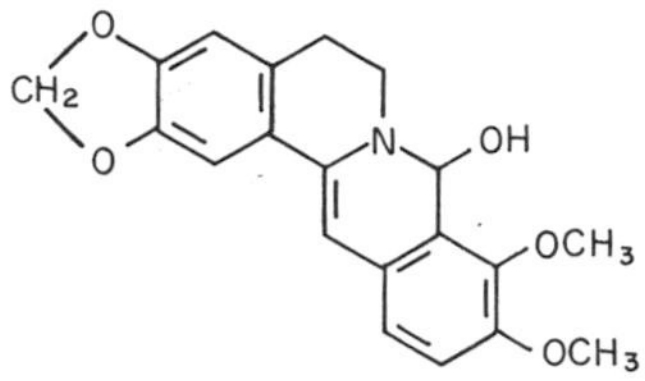

XXIV. Berberine

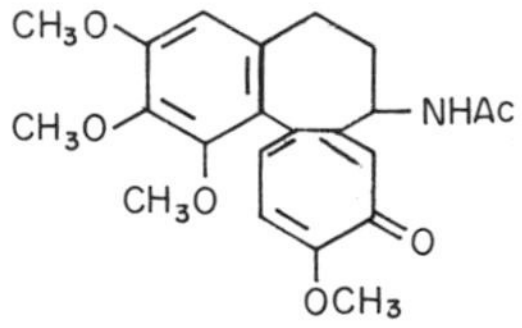

XXV. Colchicine

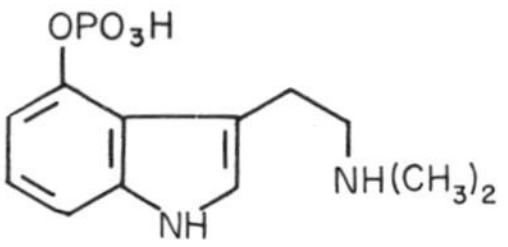

XXVI. Psilocybin

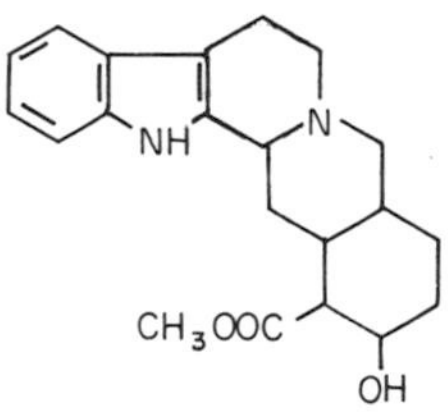

XXVII. Yohimbine

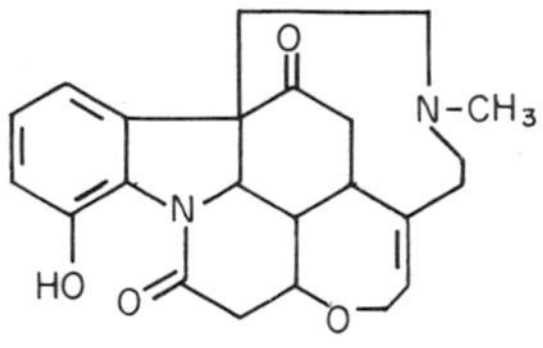

XXVIII. Vomicine

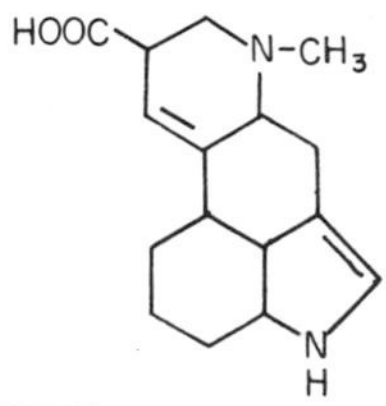

XXIX. Lysergic acid

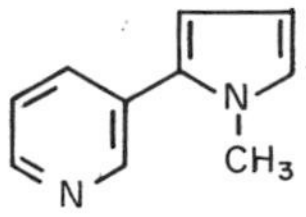

XXX. Nicotine

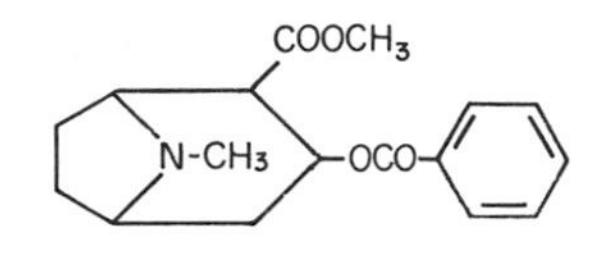

XXXI. Cocaine

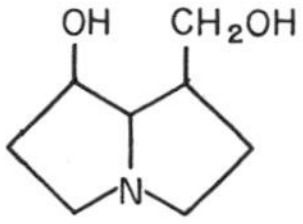

XXXII. Platynecine

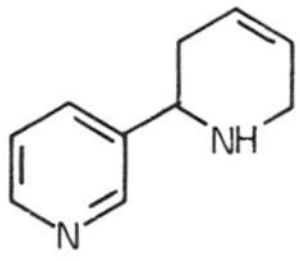

XXXIII. Anabasine

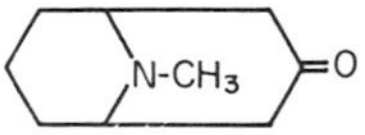

XXXIV. Pseudopelletierine

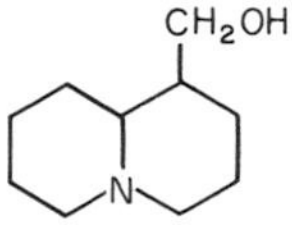

XXXV. Lupinine

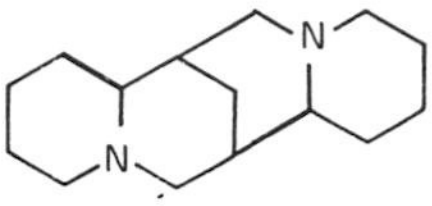

XXXVI. Sparteine

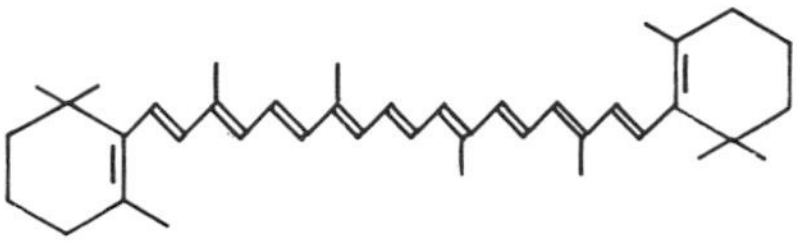

XXXVII. β-Carotene

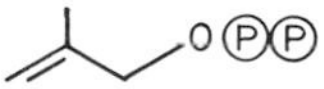

XXXVIII. Isopentenyl pyrophosphate

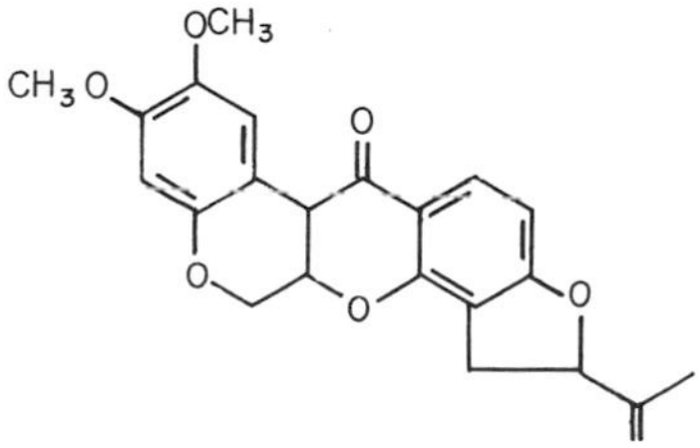

XXXIX. Rotenone

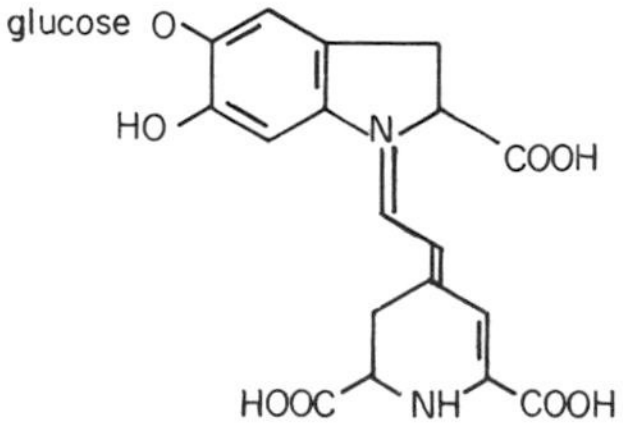

XL. Betanin

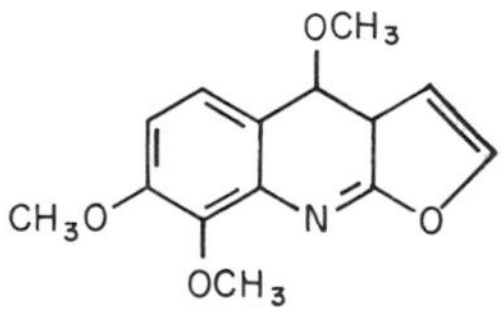

XLI. Skimmianine

Besides such common reactions specific to the different classes of compound, there are a number of other unit steps used in biosynthesis which are common to most groups. Thus, methylation of amines or of the carbon or oxygen of an enolate or phenolate union is mediated by the *S*-methyl group of methionine. Such methyl groups are often transformed into ring structures as in berberine (XXIV), the odd carbon of which arises from the *N*-methyl group of laudanosine, or congener. In enolate and phenolate ions, isopentenylations also are commonly encountered, arising from attack of isopentenyl pyrophosphate (XXXVIII). Again, the introduced isopentenyl group may be involved in further ring closures, as in rotenone (XXXIX). Reduction of carbonyl groups, and double bonds conjugated to them, is also a common step in biosynthetic reactions, as are the corresponding oxidations. Attack by hydroxyl radicals leads to simple hydroxylation of enolate anions or olefins, and to the formation of *N*-oxides, imines, eneamines, or carbinolamines when the group attacked is an amine. Other important oxidation reactions involve one-electron oxidation of enolate or phenolate anions leading to a radical which can then couple with another to give a dimer linked by *C-C* or *C-O* bonds.

Studies have shown that the individual steps in biosynthetic pathways leading to several different classes of compound are composed of a combination of several of these reactions. It is not surprising, therefore, to find that a number of different groups of plants, unrelated on other grounds, synthesize rather similar groups of compounds (26-29).

THE USE OF CHEMICAL CHARACTERS IN TAXONOMY

In spite of all the drawbacks caused by our lack of knowledge of the causes of quantitative variation in the accumulation of compounds in plants and of biosynthetic homology, there are still a vast number of substances which can be used as taxonomic characters. It has been estimated that there are about 800,000 species of plants, about half of which are angiosperms (11). The number of known simple secondary compounds, excluding macromolecules, is certainly nothing like as large as this; excluding macromolecules and simple combinations such as glycosides, probably no more than 8,000 or so have been isolated so far, and the structure and purity of many of these is still in doubt. Hence, it must be expected that many groups of biosynthetically related compounds will be found to be widespread in the Plant Kingdom and show little, if any, taxonomic value. It has also been found that, with most groups of secondary compounds which have been examined, substances which are apparently closely related occur in disparate taxa which are widely separated in all accepted taxonomic systems. Some compounds, however, are more restricted in their distribution; some classes are found only in flowering plants; others occur mainly in gymnosperms rather than angiosperms or vice versa. In yet other cases, certain orders or families have the capacity to accumulate a wide variety of biogenetically related substances which are only sporadically distributed elsewhere in the Plant Kingdom.

Thus, like many other taxonomic characters, most groups of secondary products are not sufficiently restricted in their distribution to give them a privileged position in the delineation of taxonomic categories.

In only one case, the betacyanins (e.g. XL), is a group of compounds restricted to a single Order or to closely related Orders; on the other hand, the pattern of distribution of individual compounds or their congeners has often been of use in helping to delineate genera, tribes, or families within a given Order. This is apparently because the later steps in the final biosynthetic elaboration of most compounds are often unique to specific taxa and give such substances value in determining the taxonomic divisions in the groups of plants in which they occur.

For example, the subfamily Flindersioideae of the Rutaceae contains two genera *Flindersia* and *Chloroxylon.* There has been some argument on other grounds whether this subfamily should not either better be placed in the related family Meliaceae which is in the same suborder of the Geraniales, or be elevated to separate family rank. The chemical evidence is quite clear on the first of these two alternatives. Both genera contain the tryptophan-derived furano- and chromono-quinoline alkaloids, such as skimmianine (XLI) and flindersine (XLII), and typical coumarin constituents which occur in the other members of the Rutaceae, but which are completely absent from the Meliaceae (39).

As might be expected from their disparate biosynthetic origins, the alkaloids cannot be regarded as a single class of compounds in the taxonomic sense (40). Nevertheless, it is useful to examine some general features of their distribution in plants (41).

True alkaloids are rare in the lower plants and in the gymnosperms. In the angiosperms, alkaloids are very unevenly distributed. In the monocotyledons, they are richly present in the Amaryllidaceae and Liliaceae of the order Liliflorae, but otherwise seem to occur only to a limited extent, mainly in the Gramineae. In the dicotyledons, certain Orders con-

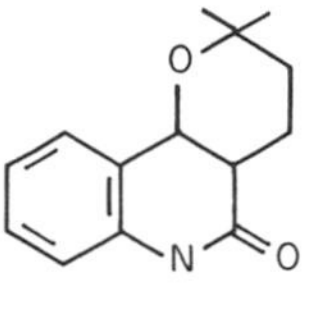

XLII. Flindersine

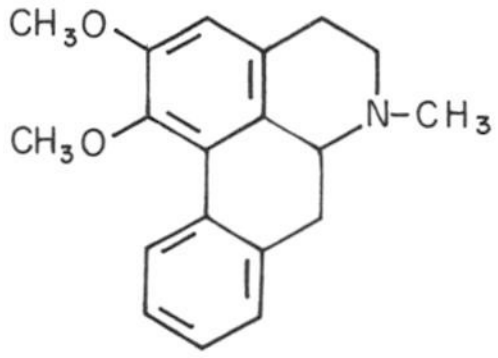

XLIII. Nuciferine

XLIV. From *Polyporus anthracophilus*

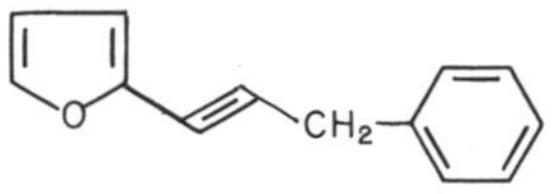

XLV. Carlina oxide

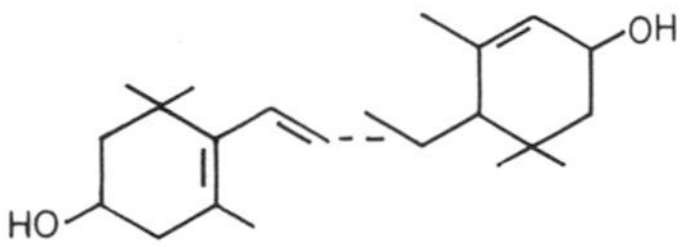

XLVI. Lutein
(chain between rings as in β-carotene, XXXVII)

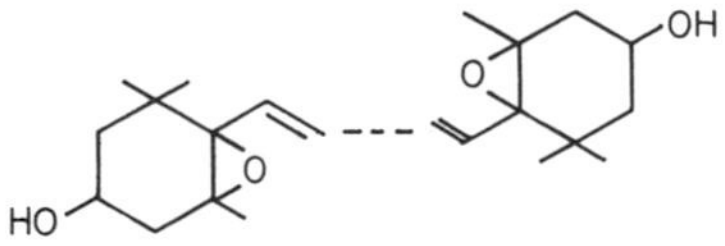

XLVII. Violaxanthin
(see XLVI)

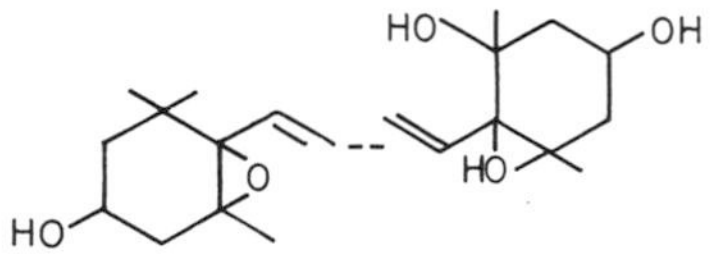

XLVIII. Neoxanthin
(see XLVI)

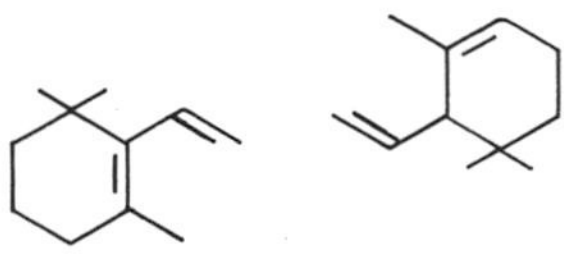

XLIX. α-Carotene
(see XLVI)

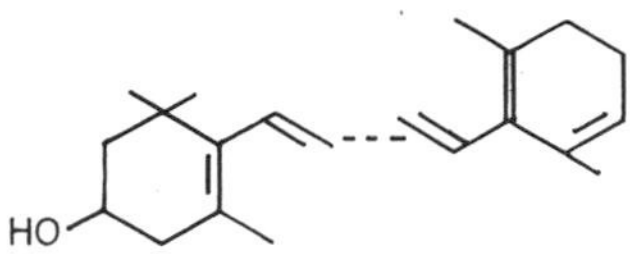

L. Rubixanthin
(see XLVI)

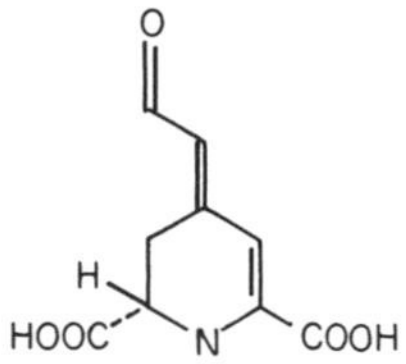

LI. Betalamic acid

The structure of neoxanthin (XLVIII) is now known to have two contiguous double bonds joining the right-hand ring.

tain large numbers of alkaloids well distributed throughout many of the families. Thus, alkaloids occur in all the 18 families of Ranales. In the Rhodeales, however, only one of the six families is conspicuous for its alkaloids, the Papaveraceae; and the same is true of the Rosales, where 95 percent of the alkaloids are found in the well-endowed Leguminosae. In the Contortae, alkaloids are present in most members of the Apocynaceae and Loganiaceae, but are rare in other members of the Order. At the family level, however, alkaloids are of help in distinguishing tribes and genera. Thus in the Apocynaceae, the three tribes—Cerberoideae, Echitoideae, and Plumeroideae—are well distinguished by the types of alkaloid that they contain; the Cerberoideae have simple pyridine or monoterpene bases, the Echitoideae contain exclusively steroidal amines, and the Plumeroideae, complex indoles (42). The separation of subfamilies in the Rutaceae has already been mentioned.

When we examine the distribution of the various groups of alkaloids in higher plants, it is found, except for clustering of certain types in some taxa, that there are many isolated examples of the main alkaloid biosynthetic types occurring throughout the angiosperms (41). The tropane group of alkaloids occurs in at least five unrelated families. Aporphine alkaloids, like nuciferine (XLIII), occur in nine angiosperm Orders. Some families elaborate a large number of different types of alkaloids; the Euphorbiaceae produce nine distinct groups, from simple tyrosine amines, such as hordenine (XXI) in *Securinaga suffruticosa,* to complex indole alkaloids such as yohimbine (XXVII), which occurs in *Alchornea floribunda.* And, of course, the presence of large numbers of alkaloids of the same class is clearly demonstrated by the indole bases in *Catharanthus roseus* (*Vinca rosea*) (42). On the whole, then, we obviously need to know more about the distribution of alkaloids and the variations in their mode of biosynthesis before we can claim that they are of use in taxonomy.

The same sort of pattern arises if we examine the distribution of any other reasonably well studied class of compounds. For example, acetylenic compounds occur in the angiosperms in eight orders unevenly spread throughout the dicotyledons, mainly in the Compositae, where they correlate with other taxonomic characters in the usually agreed subdivision of this family (43). The fact that acetylenic compounds occur also in the fungi, the Basidiomycetes, has often been put forward as a reason why chemotaxonomy is of little use in plants. But the fungal acetylenes are mainly C_9 and C_{10} compounds (e.g. XLIV), whereas those in the higher plants range from C_{10}-C_{17} (e.g. XLV). A similar example is the occurrence of the ergot alkaloids, ergosine and ergosinine (both peptides containing lysergic acid, XXIX), first isolated from the fungus *Claviceps purpurea,* in *Ipomoea argyrophylla* (44). Such co-occurrence has, quite rightly, led to much criticism of the unthoughtful use of chemical characters in taxonomy. But, in any case, the occurrence of biogenetically similar compounds in widely separated taxa is evidence of either the phylogenetic stability of certain characters or a remarkable example of convergence. In the same way, criticism that enzymic differences are of less use than morphological data in distinguishing lions from tigers, ignores the fact that no one is asking taxonomists to disregard the latter type of data. Their value in the recognition and naming of species is not in doubt, but it is surely equally true that differences in the composition or activity of enzymes might be more useful in giving a better idea of the phylogenetic relationship between categories being considered.

Some groups of substances, although regarded as secondary compounds, are so evenly distributed, at least in higher plants, as to be of little or no taxonomic value. Thus, the four carotenoids, β-carotene (XXXVII), lutein (XLVI), and smaller amounts of two β-carotene epoxides (violaxanthin (XLVII) and neoxanthin (XLVIII), are present along with chlorophylls *a* and *b* in the photosynthetic organelles of all higher plants (45). Carotenoids are present also in equivalent organelles in algae and photosynthetic bacteria. We may conclude, therefore, not only that carotenoids play an essential role in photosynthesis, but there is also a strong probability that higher plants originated from a single common ancestor. Algae, on the other hand, although always containing β-carotene or its isomer, α-carotene (XLIX), show a much wider variation in carotenoid composition, and the distribution of these less common carotenoids fits well with views on algal evolution. In flowers and fruit of higher plants, a wide variety of different carotenoids have been found. Their distribution does not appear to have any profound taxonomic significance, but certain unique compounds do occur which may be of use as taxonomic markers, such as rubixanthin (L) in *Rosa* species (45).

The distribution of the betalains, the red-violet and chemically related yellow pigments, the archetype of which is betanin (XL) itself from the red beet, like that of many other coloring matters, has been extensively studied. The betalains are all closely related biosynthetically, containing as a common unit betalamic acid (LI), which is produced probably from

L-3,4-dihydroxyphenylalanine (DOPA) by ring opening of the aromatic ring between the dihydroxy group followed by ring closure on the nitrogen atom. Betalamic acid can react with a variety of amino or imino acids to give either the red betacyanins, if the reacting moiety is 5,6-dihydroxy-2,3 dihydroindole-2-carboxylic acid (also produced from L-DOPA), or the yellow betaxanthins, if proline (LII), glutamic acid or other simple amino acids are involved. The betalains are distributed through all the families grouped in Englerian Order Centrospermae, except the Caryophyllaceae, and occur also in the related Cactaceae, but are found nowhere else in the Plant Kingdom (46). The betalain families contain a number of flavonoid compounds but, strangely, no anthocyanins (e.g. LIII), the usual red-violet flavonoid coloring matters of fruits and flowers. Conversely, the Caryophyllaceae do contain anthocyanins. On the basis of this distribution, it has been suggested that the latter family should be excluded from the Order Centrospermae (46). Taxonomists are, however, quite naturally reluctant to do this on the basis of a single character, especially since the Caryophyllaceae, with their free-central placentation, are archetypes of the group. In spite of phytochemists' cries to the contrary, it is thus undoubtedly right for the taxonomists to retain their grouping of Caryophyllaceae with the other families, until we have more information about the differences in genetic information required for plants to be able to synthesize betalains as against the other distinguishing morphological criteria for this group. It is, however, interesting to note that recent taxonomic treatments have grouped the Cactaceae with the other centrospermous families (47).

The flavonoids are another well-studied class of plant-coloring matters which show little variation in basic skeleton, that is, are all biosynthetically homologous, yet have wide differences in final structure due to variation in hydroxylation, *O*- and *C*- glycosylation, methylation, isoprenylation, and shift of the B-ring from C-2 to C-3 or C-4 of the chroman moiety. They do not occur in protists, but are found in algae, liverworts, mosses, and ferns, although the pigments present in the nonvascular plants are generally regarded as more primitive than those in higher plants. The gymnosperms are similar to ferns in the range of flavonoid that they accumulate, except that they contain a number of dimeric biflavonyls which have only otherwise been detected from two widely separated species in the angiosperms (48).

In the angiosperms, commonly occurring flavonoids are more or less ubiquitous throughout all the whole phyla. There are a number of exam-

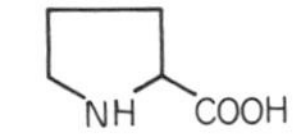

NH COOH

LII. L = Proline

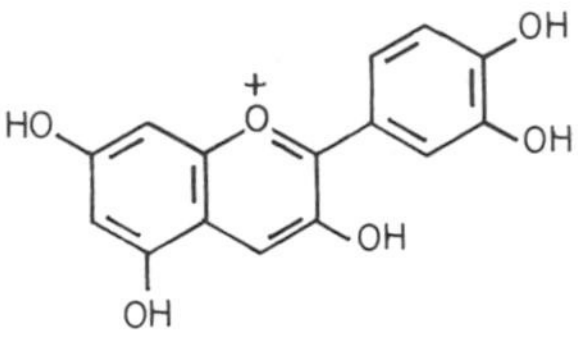

HO OH OH OH OH $\overset{+}{O}$

LIII. Cyanidin

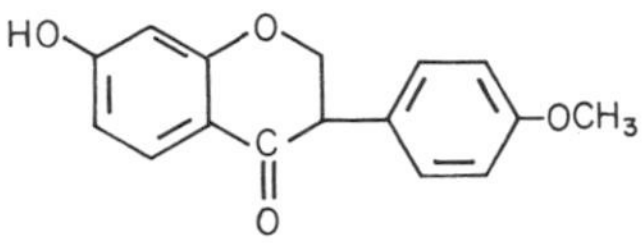

HO O OCH_3 C O

LIV. Formononetin

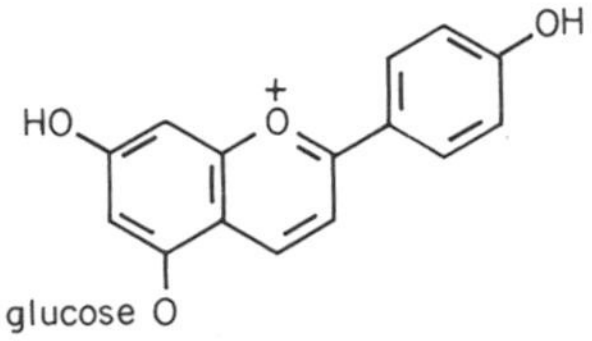

HO $\overset{+}{O}$ OH glucose O

LV. Gesnerin

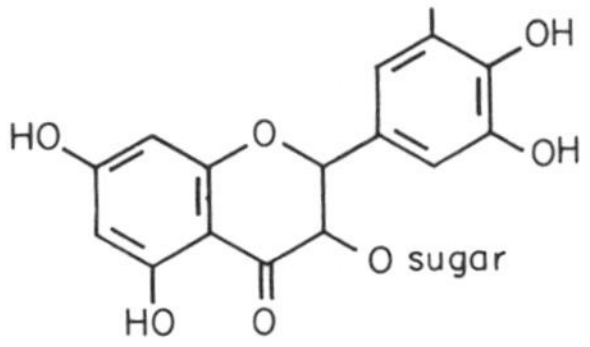

HO O OH OH HO O O sugar

LVI. Quercetin-3- glycoside

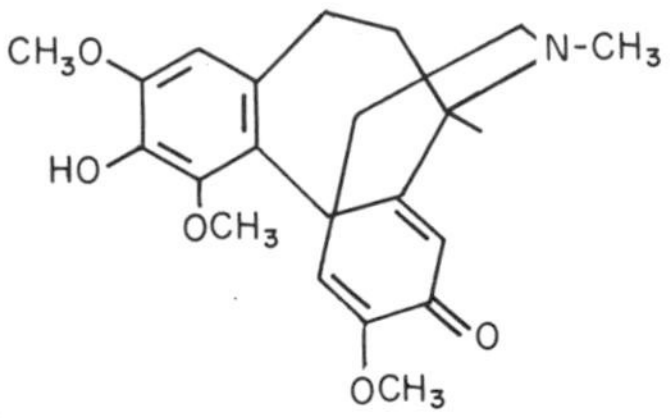

CH_3O HO OCH_3 $N-CH_3$ O OCH_3

LVII. Androcymbine

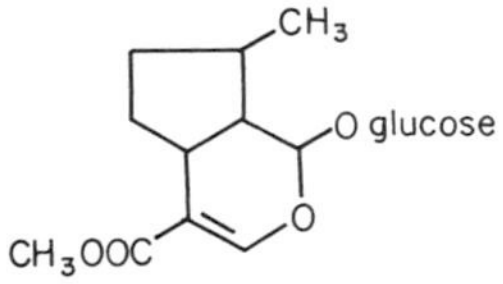

CH_3 O glucose CH_3OOC O

LVIII. Loganin

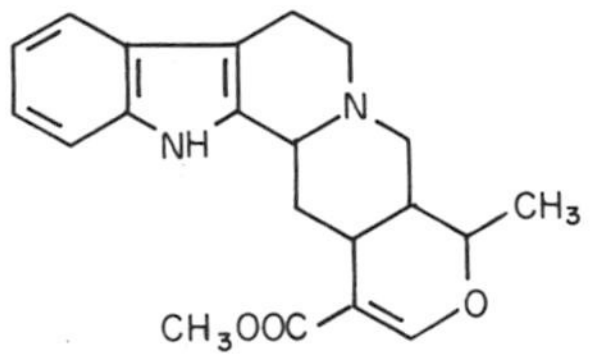

NH N CH_3 CH_3OOC O

LIX. Ajmalicine

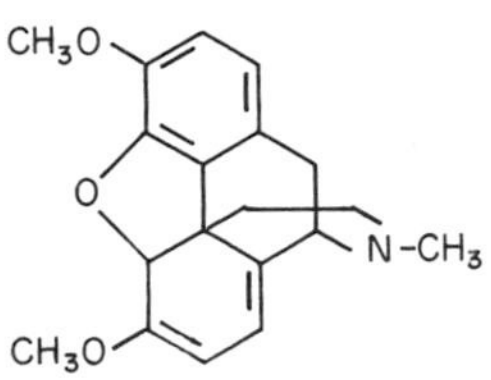

CH_3O O $N-CH_3$ CH_3O

LX. Thebaine

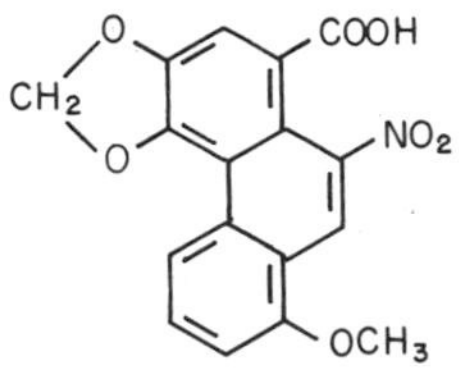

O CH_2 O COOH NO_2 OCH_3

LXI. Aristolochic acid

ples also where very disparate families contain rather rare flavonoids: thus isoflavones (e.g. LIV) are common to the Leguminosae and the Iridaceae, which are also rich in flavone *C*-glycosides.

A good example of the restricted distribution of flavonoids at the family level, however, is the occurrence of 3-deoxyanthocyanins (e.g. LV) in the Gesneriaceae, where their distribution by genera parallels the division into New World species, which contain such compounds while the Old World species do not (48).

It has also been shown that complexity of glycosidation pattern in the flavonoids, which presumably involves extra biosynthetic steps, parallels advancement in the cultivated *Pisum* species. The true wild pea, *P. fulvum,* contains the 3-glucoside derivative of quercetin (LVI); primitive Asian cultivars, *P. nepalensis,* contain the 3-diglucoside (sophoroside) and the modern pea *P. sativum,* a 3-triglucoside acylated with *p*-coumaric acid. Whether such increases in complexity are common in other cultivated plants is not known, but it indicates how useful chemical data might be in phylogenetic studies when more is known about their distribution and biosynthesis.

All these examples show that individual groups of compounds can be of use in taxonomy, at least at the family level and below. Chemical characters may have an even higher taxonomic value when further information is available on their function, stability, and the causes of quantitative variation, and, more especially, when we know more about the distribution of several biosynthetically unrelated compounds in various groups of plants. It would be unrealistic, however, to regard them as a panacea for all taxonomic ills. They are likely to be still susceptible to the effects of parallel or convergent development, inconstancy, and environmental modification.

However, chemical processes are undoubtedly essential events which distinguish certain groups of plants, and it is here that their end products, the chemical characters themselves, have a biological relevance by pointing to the occurrence of such processes. The variation in biosynthetic pathways, and the enzymes associated with them, are undoubtedly phylogenetically significant, and both the sequence of steps and their number reflect the genetic information which indicates the evolutionary history of the organism. We may conclude, therefore, that further studies on the multiple occurrence of chemical characters, and their underlying biochemistry will be of increasing use in the classification of plants and in determining their natural phylogenetic affinities.

THE USE OF COMPARATIVE PHYTOCHEMISTRY IN BIOSYNTHETIC STUDIES

The use of comparative studies in biosynthetic speculation has already been mentioned in connection with the terpenes. Other important theories on the biogenesis of alkaloids, polyketides, carotenoids, flavonoids, and lignin have also been formulated on the basis of a knowledge of the structure and distribution of these classes of compounds in plants (16).

There are also several examples where the occurrence of congeneric compounds in related taxa have yielded an essential clue to individual steps in biosynthesis. A few selected examples will suffice. In the study of the biosynthesis of colchicine (XXV) from *Colchicum autumnale,* the isolation of androcymbine (LVII) from the closely related species *Androcymbium melanthoides,* clearly showed that the likely route to ring C of colchicine was by enlargement of the equivalent ring in androcymbine (49). Similarly, the occurrence of the monoterpenoid iridoid glycoside, loganin (LVIII), and its congeners in several species of the Apocynaceae, led to the idea that the C-10 moiety in many of the complex indole alkaloids of this family, including cantharidine, ajmalicine (LIX), and vindoline, arose by the incorporation of loganin into the molecule. And, when tested, this indeed proved to be the case (50).

It has already been mentioned that the occurrence of the same compound in different taxa may be significant if it is known that it has more than one mode of biosynthesis. For example, the pyridine group in alkaloids can be formed either from glycerol and aspartate, or from ornithine, or by the transamination of a suitable polyketide. This is an example of analogous development, and whereas without knowing the pathways one might have presumed that the occurrence of pyridine alkaloids like nicotine (XXX), anabasine (XXXIII), and coniine in widely divergent taxa had arisen from convergence or parallelism, it can clearly be seen that this is not so (41).

The ability of certain plants to synthesize certain compounds at the end of a biosynthetic pathway, while other species accumulate intermediates, may be due to the variation in a single enzyme system. In the biosynthesis of morphine (I) in *Papaver somniferum,* it is known that the penultimate steps are the demethylation of thebaine (LX) to codeine and hence to morphine. Indeed, when thebaine was fed to tissue cultures of *P. somniferum,* morphine was produced; when, however, the precursor was fed to *P. bracteatum,* a species which accumulates thebaine but not morphine, no transformation occurred, as might be expected. However, tissue cultures of *Nicotiana tabacum,* which do not synthesize any of these compounds, can demethylate thebaine to morphine (51). A similar situation may occur in plants which accumulate sesquiterpenes in contrast to those which have triterpenes such as the sterols; this could be due to the lack of the enzyme system which catalyzes the dimerization of farnesyl, or nerolidyl pyrophosphate.

Such biosynthetic variations would undoubtedly not have been looked for if there was no interest in comparative studies. There can be no doubt, therefore, that comparative phytochemistry has helped both in the formulation of biosynthetic theories, and in suggesting the necessary key substances in actual biosynthetic sequences. As we learn more about biosynthetic pathways in plants, we may equally expect a fruitful feedback to comparative studies, and it is obvious that both will throw important light on the classification of plants.

It should be noted, however, that in the majority of the biosynthetic sequences which have been studied, little attention has been paid to the rate-determining steps, control mechanisms such as repression, or the relative importance of divergent pathways from the elementary precursors. We need to know more about the biological importance of individual pathways before their significance as pointers to evolution can be evaluated (16).

THE USE OF COMPARATIVE PHYTOCHEMISTRY IN THE SEARCH FOR DRUGS

From the medicinal point of view, our present knowledge of comparative phytochemistry is of little predictive value, except here and there. We can, for example, as mentioned earlier, certainly expect most members of the Apocyanaceae to contain complex indole alkaloids, but it would be difficult to predict the exact structural type present in any one species, or whether it was likely to be a rich source or not. Many studies have shown that similar alkaloids are present in closely related species but, even here, a small variation in structure, which may have little taxonomic value, may change a valuable drug into an ineffective one or even a toxic substance.

An example of the successful search for drugs in related species is the isolation of the anti-tumor agent, aristolochic acid, from *Asarum canadense.* The compound had been previously found in 12 species of the genus *Aristolochia* and in one species of *Apama,* which are in different tribes of the Aristolochiaceae family (52).

Another example is the important antimitotic agent colchicine, which has long been used for the alleviation of gout. It was found in *Colchicum autumnale* and in 15 other species of *Colchicum* and also in five species of *Merendera,* a genus in the same tribe, Melanthioideae, of the Liliaceae. An examination of selected species of nine other genera of this tribe showed colchicine to be present in each (53).

On the other hand, the well-known example of the substitution of *Citrus medica* var. *acida,* the West Indian lime, for *C. medica* var. *limetta,* the sweet lime, in the treatment of scurvy in the British Navy in the eighteenth century should point the need for caution. The former variety, which was pressed on the Navy by the desire of the Governor of Bermuda to provide a market for local enterprise, contains only one quarter of the ascorbic acid content of the true lime, and its use led to the recurrence of fatal scurvy in the 1850's (54).

We can conclude, therefore, that because of our present paucity of knowledge on the variation of both structural types and of quantities of compounds of importance in medicine in closely related genera, we can only suggest to pharmacologists that examination of related taxa might prove useful. In doing so, however, we are really little further forward than the Greeks (4), and probably less so than the Aztecs (55).

CONCLUSION

It is obvious that the recent rapid developments in comparative phytochemistry can only be welcomed. Our ignorance of the chemistry and biochemistry of most plants is an obstacle to progress in relation to both taxonomic problems and the more easily directed search for new drugs.

Some of the chemical data which are already available, however, could be made more use of as a discriminating tool to delineate certain taxa, and botanists should not be afraid of employing it. The stress so far on the use of chemical characters has been too much biased toward a static chemical speciation, and there is a definite need for biologists to correct this approach and encourage the consideration of such characters in the more dynamic context of both the living plant and plant populations. We should certainly discourage those who seek to change agreed classifications on the basis of a single simple chemical character, for there is really no evidence that such characters per se are more fundamental or privileged than any others.

On the other hand, we must not be timid in using chemistry as an aid to biological speculation. Even a single compound or biochemical operation may have more to contribute to the study of phylogenetic relationships than the most extensive morphological analysis; that is, it can be presumed that the biochemistry of extinct species can be determined from that of existing forms. Further studies on the dynamic problems in biosynthesis will undoubtedly contribute much to the clarification of general biological problems of which taxonomy is only one.

The most important contribution that medicine will make to botany will come from the widespread support that medicine gives to cell-biology and the underlying biochemistry, especially where it leads to an understanding of the role of macromolecules in the control of metabolism and growth. Botany, on the other hand, will continue to provide medicine with a range of useful compounds for the cure or alleviation of diseases both by extrinsic and intrinsic factors. Comparative phytochemistry should help by bridging the gaps in both directions and thus contribute to our better understanding of the total biological basis of all living organisms.

ACKNOWLEDGMENT

I am grateful to the CIBA Foundation, London, for generously allowing me to use their library. I am also grateful to Miss H. Miller for help in the preparation of the manuscript.

Selected References

I have been inspired by reading several recent publications by R. E. Alston, E. C. Bate-Smith, L. Constance, J. D. Bu'Lock, A. Cronquist, H. Erdtman, R. Darnley Gibbs, J. B. Harborne, E. Heftman, R. Hegnauer, J. B. Hendrickson, J. Heslop-Harrison, V. H. Heywood, A. Löve, T. Mabry, E. Mayr, N. T. Mirov, K. Mothes, R. F. Raffauf, R. C. Rollins, R. E. Schultes, O. T. Solbrig, and B. L. Turner. Only a few of their many important publications are listed below.

1. B. Holmstedt and G. Liljestrand, *Readings in Pharmacology* (London, Pergamon, 1963), pp. 1-61.
2. G. Joachimoglu, in *Hashish: Chemistry and Pharmacology*, ed. G. E. W. Wolstenholme and J. Knight, CIBA Foundation (London, J. & A. Churchill, 1965), pp. 2-11.
3. P. H. Davis and V. H. Heywood, *Principles of Angiosperm Taxonomy* (Edinburgh, Oliver and Boyd, 1963).
4. C. Heymans, "Pharmacology in old and modern medicine," *Ann. Rev. Pharmacol.* 7 (1967), 1-17.

5. G. Urdang, "Pharmacopoeias as witnesses of world history," *J. Hist. Med. 1* (1946), 46-70.
6. R. D. Gibbs, "History of Chemical Taxonomy," in *Chemical Plant Taxonomy,* ed. T. Swain (London, Academic Press, 1963), pp. 41-88.
7. J. H. Gaddum, "Pharmacologists of Edinburgh," *Ann. Rev. Pharmacol. 2* (1962), 1-22.
8. B. D. Davis, R. Dulbecco, H. N. Eisen, H. S. Ginsberg, and W. B. Wood, *Principles of Microbiology and Immunology* (New York, Harper & Row, 1968).
9. C. Djerassi, "Steroidal oral contraceptives," *Science 151* (1966), 1055.
10. R. E. Schultes, "Hallucinogens of plant origin," *Science 163* (1969), 245-254.
11. R. E. Schultes, "The Future of Plants as Sources of New Biodynamic Compounds," pp. 103-124, this volume.
12. A. L. McAlester, *The History of Life* (Englewood Cliffs, N.J., Prentice-Hall, 1968).
13. E. Mayr, "The role of systematics in biology," *Science 159* (1968), 595-599.
14. R. Hegnauer, "Chemotaxonomy, past and present," *Lloydia 28* (1966), 267-268.
15. R. Robinson, "Molecular structure of strychnine, brucine and vomicine," *Prog. Org. Chem. 1* (1952), 1-19.
16. T. Swain, "Methods Used in the Study of Biosynthesis," in *Biosynthetic Pathways in Higher Plants,* ed. J. B. Pridham and T. Swain (London, Academic Press, 1965), pp. 9-36.
17. G. G. Simpson, "Organisms and molecules in evolution," *Science 146* (1964), 1535-1538.
18. O. T. Solbrig, *Principles and Methods of Plant Biosystematics* (London, Macmillan, 1970).
19. E. Mayr, *Principles of Systematic Zoology* (New York, McGraw-Hill, 1969).
20. J. D. Watson, *Molecular Biology of the Gene* (New York, Benjamin, 1965).
21. J. G. Hawkes, *Chemotaxonomy and Serotaxonomy* (London, Academic Press, 1968).
22. R. J. Britten and E. H. Davidson, "Gene regulation of higher cells," *Science 165* (1969), 349-354.
23. M. O. Dayhoff, *Atlas of Protein Sequence and Structure,* vol. 4 (Silver Spring, Md., National Biomedical Research Foundation, 1969).
24. R. J. L. Canovas, L. N. Ornston, and R. Y. Stanier, "Evolutionary significance of metabolic control systems," *Science 156* (1967), 1696-1702.
25. T. A. Geissman, *Principles of Organic Chemistry,* 3rd ed. (San Francisco, Calif., Freeman, 1968).
26. T. Swain, ed., *Chemical Plant Taxonomy* (London, Academic Press, 1963).
27. T. Swain, ed., *Comparative Phytochemistry* (London, Academic Press, 1966).
28. R. E. Alston and B. L. Turner, *Biochemical Systematics* (Englewood Cliffs, N.J., Prentice-Hall, 1963).
29. R. Hegnauer, *Chemotaxonomie der Pflanzen,* vols. I-V (Basel, Birkhauser, 1962-1969).
30. E. F. Woodward and E. Smith, "Acquisition and variability of investigated plant material," *Lloydia 25* (1962), 281-284.
31. E. S. Mika, "Selected aspects of the effect of environment and heredity on the chemical composition of seed plants," *Lloydia 25* (1962), 291-295.

32. A. H. Williams, "Dihydrochalkones" in ref. 27 above, pp. 297-302.
33. W. E. Hillis, "Variation in polyphenol composition within species of *Eucalyptus*," *Phytochemistry 5* (1966), 541-551.
34. N. T. Mirov, "Chemistry and plant taxonomy," *Lloydia 26* (1963), 117-124.
35. E. Heftman, "Biochemistry of steroidal saponins and glycoalkaloids," *Lloydia 30* (1967), 209-223.
36. R. E. Alston, "Chemotaxonomy or Biochemical Systematics," in ref. 27 above, pp. 33-56.
37. L. H. Fikenscher and M. R. Gibson, "Investigation of *Dryopteris* Species," *Lloydia 25* (1962), 196-200.
38. J. B. Hendrickson, *The Molecules of Nature* (New York, Benjamin, 1963).
39. J. R. Price, "The Distribution of Alkaloids in the *Rutaceae*" in ref. 26 above, pp. 429-452.
40. R. Hegnauer, "The Taxonomic Significance of Alkaloids," in ref. 26 above, pp. 389-427.
41. R. Hegnauer, "Comparative Phytochemistry of Alkaloids," in ref. 27 above, pp. 211-230.
42. R. F. Raffauf, "Some chemotaxonomic consideration of the Apocynaceae," *Lloydia 27* (1964), 286.
43. N. A. Sørensen, "The Taxonomic Significance of Acetylenic Compounds," in *Recent Advances in Phytochemistry*, ed. T. J. Mabry (New York, Appleton-Century-Crofts, 1968), I, 187-228.
44. A. Hofmann, "Ergot: a Rich Source of Pharmacologically Active Substances," pp. 235-260, this volume.
45. T. W. Goodwin, "The Carotenoids," in ref. 27, pp. 121-138.
46. T. J. Mabry and A. S. Dreiding, "The Betalains," in *Recent Advances in Phytochemistry*, ed. T. J. Mabry (New York, Appleton-Century-Crofts, 1968), I, 145-160.
47. A. Cronquist, *The Evolution and Classification of Flowering Plants* (Boston, Mass., Houghton-Mifflin, 1968).
48. J. B. Harborne, *Comparative Biochemistry of the Flavonoids* (London, Academic Press, 1966).
49. A. C. Baker, A. R. Battersby, E. McDonald, R. Ramage, and J. H. Ulment, "Biosynthesis of colchicine," *Chem. Comm.* (1967), 390.
50. A. R. Battersby, R. S. Kapil, J. A. Martin, and M. Lugt "Loganin as a Precursor of indole alkaloids," *Chem. Comm.* (1968), 133.
51. K. Mothes, "Biogenesis of alkaloids and problems of chemotaxonomy," *Lloydia 29* (1966), 158-171.
52. R. W. Doskotch and P. W. Van Evenhoven, "Isolation of aristolochic acid from *Asarum canadense*," *Lloydia 30* (1967), 141-144.
53. K. R. Fell and D. Ramsden, "Colchicum," *Lloydia 30* (1967), 133-135.
54. A. Horden, "Some Historical Considerations," in *Psychopharmacology*, ed. C. R. B. Joyce (London, J. B. Lippincott, 1968).
55. E. C. Del Pozo, "Aztec pharmacology," *Ann. Rev. Pharmacol. 6* (1966), 9-15.

Medicinal Plants and Empirical Drug Research

RUDOLF HÄNSEL
Institut für Pharmakognosie
Freie Universität, Berlin

INTRODUCTION

There are a number of large pharmaceutical firms which maintain a so-called "natural products program" in which they conduct, with the assistance of ethnobotanists, a systematic examination of the exotic flora for compounds of pharmacological interest. There are other firms which view this sort of plant screening as unproductive, the results obtained being not proportional to the effort involved. This paper examines the relationship between botany and the total field of drug research. The subject cannot, of course, be exhaustively considered in the space available and, in this introductory section, I will try to outline the limitations I have imposed.

In the realm of science, drug research belongs to the "interdisciplinary group of sciences," since it involves a number of heterogenous technical field which are united in their application to a common object. The science of botany can be applied to "drugs" in a number of different ways. We may begin with plant physiology. Here the plant is considered as a living system, as a model system, which responds to chemical stimulation; and we can consider those chemical compounds which influence this system not only because of their biological activity but because they may be potential pharmaceuticals in the broadest sense. For example, the cytostatic activity of urethane, a compound used medicinally in the treatment of leukemias and lymphogranulomas, was first discovered by the observation that certain arylcarbamic acid esters inhibited the germination of wheat and oats (1).

We can also view plants form the ecological-sociological standpoint as a part of a living community. Studies of the phenomena of parasitism, symbiosis, and antibiosis, as well as the chemical basis of such phenomena, have led to the discovery of drugs which are active as chemotherapeutic agents. But since my theme is "Medicinal Plants and Drug Research" and not "Botany and Drug Research," I do not intend to consider such interrelationships of the plant sciences and drugs.

One can, on the other hand, view the plant world merely as the source of new types of chemical compounds, frequently of complex structure. Seen in this way, the plant is a kind of factory, a mere producer of individual chemicals of often unknown constitution and, therefore, of unknown, often novel, pharmacodynamic activity. I interpret the term "Medicinal Plants" used in the title specifically from this viewpoint and thus will limit my theme to that of describing natural products of use as drugs.

The importance of natural products in drug research has been repeatedly stressed in several recent reviews. It will be sufficient, therefore, if I deal here with the three most important applications: (a) Constituents isolated from plants which are used directly as therapeutic agents. Examples of such natural substances are digitoxin, strophanthin, morphine, and atropine, which are all still unsurpassed in their respective fields. (b) Plant constituents which are used as starting materials for the synthesis of useful drugs. For example, adrenal cortex and other steroid hormones are normally synthesized from plant steroidal sapogenins and the "synthetic" penicillins from natural penicillin. (c) Natural products which serve as models for pharmacologically active compounds in the field of drug synthesis. There are numerous reasons why plant constituents which are potentially useful as drugs cannot be employed directly. The plant material may be either unavailable or available only in limited quantity; furthermore, the plant may not lend itself to cultivation. Frequently, the side effects of a natural product often prevent its use in medicine and can be resolved only by preparation of a synthetic derivative—for example, cocaine, which led to the development of modern local anesthetics, the coumarins as precursors of modern antithrombins, and the modification of colchicine and of podophyllotoxin to obtain anti-tumor preparations.

To be sure, without naturally occurring active principles, it seems probable that neither the principle nor the activity would otherwise have been discovered. Put yourself in the place of a chemist who would like to develop a remedy for cardiac insufficiency; methods currently available would not lead him to synthesize a digitoxin-like molecule without knowledge of the natural prototype.

In explaining the ways in which plant constituents have significance as medicinal agents, I took for granted the existence of such active principles in plants. Now there are more than 300,000 species of higher plants alone. An extremely small proportion of the total has been investigated phytochemically; even smaller is the proportion which yield well-defined drugs. What does this mean? Do all species of higher plants contain substances which, when given to the animal or human organism, display some pronounced pharmaceutical activity? Or is the occurrence of active principles restricted to a small number of species? Furthermore, how can active plants be distinguished from harmless, non-toxic ones?

These questions are of course misleading, since they imply that there is some kind of deductive regularity which permits the finding of new medicinally useful compounds. Pharmacological investigations are decisive in seeking both for new synthetic compounds and for new natural products. Such investigations encompass not only systematic pharmacological screening of plant extracts, but also chance observations. Here are differences from the study of new synthetic compounds, for, in the discovery of active plant constituents, the experience of folk medicine plays a large part. Not only does folk medicine frequently point out the type of pharmacological action which can be expected and which can later be verified by animal testing. It has, it seems to me, an even greater psychological importance, because it shows that certain plants are still used by man. As the history of drug research shows, the often highly colored ethnological observations from exotic countries can still stimulate the need to investigate the plants mentioned.

Thus, there are really no strictly deductive principles of drug discovery which lead to the easy detection of active principles. There are, however, certain rules of experience which have a limited validity. In the following sections, I wish to outline two specific working hypotheses for medicinal plant research: a) that involving chemotaxonomy, and b) that of correlation between pharmacological activity and physical properties.

THE USE OF CHEMOTAXONOMY

We observe in nature a great multiplicity of contrasting forms. The multiplicity of forms in the plant world does not consist so much of a continuous change from type to type (for example, from species to species), but instead is discontinuous or, more exactly, hierarchical. Botanical taxonomy is concerned with the hierarchical organization of the plant world in the different taxa. Despite all differences in form or in growth, the biochemical pathways of primary metabolites do not seem to differ from species to species. Assimilation and dissimilation of sugars occur in all green plants basically in the same way. There are, however, very great differences in the so-called secondary metabolism of higher plants; that is, in the synthesis and storage of such compounds as alkaloids, terpenoids, and glycosides. Chemotaxonomy (2-4) is concerned with these similarities and differences in the secondary metabolism of plants; or, expressed somewhat differently, with the distribution of secondary constituents in the Plant Kingdom. The most important principle of this relatively new branch of plant biochemistry is that *the secondary constituents of plants which are closely related taxonomically resemble one another more than those of plants which are more distantly related.* The "law" on which this principle rests does not, in my opinion, imply a relationship between the form of a plant and its chemistry. Rather, the state of affairs appears to be due to variations in acquired mechanisms of excretion and storage of metabolites (some of which are toxic), which has led to deviations from the original condition of similarity. In this connection, it must be remembered that, unlike the animal, the plant retains all its metabolic products within its organs (5). Whatever the theoretical biological background of chemotaxonomy may be, the import consideration is that morphologically related plants often contain chemically similar secondary constituents with pharmacodynamically similar actions. For drug research, this principle can be modified as follows: if I seek variants of a certain structural type, there is greater likelihood that the desired compounds will be found among near-relatives of the plant from which the basic type had been discovered than in species which botanically are remotely related.

Now for a few examples. The foxglove (*Digitalis purpurea*) was introduced into modern medicine by Withering on the basis of its use as a heart stimulant in folk medicine. Research workers were soon seeking analogs of the active principle, digitoxin, which had shorter latent periods of action and lower cumulative effects. Synthetic variants of the digitoxin, however, gave less prompt results than did the search for suitable natural variants of the drug in plants related to *D. purpurea.* A species native to southeastern Europe, *D. lanata,* proved to be a lucky find, even from the technical point of view. It is easily adapted to field cultivation and contains a three- to fivefold greater concentration of active principles than the foxglove, and the compounds could be easily obtained in crystalline form.

A second example is *Rauwolfia serpentina,* which is native to India. The usable parts of this long-lived plant are the underground roots. When the world demand for Radix Rauwolfia for use as a hypotensive agent became so great that it appeared that the whole area of India in which the plant was grown would be exhausted, the Indian government proclaimed an embargo on exports. It was immediately obvious that, in the search for new sources of *Rauwolfia* alkaloids, investigators would limit their studies to other species of the same genus. Sure enough, the African species, *R. vomitoria,* which is a much larger plant and therefore potentially more valuable as a drug source, was found to be a suitable substitute drug.

A third example can be taken from the search for *steroidal sapogenins.* Sapogenins are widely distributed in the Plant Kingdom. Two types are recognized: the triterpene saponins, with a C-30 compound as aglycone, and the steroidal saponins, with a C-27 skeleton as the genin. Steroidal saponin-containing plants are eagerly sought after as sources of starting material for the synthesis of certain hormones, principally those of the cortisone type. In this case, as before, phytochemical screening was restricted to certain plant families in which steroid-containing species were known to occur. As a result, we have today a reasonably good knowledge of the distribution of triterpene and steroidal saponins in the Plant Kingdom (2). Steroidal saponins are found in several families of the monocotyledons; in the dicotyledons, their occurrence is limited to plants related to the genus *Digitalis.*

A fourth example is the *distribution of alkaloids and cardenolides in a single family* (2, 6). The Apocynaceae is the best example, for it contains numerous medicinal plants. Especially notable are the genera *Rauwolfia* and *Catharanthus,* which contain alkaloid-producing plants, and *Strophanthus, Thevetia, Nerium* with cardenolide-containing species. How are alkaloids and cardenolides distributed in this family? We can examine the subdivisions of the family as they are conceived by the taxonomist on the basis of certain morphological characteristics (e.g. type of fruit) and see whether or not there are relationships between this taxonomic classification, on one hand, and the distribution of plant constituents, on the other. The Apocynaceae is usually divided into three subfamilies: the Cerberoideae, the Echitoideae, the Plumerioideae, each of which is further divided into tribes and genera. A superficial examination shows that cardio-active glycosides are found only in species of the subfamily Echitoideae; representatives with typical indole alkaloids occur only in species of the subfamily Plumerioideae; and, finally, certain nitrogenus compounds derived from piperidine are found only in species of the subfamily Cerberoideae. Many of the Apocynaceae are more or less toxic. If we know from its botanical characteristics to which subfamily a given species belongs, we can infer whether the toxicity is caused by N-containing indole bases or by glycosidic cardenolides. Let us now examine more closely the distribution of indole bases in the Plumerioideae. Nearly 600 alkaloids have been isolated from species of this subfamily, but the chemical constitution of only about one-half of these is known. They are, without exception, indole bases in which the non-indole portion of the molecule is coupled primarily with the α-carbon of the indole moiety. We can classify the 300 known apocynaceous alkaloids into eight chemically related types (Fig. 1). However, with respect to their distribution in the various tribes, it is not possible to establish strict conformity with any specific type (Table 1).

In these four simple examples from chemotaxonomy, I have shown that it is possible, with limitations, to make certain predictions concerning the occurrence of secondary plant constituents from the botanical position of a given plant in the plant system. Now let us turn to the second working hypothesis which is useful in medicinal plant research: specifically, the use of correlations between pharmacological activity and physical properties.

CORRELATIONS BETWEEN PHARMACOLOGICAL ACTIVITY AND PHYSICAL PROPERTIES

In a biological sense, correlation is a term usually restricted to describing the connection between two or more observations, and attempts are usually made to determine their correspondence. This does not—and certainly not in every stage of an investigation—always allow one to distinguish whether there is a linear causal relation between them or whether their relationship is much more complex. Let me exemplify what I mean. Certainly, only a pharmacological or a physiological investigation can decide whether a given plant constituent is pharmacologically active. However, pharmacodynamic activity, in some cases, may be coupled with other easily established properties of the substance: for example, with its basicity, with its lipophilic nature, with

Fig. 1 Alkaloid types of the Plumerioideae.

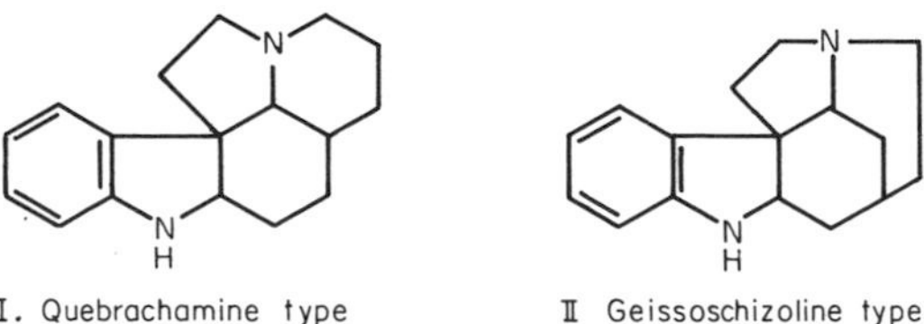

I. Quebrachamine type

II Geissoschizoline type

III. Eburnamine type

IV. Catharanithne type

V. Voacorine type

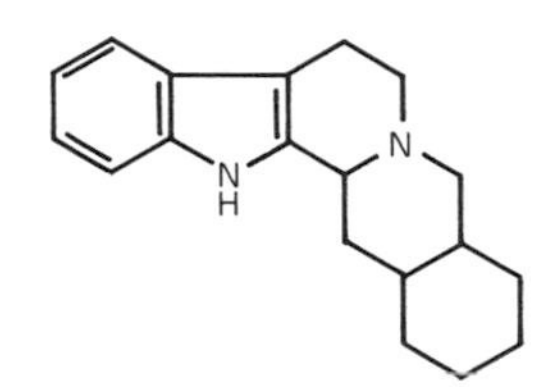

VI. Reserpine type

VII. Alstonine type

VIII. Ajmaline type

its bitter taste, with certain peculiarities of the chemical constitution, and so on. Again, let me give examples. Sertürner's well-known discovery of morphine serves as an extraordinarily good example in drug research. The real value of Sertürner's discovery consisted in the fact that the active principle of a drug can possess basic properties. Therefore, without the help of scientific pharmacology, it was possible, in Sertürner's case, to utilize a purely physical method for the fractionation of a drug extract to obtain the active principle: in this case the alkaloid. In rapid succession, the active alkaloids of numerous drug plants were subsequently discovered. The importance of Sertürner's

Table 1. Distribution of Alkaloid Types in the Plumerioideae

	Alkaloid-type*							
Tribe and Genus	I	II	III	IV	V	VI	VII	VIII
Carisseae								
Melodinus	x							
Pleiocarpa	x	x	x					
Hunteria		x	x					
Picralima		x	x					
Tabernaemontaneae								
Rejoua				x				
Ervatamia				x	x			
Tabernaemontana			x	x	x			
Stemmadenia	x			x				
Voacanga			x	x	x			
Alstonieae								
Gonioma	x							
Alstonia		x	x			x	x	
Paladelpha								
Blaberopus								
Diplorrhynchus		x	x			x		
Aspidosperma	x	x	x			x	x	x
Geissospermum		x	x					
Haplophyton	x		x					
Rhazva	x		x					
Amsonia	x					x		
Catharanthus	x	x	x	x	x		x	
Vinca	x	x	x				x	x
Rauwolfieae								
Rauwolfia			x			x	x	x
Ochrosia					x	x	x	
Vallesia	x					x		
Kopsia	x							

*See Fig. 1.

discovery, consequently, lies in the fact that the basic character of plant constituents is frequently correlated with pronounced pharmacological activity in men and animals. The implication of the presence of pharmacological activity is directly bound up with the definition of the term alkaloid.* We do not understand the theoretical basis for this correlation between the basicity of plant constituents and their action on the nervous system; it is purely an empirical observation.

One could, however, argue in the following way. In physiological pH-ranges, alkaloids are soluble both in water and in lipids. These are the physicochemical conditions for them to be transported passively to a potential place of action. In other words, perhaps the distribution coefficient is one of the main reasons for which there are an unusual number of pharmacodynamically active substances in this group of plant constituents.

Let me give a second example of similar correlative relationship: distribution coefficient and pharmacodynamic activity. Plant constituents may be classified in a number of different ways: that which is of interest in relation to the previous example is a classification according to the distribution coefficient of the compound between lipid and water (8, 9). In order to remove certain metabolic end products from active circulation, the plant has developed a series of excretion mechanisms. Two of the more important of such mechanisms are (a) coupling of the product with a sugar residue, which thereby renders it more soluble in the cell sap, and aids its removal into the vacuole; (b) the metabolic end product is lipophilic and can be stored in special excretory organelles, which exist in numerous morphological forms. We can see that two main types of secondary plant metabolites will exist; type *a* is more strongly hydrophilic, and type *b* more strongly lipophilic. With reference to pharmacodynamic activity, it appears that substances with the same or similar basic structure are more active the greater their lipid-water distribution coefficient. Possibly, this depends, in a number of cases, on absorption and penetration phenomena (10). Once again, a few examples will illustrate this principle:

(i) *Coumarins.* Coumarins occur in the Plant Kingdom either as simple hydroxy- and methoxy-derivatives or their corresponding glycosides, all of which are soluble in cell sap, or as non-glycosidic compounds, polysubstituted with lipophilic groups such as isoprene residues and furano-linked rings all of which are lipophilic. There are several of these latter coumarin derivatives known which have pronounced activity in men and animals. One can begin with the photosensitizing action of the furano-

*Alkaloids are basic plant constituents containing amine nitrogen, usually in heterocyclic form, which produce a pronounced, mostly quite specific action on different areas of the nervous system (7).

coumarins. There are also lipophilic coumarins which have a distinct effect upon the central nervous system. For example, 7-methoxy-coumarin produces stimulation of the CNS, as do the visnagins from the Egyptian plant *Ammi visnaga.* Numerous isoprene-substituted coumarins exhibit vasodilating and spasmolytic properties. There are, on the other hand, other furanocoumarins, which have a central-stimulating action, e.g. osthol (11). Even the type compound of the whole series, coumarin itself, is active as a CNS-sedative. More than one hundred derivatives of coumarin have been prepared synthetically. Substitution of the coumarin nucleus with hydroxy groups results in a loss of activity, but hydrogenation and ring opening, reduction of the carbonyl group, or other synthetic changes which do not essentially alter the distribution coefficient, produce no reduction in activity (Fig. 2) (12). Umbelliferone is active as an antibiotic against brucellosis, whereas glycosidation results in a complete loss of activity (13).

(ii) *Kawa-lactones* (14). The active compounds of the famous Polynesian kawa drug, the so-called kawa lactones, also belong to the *a*-pyrone group of plant constituents. Within the plant, these pyrones are localized in lipoidal cells as can be easily proved histochemically by putting a section under the microscope and adding a few drops of concentrated sulfuric acid. The oil cells are stained deeply red—a color reaction, which is characteristic for the kawa-lactones (e.g. methysticin, dihydromethysticin, kawain). Hence, in this case also, we are dealing with lipophilic plant substances. Thorough pharmacological investigations have shown that these lipophilic kawa-lactones are highly active on the central nervous system.

Fig. 2 Coumarin as a model for synthetic sedatives (12).

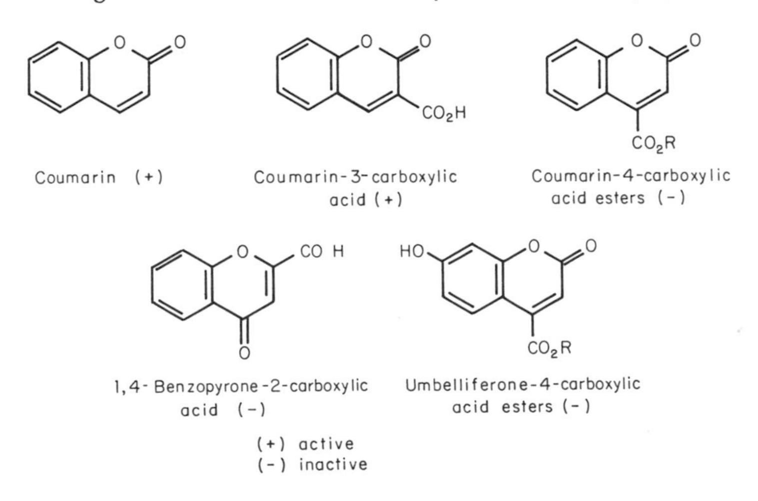

(iii) *Flavonoids.* Like the coumarins, the flavonoids occur in plants either as sap-soluble glycosides or alkylated lipophilic derivatives. This is very nicely demonstrated in citrus fruits (15). If the juice is expressed from any citrus fruit flesh and the flavonoids isolated, glycosides of the hesperidin-eriodictyol type are obtained. On the other hand, separation of the so-called flavedo-layer, which contains excretory cells, and expression of the volatile oil followed by isolation from the lipophilic fraction yields highly methylated flavonoids of the nobiletin-tangeretin type. How both groups of flavonoids act as vitamin P factor, for example, with respect to their anti-inflammatory activity has been clarified by the investigation of Freedman and Merritt (16). The glycosides possess no activity, but the lipophilic flavonoids attain the activity of hydrocortisone (Table 2). As in the other examples, we see a correlation between the lipoic

Table 2. Anti-inflammatory Activity of Flavonoids (15)

Substance	Structure	Potency (A.I units/g)
Hesperidin	5,7,3'-Trihydroxy-4'-Methoxyflavonone	0
Naringin	5,7,4'-Trihydroxyflavanone-7-neo-hesperidoside	0
Tangeritin	5,6,7,8,4'-Pentamethoxyflavone	50
Nobiletin	5,6,7,8,3',4',-Hexamethoxyflavone	50
Hydrocortisone phosphate		74
RS-1		333

nature of the compounds and their pharmodynamic activity. This is also true of their toxicity, as tested against insects and fish. The flavonoids are commonly considered as completely harmless substances from the toxicological point of view. This does not apply, as Indian chemists (17, 18) showed some time ago, to the lipophilic members of the series. The unsubstituted derivatives, the methoxy-substituted chalkones and flavones, are active as insecticides and as fish poisons. When hydroxy-groups are attached to the skeleton, the chalkones and flavones become more readily water-soluble and these hydroxy-derivatives turn out to be harmless (Fig. 3).

(iv) *Constituents of Colchicum.* As a final example, I have chosen the characteristic constituents of various *Colchicum* species. It is well known that colchicine (IX, Fig. 4), which is used in the treatment of gout, has a high general toxicity. When a methoxy-group is replaced by a hydroxy-

Fig. 3 Toxicity of flavanoids.

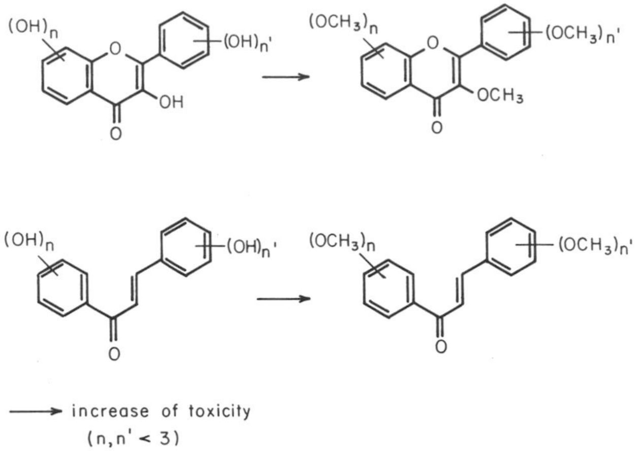

→ increase of toxicity
($n, n' < 3$)

Fig. 4 Colchicine and derivatives.

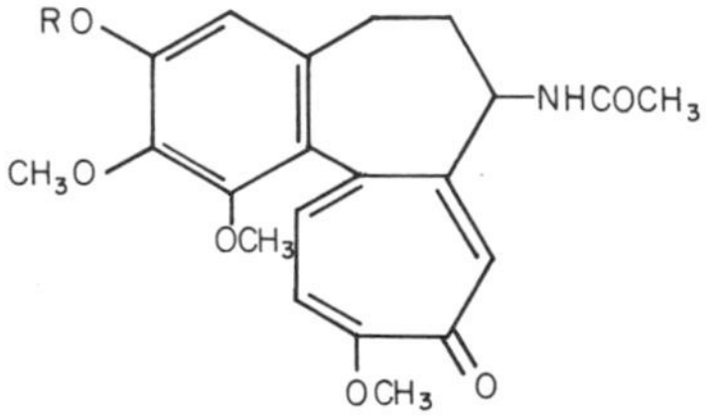

IX. Colchicine: $R = CH_3$
X. Substance C: $R = H$
XI. Colchicoside: R = glucosyl

group in the colchicine molecule, the so-called Substance C (X) is obtained; when this hydroxy-group is linked to a sugar molecule, a compound called colchicoside (XI) results. The polarity of these molecules increases in the order: colchicine-substance C-colchicoside. A striking decrease in toxicity is associated with this increase in polarity: colchicoside is only slightly poisonous.

I need not give any more examples of the correlation between activity and physical properties in discovering drugs. We saw how a chemist could isolate an active principle from a complex plant extract by systematic stepwise enrichment, without needing to carry out experiments on animals. It is possible, in many cases, to plan the best way to isolate active principles merely by proper formulation of a question or by a suitable working hypothesis. To state it in another way: it is well worthwhile, when investigating active plant principles, to utilize, as far as possible, all the experience of previous investigators.

CONCLUSION

This survey, entitled "Medicinal Plants and Empirical Drug Research," has been essentially a review of the literature. The term "Medicinal Plants" has been used as a synonym for "plants and plant constituents with pronounced activity in men and animals." The active principles were found empirically, that is, by means of practical research based on experience, rather than as a result of theoretical deductive reasoning. In certain stages during the search for active principles, it is possible to forego animal experiments ordinarily needed to evaluate preliminary results and to formulate hypotheses. Examples of such hypotheses have been given and labeled *chemotaxonomic* and *correlative* working hypotheses. Of course, all of the empirically established rules may fail to apply when fundamentally new problems arise. There are no working hypotheses enabling us to find plant constituents with certain specific predetermined kinds of activity.

The number of compounds isolated from plants so far is much smaller than the number of synthetic chemicals; the ratio is probably 1:100. The question naturally arises whether the search for new types of plant constituents is coming to a close. However, the great number of new discoveries, such as the antibiotics and cytostatic agents, which have arisen as a result of various systematic plant screening programs–phytochemical as well as pharmacological–seems to indicate that the pharmacochemical investigation of the Plant Kingdom is nowhere near its end. Many scientists active in drug research are dissatisfied with the empirical methods used so far and seek to place drug research on a more theoretical basis. Still, the empirical approach has proved itself from the pragmatic viewpoint–the viewpoint of both the patient and the physician. This is evident from the results already obtained, and so long as empirical drug research continues, medicinal plants will continue to play an important role.

ACKNOWLEDGMENTS

I would like to thank Professor V. E. Tyler, Dean, School of Pharmacy, Purdue University, Lafayette, Indiana, for his great help in translating the German manuscript.

I am indebted to Prof. R. E. Schultes, Director and Curator of Economic Botany, Botanical Museum of Harvard University, for valuable suggestions and advice during the preparation of this manuscript.

References

1. W. A. Sexton, *Chemische Konstitution und biologische Wirkung* (Verlag Chemie, 1958), p. 351.
2. R. Hegnauer, *Chemotaxonomy der Pflanzen* (Basel-Stuttgart, 1963-69), Vols. I-V.
3. R. Hegnauer, "Chemotaxonomy: past and present, *Lloydia 28* (1965), 267.
4. H. Erdtman, "Organic Chemistry and Conifer Taxonomy," in *Perspectives on Organic Chemistry,* ed. A. Todd (New York-London, 1956), p. 453.
5. H. Reznik, "Vergleichende Biochemie der Phenylpropane," *Ergebn. Biol. 23* (1960), 14.
6. R. F. Raffauf, "Some chemotaxonomic considerations in the Apocyneceae," *Lloydia 27* (1964), 286.
7. Mothes' definition (1950), in H. Römpp, *Chemielexicon,* 5th ed. (Stuttgart, 1962), I, 147.
8. R. Hänsel, "Flavonoide, Endausgestaltung und Verteilung über das Pflanzensystem," *Planta medica 14* (1962), 453; "Flavonoide, chemische Wertbestimmung und therapeutische Wirkung," *Pharm. Weekbl. 100* (1965), 1425.
9. H. Rimpler, "Zur chemotaxonomischen Wertigkeit chemischer Merkmale: Exkretionsmechanismen und Exkrete," *Planta medica 13* (1965), 412.
10. K. Münzel, "Einfluss der Formgebung auf die Wirkung eines Arzneimittels, in *Progress in Drug Research* (Basel-Stuttgart, 1966), X, 204.
11. J. F. Shvarov, "Pharmacological examination of osthol. ref.," *Chemical Abstracts 62* (1965), 2140.
12. F. v. Werder, Über Abkömmlinge der Cumarin-3-carbonsäure, einer neuen Klasse synthetischer Heilmittel," in *E. Merck's Jahresberichte 50* (1936), 88.
13. M. P. Duquenois, M. Haag-Berrurier, and E. Greib, "Metabolisme et chez l'homme du 7-β-glucoside d'ombelliferone," *Bull. Acad. nat. Med.* (Paris) *149* (1965), 451.
14. R. Hänsel, "Characterization and physiological activity of some Kawa constituents," *Pacific Science 22* (1968), 239.
15. L. Freedman and A. J. Merritt, "Citrus flavonoid complex, chemical fractionation and biological activity," *Science 139* (1963), 344.
16. K. Walther, H. Rimpler, and C. Leuckert. "Vorkommen von Falvonoiden in den Exkreträumen von Citrusfrüchten," *Planta medica 14* (1966), 453.
17. V. V. S. Murti, N. V. S. Rao, and T. R. Seshadri, "Insecticidal properties and chemical constitution," *Proc. Indian Acad. Sci. 25A* (1947), 22.
18. T. R. Seshadri and N. Viswanadham, "Insecticidal properties and chemical constitution," *Proc. Indian Acad. Sci. 25A* (1947), 337.

Biodynamic Agents from Microorganisms *

NESTOR BOHONOS

Children's Cancer Research Foundation Inc. and Department of Pathology,
Harvard Medical School at Children's Hospital,
Boston, Massachusetts

Microorganisms have long been recognized as being involved in plant and animal diseases; the preparation of food or feed products; the production of alcoholic beverages, organic solvents and acids; nitrogen fixation; decomposition of organic matter; production of toxins, and so on. It is, however, only in the last thirty years that they have become important in the pharmaceutical and animal feed industries because of their sophisticated chemical products or processes.

Drugs made by microbial processes (antibiotics, corticoids, and vitamin B_{12}) constituted approximately 16 percent ($654 million) of the total manufacturers' shipments of ethical pharmaceutical preparations ($4,143 million) made in the United States in 1967 (1). Antibiotics are the major commercial fine chemicals produced by microorganisms and because of their phenomenal impact on the practice of medicine, the term antibiotics has in many circles become synonymous with microbial products. This has in some instances prejudiced the evaluation of microbial compounds for biological activity not related to the observed phenomenon of antibiosis, and it is unfortunate that chemotherapists were not initially more receptive to the potential of microbial products as biodynamic agents for many types of activities.

*The preparation of this review was supported in part by Research Grant No. FR05526 from the Division of Research Facilities and Resources, National Institute of Health.

Other papers in this symposium discuss some aspects of microbial products. The objective of this presentation is to broadly review the many potentials and characteristics of natural product research with microorganisms. Since it is impossible to include all important observations, examples will be selected to illustrate the approaches which have been used, the versatility of microbes as chemical synthesizers, and the types of biological activities which have been observed. It is understandable that a preliminary observation of activity does not necessarily indicate a clinically useful product. Such a compound may, however, be a candidate for further evaluation, a structural model for chemical studies, or it may suggest testing of related products.

In contrast to natural product research involving animals or higher plants, that with microorganisms has the following characteristic advantages:

1. Cultures are readily available from collections or by isolation.
2. Many cultures may be grown with little effort in relatively simple equipment.
3. The short generation time facilitates rapid multiplication of cells or utilization of substrates.
4. Much technology has already been developed for cultivation of the more complex microorganisms—e.g. Basidiomycetes and tissue cultures, as well as for the scale-up of fermentations and isolation procedures.
5. Modification of fermentation conditions and/or mutations may result in higher yields and/or new products.
6. The large spectrum of products and activities already discovered indicates good potential for further research.

In some instances new products or processes have been discovered as a consequence of astute observations in laboratories, but most microbial product knowledge has developed as a consequence of organized search or screening programs. These programs have been based on such characteristics as the following:

1. Physical or chemical properties of compounds such as crystalline deposits in culture growths; color; u.v. absorption; color reactions; paper or thin-layer chromatography; carbon, halogen, sulfur or phosphorous balances; isotope tracer techniques, and so on.

2. Growth stimulation, growth inhibition or toxicity tests with bacteria, yeasts, fungi, algae, protozoa, higher plants, animals, insects, tissue slices, tissue cultures, and so on.
3. Induction of morphological changes in microbial, plant or animal cells.
4. Effects on glucose, protein, nucleic acid, or fat metabolism; inhibition of specific enzyme activities or viral multiplication.
5. Pharmacological activities such as those involving the central nervous system, anti-inflammatory effects, prolongation of phenobarbital sleeping times, contraction or relaxation of smooth muscles, diuretic action, effect on blood pressure or blood clotting, and so on.
6. Investigation of toxic principles in food poisoning or plant diseases.
7. Folklore indications—e.g. biologically active agents in ergot producing organisms and Basidiomycetes.

Some screening procedures, such as the microbial inhibition tests, are so sensitive and nonspecific that it is relatively easy to observe activities. Frequently an organism may produce several compounds which will respond in a test system, and in most screening programs an early objective is to resolve activities as well as to differentiate them from those previously observed. This may involve a gross examination for culture types (on the premise that different types of cultures may favor a diversity of products); expansion of activity tests; simple extraction, adsorption and stability tests; chromatography; absorption spectra; and so forth. If such preliminary observations indicate a potentially new product, it is common practice—particularly in laboratories in which the objective is new useful drugs—to make a crude concentrate for *in vivo* testing. Those products which exhibit interesting *in vivo* activity may be further studied in a more intensive identification program before chemical isolation efforts are initiated. An identification program is greatly facilitated by a collection of reference compounds and an awareness of the characteristics of known products. The complexity of this work is probably best demonstrated by the fact that more than one thousand antibiotics have been reported from *Streptomyces* alone. Some of these have been obtained in such limited amounts that they have not been available for even limited distribution to the more active cooperating laboratories, and others have not been adequately characterized for absolute identification. Confusions in the literature exist because natural products are frequently announced prior to the determination of their structures with inadequate or erroneous description of characteristics, errors in purity or analysis, essentially simultaneous announcements, or disregard for priority publications.

STRUCTURALLY RELATED PRODUCTS

When a compound is found to have interesting properties, chemists generally prepare related products in order to improve on activity, minimize side effects or to develop an understanding of structure-activity relationships. In several series of compounds, the variations in structurally related products of microbial origin are as diverse and unique as might be expected from the most sophisticated synthetic program. Generally, different species of a genus are involved in producing major variants but in some cases, particularly with the fungi imperfecti, organisms belonging to diverse genera have been found to produce related products.

The purine nucleotides in Fig. 1 illustrate the diversity of microbial products which are related to adenosine or nebularine (purine riboside). 3'-Deoxyadenosine (cordycepin), 3-amino-3'-deoxyadenosine, puromycin, lysylaminoadenosine, homocitrullylaminoadenosine, angustmycin A (decoynin) and angustmycin C (psicofuranine) have unusual modifications of the ribose moiety (2, 3). In aristeromycin the ribofuranoside unit is replaced by a substituted cyclopentyl radical. The antibiotic nucleocidin (4) is a sulfamic acid ester of an adenine C6 glycoside (5). In addition to S-adenosylmethionine, an important biological methyl donor, S-adenosylethionine (6), 5'-methylthioadenosine (7-10), and 5'-ethylthioadenosine (11) have been obtained from microbial systems. Tubercidin, toyacomycin, and sangivamycin are 7-deaza derivatives—the latter two compounds containing unusual purine substituents on the 7 carbon. In the formycins, a pyrazole replaces the imidazole ring and the ribose moiety is attached by a C-C linkage. The 6-amino position in puromycin and N^6-(Δ^2-isopentenyl)-adenosine is alkylated, and in septacidin the glycosidic linkage is in this 6 position. Of the compounds indicated in Fig. 1, 3'-deoxyadenosine (12), lysylaminoadenosine (13), and homocitrullylamenoadenosine (14) were isolated from *Cordyceps militaris;* the isopentenyladenosine from *Corynebacterium fascians* (15) and tRNA of yeast (16); and the other compounds were isolated as antibiotics from *Streptomyces* (2, 3). A methylthio-derivative, N^6-(Δ^2-isopentenyl)-5'-methylthioadenosine occurs in tRNA of

Fig. 1

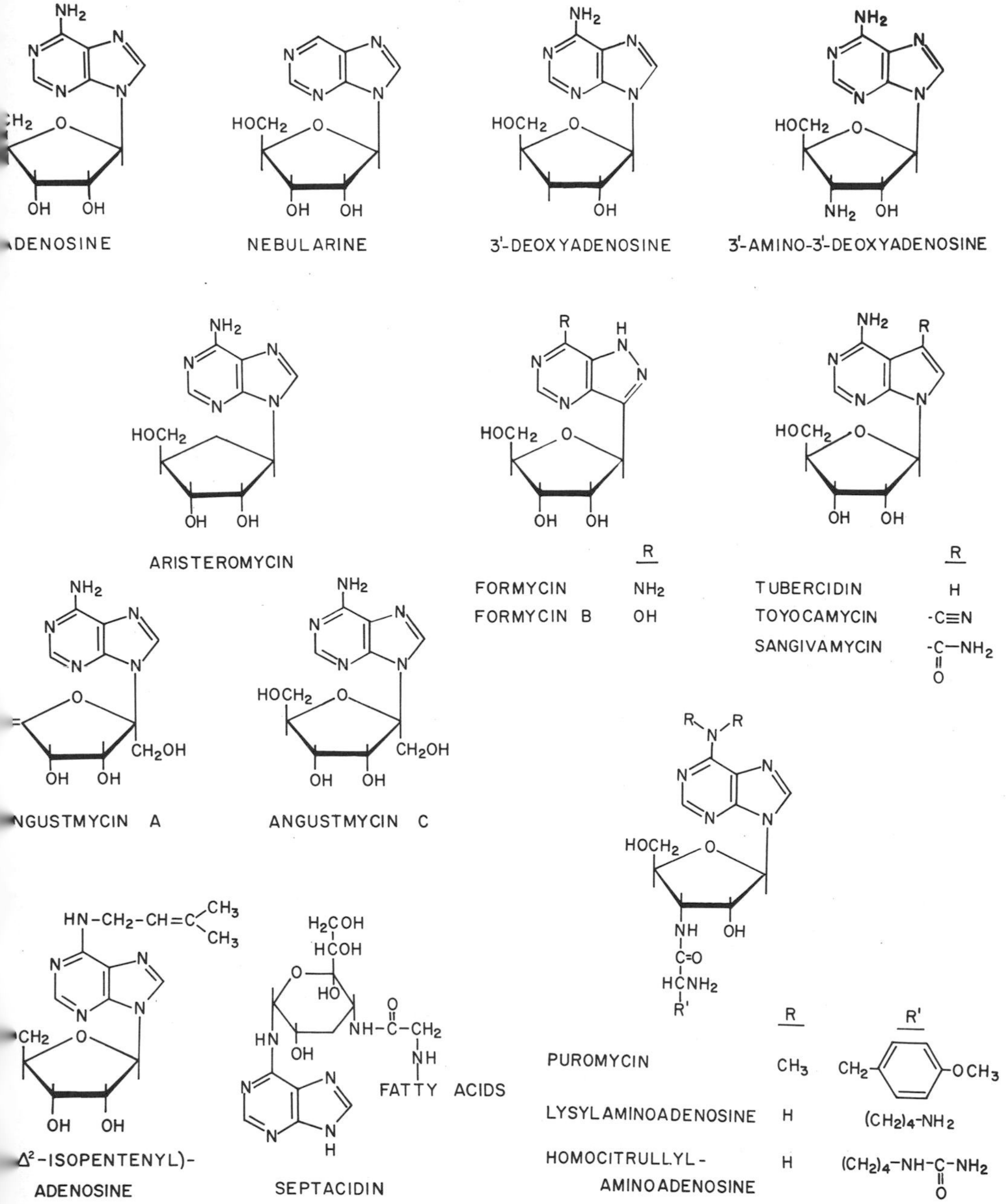

DENOSINE

NEBULARINE

3'-DEOXYADENOSINE

3'-AMINO-3'-DEOXYADENOSINE

ARISTEROMYCIN

	R
FORMYCIN	NH_2
FORMYCIN B	OH

	R
TUBERCIDIN	H
TOYOCAMYCIN	-C≡N
SANGIVAMYCIN	-C(=O)—NH_2

NGUSTMYCIN A

ANGUSTMYCIN C

	R	R'
PUROMYCIN	CH_3	CH_2–C_6H_4–OCH_3
LYSYLAMINOADENOSINE	H	$(CH_2)_4$-NH_2
HOMOCITRULLYL-AMINOADENOSINE	H	$(CH_2)_4$-NH-C(=O)-NH_2

Δ^2-ISOPENTENYL)-ADENOSINE

SEPTACIDIN

Escherichia coli (17, 18). A hydroxy-isopentyladenosine, zeatin riboside, has been isolated from the fungus *Rhizopus roseolus* (19).

The recently reported antiviral antibiotic pyrazomycin is 3(5)-ribofuranosyl-4-hydroxypyrazole-5(3)-carboxamide (20). This compound has a striking similarity to the 5-amino-4-imidazolecarboxamide ribonucleotide intermediate in purine nucleotide biosynthesis and to the formycins. It is tempting to speculate that if the organism which produces this antibiotic did not have a genetic block in the pyrimidine ring closure, the end product would have been a formycin.

An equally impressive example of the production of structurally related products is found in the actinomycins. More than twenty actinomycins have been characterized and found to contain different combinations of amino acids (2, 3). The type of actinomycin produced depends upon the strain, the composition of the medium, the stage of growth, and the possible assimilation of compounds added to the medium—e.g. sarcosine, piperidine-2-carboxylic acid, and isoleucine. All of the actinomycins are produced by *Streptonyces* but the phenoxazine chromophore has also been observed in the pigments cinnabarin (polystictin) (21-23) and tramesanguin (24) produced by *Trametes cinnabarina.*

Many other antibiotics, pigments, and fungal metabolites are found in structurally related forms.

GROWTH FACTORS AND VITAMINS

The first microbial biodynamic agents to be discovered were the macromolecular toxins, and much is known about the production and characteristics of some of these endotoxins and exotoxins. They are, however, compounds with limited utility except in the production of biologicals, and they present no leads for chemical studies.

The first really important group of microbially produced pharmacologically active agents to receive intensive study and to be significant contributions to medical science were the growth factors, or vitamins. Studies initiated in the late thirties revealed that many of the factors required for the growth of some fastidious organisms on chemically defined media were frequently the same as some of the water-soluble vitamins required for animal growth. It also became apparent that some microorganisms which did not require these growth factors were indeed excellent producers of these products; hence they were used for the preparation of nutritional supplements and in some cases of pure vitamins. Because of their adaptability and convenience, microbial assays were developed for some vitamins. Several compounds were first discovered and isolated as microbial growth factors before they were recognized as vitamins. Microbial studies have been of significance in the discovery, development, or production of ascorbic acid, biotin, β-carotene, cobalamin, folic acid, nicotinamide, pantothenic acid, pyridoxine, riboflavin, thiamin, vitamin D, and vitamin K. Highly efficient commercial fermentation processes have been developed for the production of riboflavin and β-carotene. All of the commercial cobalamin (vitamin B_{12}) is produced microbiologically. A fermentation step is critical in the production of L-sorbose for the synthesis of ascorbic acid.

Natural product research abounds in examples of serendipity, and the folic acid saga has several examples of this. Folic acid was first reported to be a growth factor for bacteria (24) and subsequently (25) it was suggested that it was the same as Factor U, a chick growth factor. Although liver and yeast were relatively rich sources of folic acid, the isolation of this factor was difficult and laborious because of the trace quantities in these materials. In the course of investigating superior sources of this vitamin, a *Corynebacterium* isolated as a contaminant in a riboflavin fermentation was found to produce relatively high yields, and sufficient crystalline material was isolated (26) for chemical and biological studies. Subsequently it became apparent that there was a series of folic acid compounds and the first product isolated from liver was characterized as pteroylglutamic acid (Fig. 2). The product isolated from the *Corynebacterium* was identified as pteroyltriglutamic acid (27). An investigation of comparative vitamin contents in malignant and normal tissues revealed the largest increase in folic acid in tumors (28); thus, when the existence of different folates was established, it was of interest to determine if any of these compounds affected tumor growth. Tests with spontaneously developed mammary tumors in rats revealed antineoplastic activity of pteroyltriglutamic acid (29, 30), whereas the pteroylglutamic acid was inactive (31). Pteroyltriglutamic acid was the first isolated microbial product to be found active in inhibiting the growth of malignant tissues. A program of synthesis of folic acid antimetabolites was initiated, and this led to the development of the first clinically effective anticancer agents aminopterin (32) and methotrexate (amethopterin) (33).

The development of the iron complexing agent "desferrioxamine B" is another example of the modification of a growth factor to a new, clinically effective agent. Following the isolation of coprogen (34), a

Fig. 2

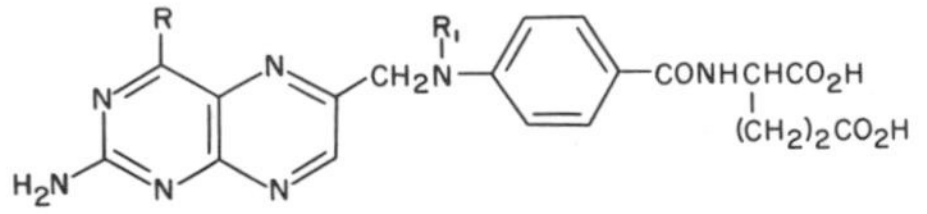

PTEROYLGLUTAMIC ACID (FOLIC ACID), R=OH, R_1=H
4-AMINOPTEROYLGLUTAMIC ACID (AMINOPTERIN), R=NH_2, R_1=H
4-AMINO-N^{10}-METHYLPTEROYLGLUTAMIC ACID (METHOTREXATE), R=NH_2, R_1=CH_3

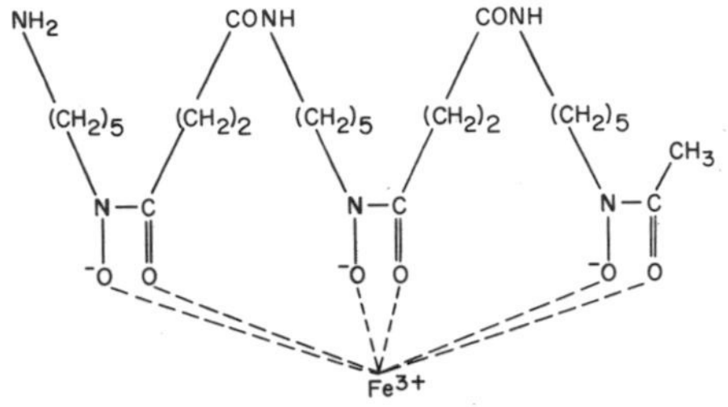

FERRIOXAMINE B

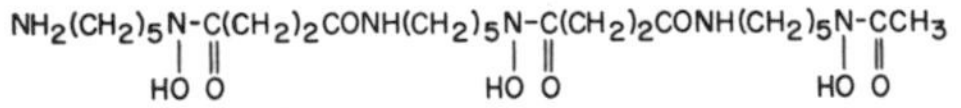

$NH_2(CH_2)_5N(OH)\text{-}C(=O)(CH_2)_2CONH(CH_2)_5N(OH)\text{-}C(=O)(CH_2)_2CONH(CH_2)_5N(OH)\text{-}C(=O)CH_3$

DESFERRIOXAMINE B

number of iron chelate growth-promoting agents were isolated (35). The iron could be removed from some of these products, and from ferrioxamine B, desferrioxamine B (Fig. 2) was obtained (36). *In vivo* testing indicated the desferro compound to be a very effective iron scavenger, and it was developed for the clinical treatment of primary and secondary hematochromatosis (37) and acute iron poisoning (38).

ANTIBIOTICS

The recognition of the clinical efficacy of antibiotics virtually caused an explosion of research for new antibiotics and, with this, research on fermentation processes, microbial physiology, and microbial products. The collaborative efforts of microbiologists, chemists, and chemotherapists resulted in some sixty antibiotics which were of sufficient efficacy to be produced commercially. One of the great developments from penicillin research was the refinement and scale-up of aerobic fermentations to the 60,000- and 100,000-gallon capacities now used by some manufacturers. The original penicillin-producing cultures were poor producers but mutations and media were developed with increased yields

TABLE 1

Properties of Some Penicillins and Cephalosporins

	Activity against				
Compound	Pen'ase-neg.[a] *Staph. aureus*	Pen'ase-pos.[a] *Staph. aureus*	Other cocci	Gram-neg. bacilli	Oral absorption
Penicillins					
Penicillin G	High	None	High	None	Poor
Penicillin V	High	None	High	None	Good
Phenethicillin	High	Low	Variable	None	High
Oxacillin	High	High	Variable	None	High
Nafcillin	High	High	Variable	None	Medium
Cloxacillin	High	High	Medium	None	High
Dicloxacillin	High	High	Medium	None	High
Ampicillin	High	None	High	Medium	High
Cephalosporins					
Cephalothin	High	High	High	Medium	Poor
Cephaloridine	High	High	High	Medium	Poor
Cephalexin	High	High	High	Medium	High

[a]Pen'ase = Penicillinase

more than a hundredfold. Initial penicillin fermentations produced mixtures of penicillins with characteristic biological and chemical properties, but by selection of cultures and the use of precursors, fermentations could be directed to produce primarily one specific product. This technology was readily adapted to other fermentations.

With extensive use of the penicillins it became apparent that sensitization to these antibiotics and the development of resistant strains were becoming serious problems. New semisynthetic penicillins (Table 1) which had increased antibacterial activity and/or improved resistance or pharmacological characteristics were produced by acylation of the 6-aminopenicillanic acid degradation unit of penicillin (Fig. 3). The development of fermentations to produce 6-aminopenicillanic acid in good

Fig. 3

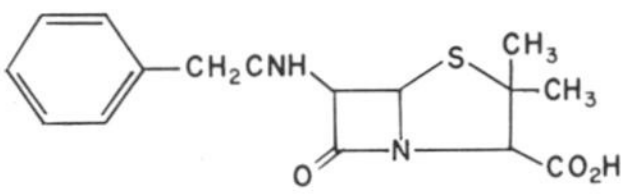

BENZYLPENICILLIN (PENICILLIN G)

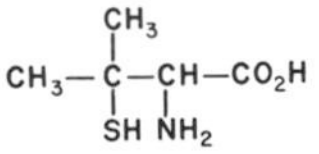

PENICILLAMINE

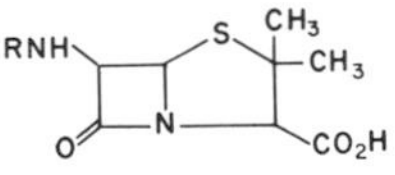

6-AMINOPENICILLANIC ACID (R=H)

SEMI-SYNTHETIC PENICILLINS (R=ACYL)

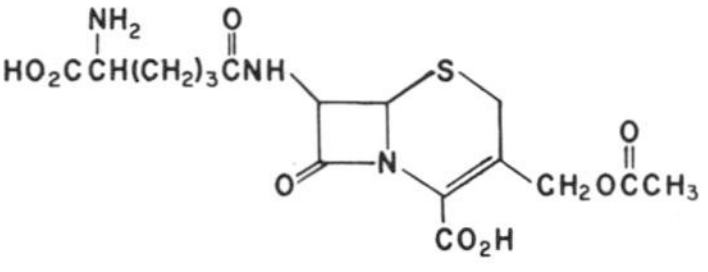

CEPHALOSPORIN C

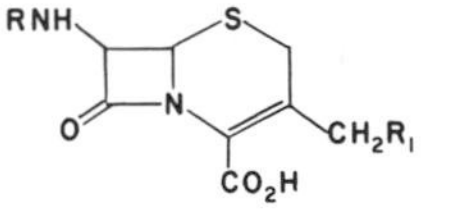

7-AMINOCEPHALOSPORANIC ACID ($R=H, R_1=CH_3\overset{O}{\overset{\|}{C}}O-$)

SEMI-SYNTHETIC CEPHALOSPORINS (R=ACYL)

yields facilitated the commercial production of the new semisynthetic penicillins (3, 39, 40). With the somewhat chemically similar antibiotic cephalosporin C (Fig. 3), the conversion of 7-aminocephalosporanic acid has made possible the production of more efficacious cephalosporin antibiotics (40).

Another spin-off from penicillin research has been the degradation product penicillamine (β,β-dimethylcysteine). The diversity of pharmacological properties of this compound are attributed to its chelating capacity, sulfhydryl reducing activity, or its vitamin B_6 antagonism. D or DL-Penicillamine has been reported active in the treatment of Wilson's disease (41), lead poisoning (42), cysteinuria (43), macroglobinemia (44), and schizophrenia (45). It can influence immune response (46), the level of circulating rheumatoid factor (47), and collagen metabolism (48). Although this drug is being used clinically, it is implicated in severe side reactions, and there is considerable controversy regarding its use.

The tetracyclines are the most widely used broad spectrum antibiotics. The five most important commercial tetracyclines (Fig. 4) differ in substituents in the 5, 6, and 7 positions and in pharmacological properties. 7-Chlortetracycline, tetracycline, and 6-demethyl-7-chlortetracycline are produced by *Streptomyces aureofaciens.* Tetracycline was discovered by hydrogenolysis of chlortetracycline (49, 50) and in fermentations (51, 52). Because the chlortetracycline process was developed to a very high efficiency, it was more feasible to produce tetracycline by the chemical conversion procedure. Commercial fermentation procedures have now been developed for the production of tetracycline, but it is believed that the bulk of this antibiotic is still made from 7-chlortetracycline. The dis-

Fig. 4

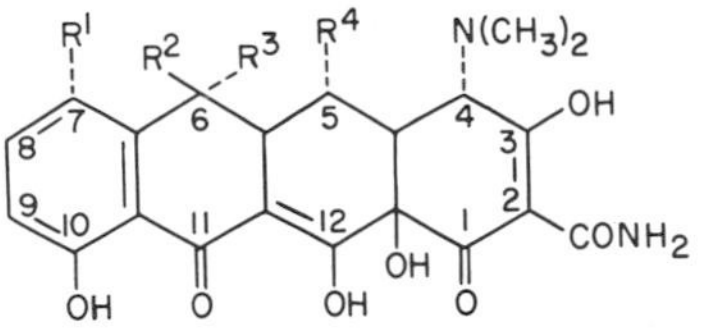

	R^1	R^2	R^3	R^4
TETRACYCLINE	H	OH	CH_3	H
7-CHLORTETRACYCLINE	Cl	OH	CH_3	H
6-DEMETHYL-7-CHLORTETRACYCLINE	Cl	OH	H	H
5-HYDROXYTETRACYCLINE	H	OH	CH_3	OH
6-METHYLENE-5-HYDROXYTETRACYCLINE	H	–	$=CH_2$	OH
α-6-DEOXY-5-HYDROXYTETRACYCLINE	H	H	CH_3	OH

covery of 6-demethyl-7-chlortetracycline arose from the observation that one of the mutants from the *Streptomyces aureofaciens* strain development program lacked the capacity to introduce the methyl radical in the 6 position (53). 5-Hydroxytetracycline was first obtained from a *Streptomyces rimosus* culture filtrate (54), and this product serves as starting material for chemical conversion to 6-methylene-5-hydroxytetracycline (55) and 6-*a*-deoxy-5-hydroxytetracycline (56). 7-Chlortetracycline and 5-hydroxytetracycline have individual superior characteristics, and it was hoped that a drug such as 5-hydroxy-7-chlortetracycline might prove superior. This product was prepared using procedures developed in biogenetic studies (57) but its extreme instability precluded any therapeutic potential. A new tetracycline, 7-dimethylamino-6-deoxy-6-demethyltetracycline, has been synthesized and found to have an antibacterial spectrum quite different from the other tetracyclines (58). It has been introduced for the treatment of some bacterial infections which are refractory to the older tetracyclines.

The clinically important macrolide, N-glycoside, and peptide antibacterial antibiotics were all discovered in direct antibacterial screening programs. Chloramphenicol is the only antibiotic which in commerce is made exclusively by organic synthesis.

Griseofulvin, a clinically effective antifungal antibiotic, was developed as a consequence of an observation that conifers would not grow in a certain heath soil. The predominating fungus in this soil was found to produce "curling factor," a product so designated because it caused a distortion in the hyphae of *Botrytis allii* (59). After isolation and chemical characterization, this product was found to be identical with griseofulvin, first isolated in a chemical screening program on fungal metabolites (60). Experimental quantities were made available for biological testing, and observations of systemic antifungal activity in plants were soon followed by reports that orally administered griseofulvin was effective in the treatment of ringworm in guinea pigs (61) and cattle (62). Nineteen years after its discovery as a chemical curiosity, it was found effective in human therapy (63, 64). A number of species of *Penicillia* produce this antibiotic, and both the dechloro- and bromo- analogues have been isolated from fermentations (65, 66). Griseofulvin and a large number of related products have been synthesized, but to date none of these has been accepted as clinically superior for the treatment of dermatomycoses. Anti-inflammatory activity was observed when this product was used for the treatment of man and guinea pigs (61, 67). This activity was confirmed with the conventional cotton pellet and tuberculin skin sensitivity tests, and is now attributed to the metabolism of griseofulvin to 3-chloro-4, 6-dimethoxysalicylic acid (68). This antibiotic has also been reported active in the treatment of gout (69), peripheral diseases (70), and coronary dyscrasias (71). Griseofulvin can, however, disturb porphyrin excretion in man (72), alter prothrombin time of patients on warfarin therapy (73), and exhibit hepatocarcinogenicity in mice (74, 75).

A number of microbial products have been found with structural similarities to griseofulvin (Fig. 5) but few have had any biological testing. One of these, geodin (76), is inhibitory to fungi, but no systemic activity has been reported for it. Another closely related product, trypacidin, was isolated in a screening program for antiprotozoal antibiotics (77). It was found to be highly effective in the control of *Toxoplasma gondii* infections in mice (78).

Azomycin (Fig. 5) was first reported as an antibacterial antibiotic of high nitrogen content (79), and three years later it was stated to have antitrichomonal activity (80). An independent discovery of azomycin as an antitrichomonal antibiotic (81) led to a synthetic program on nitroimidazoles and the discovery of the clinically important drug metronidazole (82), which possesses higher antitrichomonal activity and lower toxicity.

Fig. 5

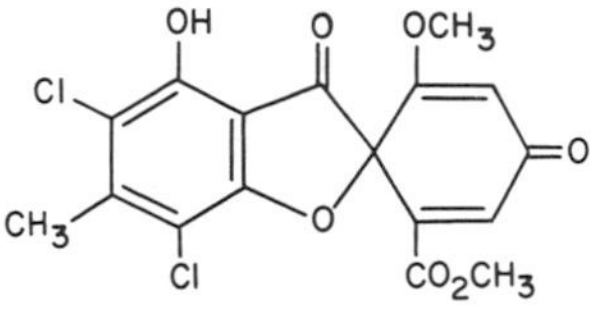

GEODIN

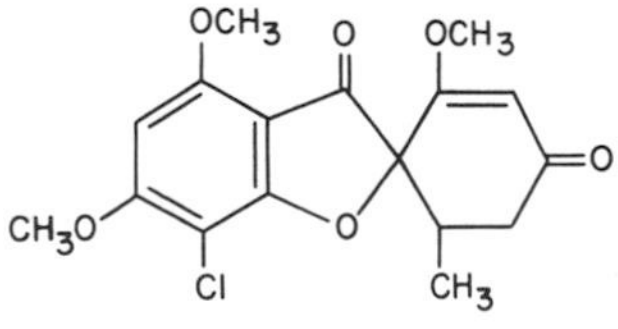

GRISEOFULVIN

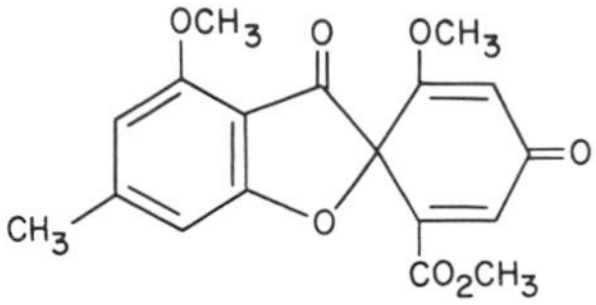

TRYPACIDIN

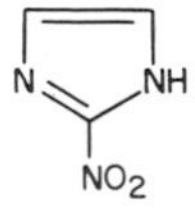

AZOMYCIN

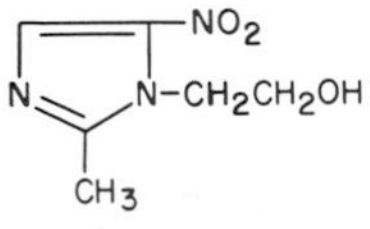

METRONIDAZOLE
(8823 R.P., FLAGYL)

Fig. 6

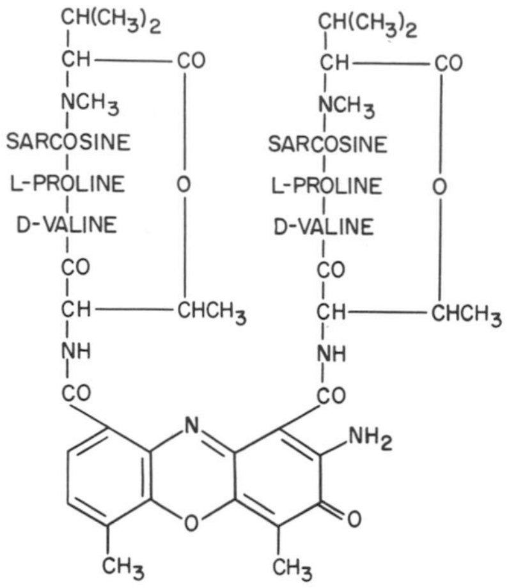

ACTINOMYCIN D

A tremendous amount of work has been done in screening for cytotoxic antibiotics with the expectation that such products would be antineoplastic agents. Antibiotics obtained by other screening procedures have also been tested in conventional animal tumor test systems. That these efforts have been fruitful is indicated by the number of antibiotics with accepted clinical efficacy and those compounds currently in clinical testing. Among the more important microbial products exhibiting antitumor activity are the actinomycins, adriamycin, bleomycin(s), chromomycin(s), daunomycin, diazomycins, mithramycin, the mitomycins, nogalomycin, olivomycin, puromycin, streptonigrin, steroid hormones, asparaginases, polysaccharides, and other macromolecules.

Actinomycin D (Fig. 6), the first antibiotic clinically accepted for the treatment of malignancies (83), was isolated because of the antibacterial activity it displayed (84). Very few of the actinomycins have had adequate antitumor testing to warrant any conclusions on structure activity relationships. In addition to the actinomycins there are other series of antineoplastic antibiotics in which the individual products are difficult to separate and have not had sufficient evaluation. It is of interest that 8-azaguanine and 5-azacytidine were first synthesized chemically as prospective antitumor agents but subsequently isolated as antibiotics (85, 86). 6-Azacytidine, 6-azauridine, and several other uridines have been obtained by microbial incorporation of the appropriate synthetic pyrimidines added to the fermentation medium (87, 88).

The activities of the antitumor antibiotics may be attributed to their effects on ribonucleic acid, deoxyribonucleic acid and protein metabolism; it is therefore not surprising that many of them exhibit antiviral activity in experimental systems. In recent years several investigators have reported on the antiviral activities of macromolecular products and of disulphide-bridged diketopiperazine compounds related to gliotoxin, the first microbial product isolated because of its antifungal or antibiotic properties (89).

The antitumor antibiotics actinomycin D, puromycin, and mitomycin C decrease immune response in organ transplants (90, 91, 92). Chloramphenicol, actinomycin D, puromycin, and cycloheximide are commonly used inhibitors in biochemical metabolism studies. The latter two compounds have also been useful in studies on memory processes (93, 94).

Other activities observed with compounds primarily classified as antibiotics include effects on neuromuscular activity, lipolysis, liver regeneration, contraction of smooth muscles, and changes in vascular permeability.

ENZYMES

Asparaginases from several bacterial sources are currently showing considerable promise in the treatment of specific malignancies. Such microbial enzymes as amylases, cellulases, hemicellulases, pectinases, pentosanases, invertases, lactases, glucose oxidases, dextranases, catalases, peroxidases, proteases (including streptokinase, fibrinolytic enzymes, keratinase and kininase), clearing factor lipase and uricase are useful or of interest in systemic, topical, or oral therapy; for diagnostic or for nontherapeutic purposes.

TOXINS

Many excellent reviews detail the occurrence, chemistry, and biological activity of microbial toxins found in food or feed products (95-98), or as phytotoxic agents in plant diseases (99, 100).

Ergot, a mixture of peptide-type ergot alkaloids was probably the first fungal toxin to be recognized and used therapeutically. Traditionally it consisted of the dried sclerotia of the fungus *Claviceps purpurea* grown on rye plants, but it is now possible to produce these alkaloids by submerged fermentations of specific strains (101). The alkaloids are clinically useful as oxytocic agents.

The orally effective anticoagulant dicoumarol was discovered as the active agent in spoiled hay which could cause capillary hemorrhaging of

Fig. 7

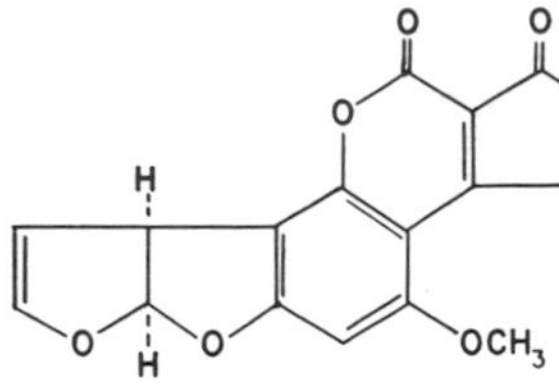

AFLATOXIN B_1

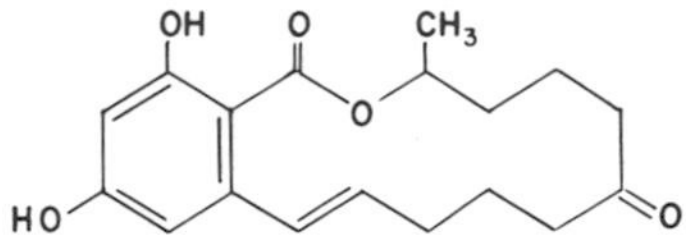

ZEARALENONE

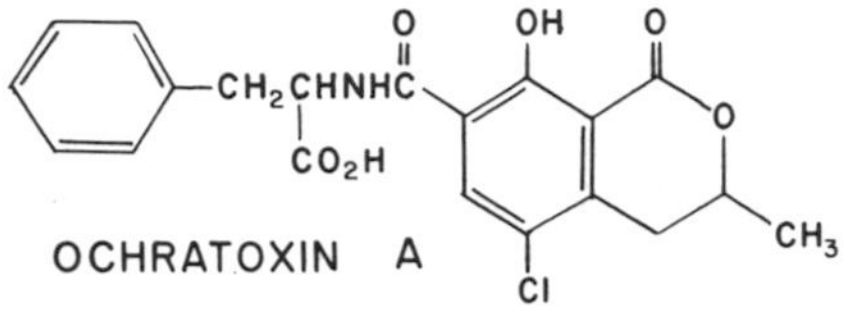

OCHRATOXIN A

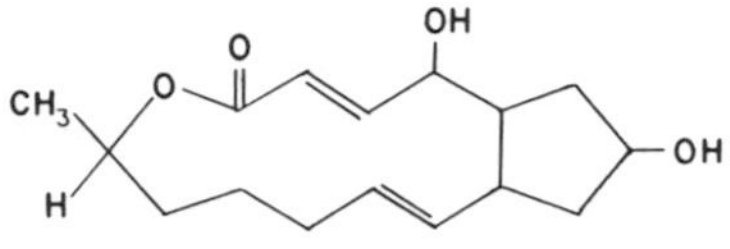

DECUMBIN

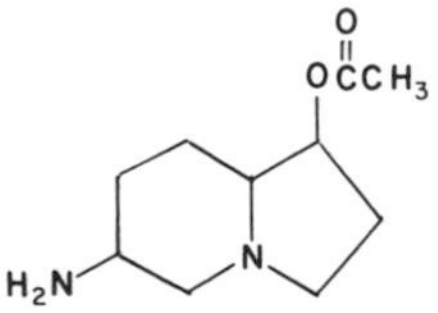

SLAFRAMINE

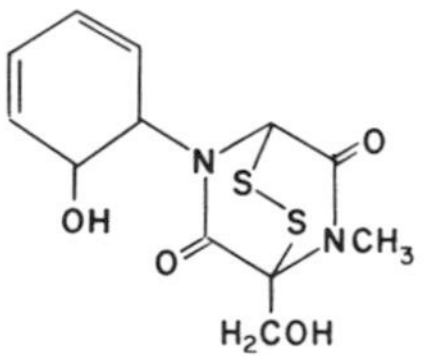

GLIOTOXIN

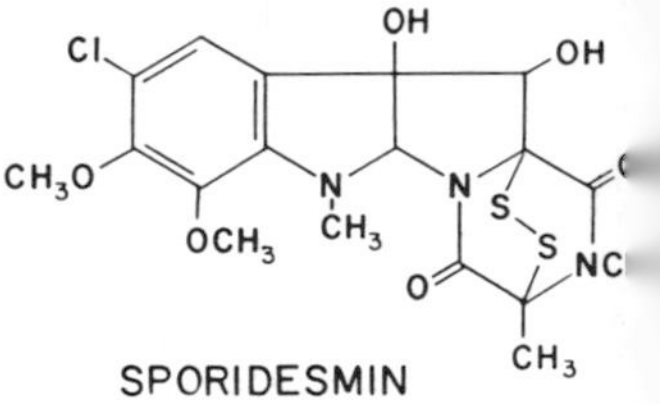

SPORIDESMIN

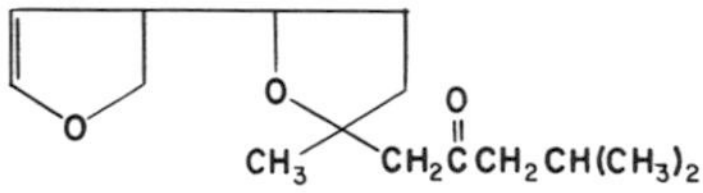

IPOMEAMARONE

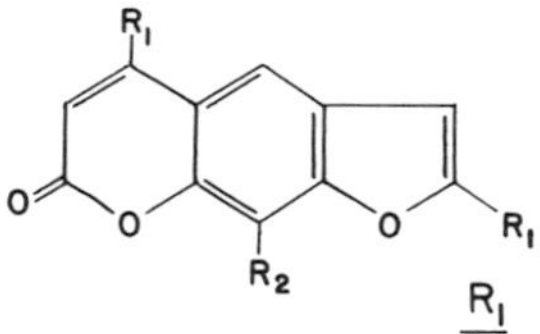

	R_1	R_2
8-METHOXYPSORALEN	H	OCH_3
4,5′,8-TRIMETHYLPSORALEN	CH_3	CH_3

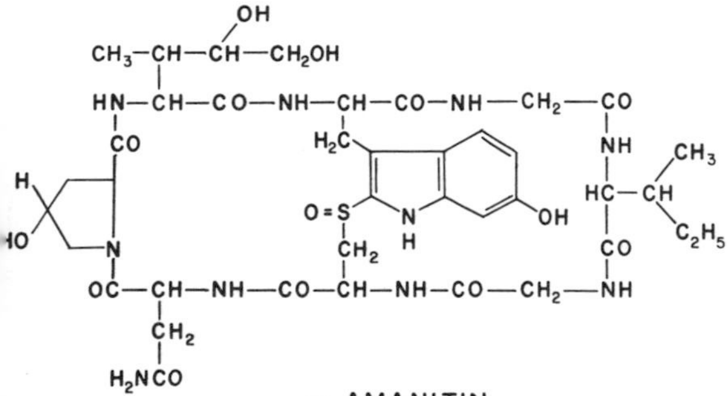

α-AMANITIN

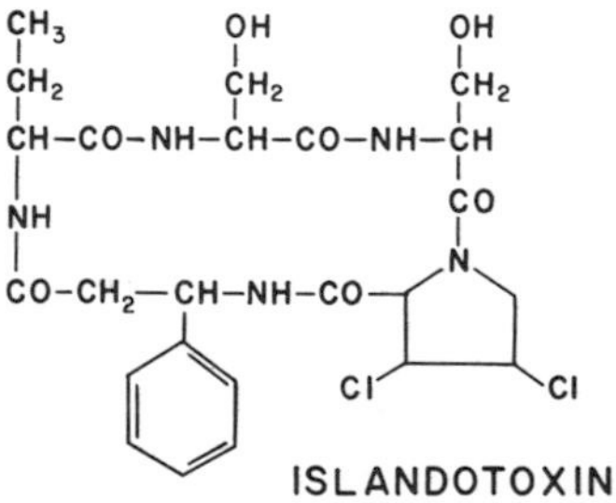

ISLANDOTOXIN

cattle (102). This product is believed to be formed by microbial conversion of the plant component o-coumaric acid to 7-hydroxycoumarin, which on reaction with formaldehyde is converted to dicoumarol (103).

The aflatoxins are a group of mycotoxins which first received attention in England because of serious losses in poultry fed peanut meal infected with *Aspergillus flavus*. This class of toxins has now been detected in many food and feed products. Aflatoxin B_1 (Fig. 7), the most toxic member, has an LD_{50} in ducklings in the order of 0.3 mg/kg (104). These compounds are highly carcinogenic, and an awareness of their presence, activity, and condition for their synthesis could have a significant impact on the epidemiology of cancer. Ochratoxin A, produced by *Aspergillus ochraceus,* has the same order of toxicity to ducklings as aflatoxin B_1 (105). It causes fatty infiltration of the liver (106) and inhibits bovine carboxypeptidase (107). Ochratoxin B (dechloro-ochratoxin A) and ochratoxin C (the ethyl ester of ochratoxin A) are relatively nontoxic (105).

Zearalenone was the causative agent in *Gibberella zeae* contaminated feeds which resulted in disturbances of estrogenic cycles in swine (108). This unique structure (Fig. 7) was amenable to synthesis, and a large number of related compounds have been made with estrogenic and/or anabolic activities. Zearalenone or its derivatives may become useful in animal husbandry.

Slaframine was isolated from *Rhizoctonia leguminocola* infected red clover which caused excessive salivation, diarrhea and, at times, death of cattle (109, 110). The administration of this compound to hepatectomized rats does not produce salivation; thus it appears that slaframine is transformed by the liver to another active product (111). Because this compound increases pancreatic flow it is being investigated for use in cystic fibrosis.

Decumbin, first isolated from a corn contaminant, can cause diarrhea and death (112). This compound was subsequently reported as cyanein (113) and brefeldin A (114). It exhibits a range of activities–e.g. antifungal, cytostatic, antimitotic, antinematocidal, antiviral, and inhibition of seed germination (115). These toxic manifestations may be associated with the α,β-unsaturated lactone structure.

Sporidesmin, with structural similarity to gliotoxin, is formed by the growth of *Pithomyces chartarum* on pasture grasses. It causes heptotoxicity and facial eczema of cattle and sheep (116).

8-Methoxypsoralen and 4,5′,8-trimethylpsoralen produced by *Sclerotia scleriotorum* (117) cause dermatosis in celery workers (118).

Other hepatotoxins include the amatoxins (119), the phallotoxins (119), islandotoxin (120), luteoskyrin (121), ipomeamorone (122), and xanthocillin (123). Citrinin (124) and citreomycetin (125) cause nephrotoxicity (126). Citreoviriodin causes paralysis and respiratory failure (127). Other mycotoxins include the spireopoxy compounds such as diacetoxyscirpenol, trichodermol, verrucarol, nivalenol, etc.; cyclopiazonic acid; patulin; maltorzine and muscarine (128). Unidentified products in contaminated feeds have been reported to include tremorigenic toxins (129, 130), emetic material (131), and agents which cause prolonged prothrombin times, hemorrhagic enteritis, subdural hemorrhaging and central nervous system disturbances (132, 133).

PHYTOTOXINS AND PLANT GROWTH REGULATORS

Toxic manifestations of many plant diseases are attributed to toxins produced by phytopathogens. Although relatively few toxins have been thoroughly investigated, it has become apparent that they represent a variety of structures and modes of actions (99, 100).

Wildfire toxin, isolated from *Pseudomonas tabaci,* can induce chlorotic lesions in tobacco and other plants (134). A methionine antimetabolite theory for the mode of action has been suggested because of reversal of toxicities to *Chlorella vulgaris* by methionine (135) and similar toxic manifestations of methionine sulfoximine in tobacco plants (136), algae (136), and rats (137). Both wildfire toxin and methionine sulfoximine inhibit glutamine synthetase prepared from rat brain (137) and from peas (138). Although the structure of this toxin is still not resolved (139) this compound does not appear to have any similarity to methionine.

Rhizobitoxinine, obtained from *Rhizobium japonicum,* causes chlorosis of many plants (140, 141). This toxin of unknown structure is believed to inhibit the enzymatic conversion of cystathionine to homocysteine (142).

Peptide and glycopeptide plant toxins have been associated with diseases of grains, potatoes, tomatoes, and digitalis.

Fusaric acid, dehydrofusaric acid, α-picolinic acid, phytotonivein, novarubin, and lycomarasmin have been implicated in *Fusaria* wilt diseases. These toxins are believed to affect electrolyte or water transport. Other types of fungal pathogens have been found to produce toxic quinones, hemiquinones, and glucosides.

Of particular interest are those compounds which have dose-dependent

Fig. 8

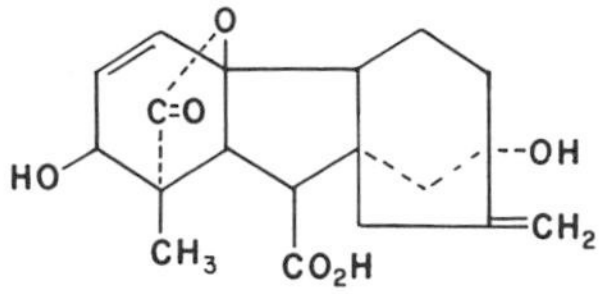

GIBBERELLIC ACID

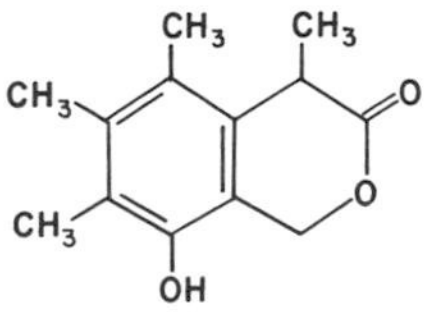

SCLERIN

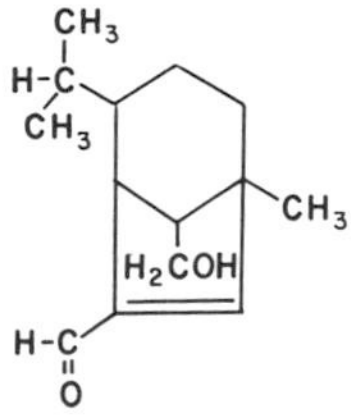

HELMINTHOSPOROL

toxic or growth-stimulating properties or which, on modification, can be converted from toxins to growth stimulators (Fig. 8). Gibberellic acid was first recognized as the active principle in "Bakanae byo," or "foolish seedling," disease of rice, a disease in which plants grow unusually tall and die. To date, twenty-four gibberellins have been identified from fungal and plant sources (143). These compounds are regarded as growth hormones because of their stimulation of growth in many plants at high dilutions. Gibberellic acid stimulates amylase formation and thus reducing sugars in germinating grains (144). In addition to its activity on plants, gibberellic acid can influence insect metabolism (145-147), mammalian resistance to ascites tumors (148), and the growth of *Mycobacterium tuberculosis* (149).

Helminthosporol, helminthosporal, and helminthosporic acid have some structural similarities to the gibberellins. The dialdehyde helminthosporal is a phytotoxin (150), whereas the reduction product helminthosporol can under specific conditions enhance plant growth (151). Both helminthosporol and helminthosporic acids stimulate the release of reducing sugars from de-embryonated barley (152).

The oxychromones sclerin, sclerotonin A, and sclerotonin B produced by a *Sclerotinia* fungus represent still another type of plant growth regulator (153, 154). Sclerin was found to stimulate amylase and lipase production in the organism from which it was isolated (153).

Indoleacetic acid (155), L-prolyl-L-valine anhydride (156, 157), and L-prolyl-L-leucine anhydride (157, 158) exhibit concentration-dependent stimulation on inhibition of plant growth.

6-Methoxymellein (159) and the polypeptide monilicolin A (160) are considered phytoalexins—a class of fungal induced host metabolites which are believed to account for host resistance to fungal infection.

INSECTICIDES

Bacterial, fungal, protozoal, and viral pathogens, as well as microbial products which exhibit insecticidal properties, have been recently reviewed (161). Compounds 379X (162), 379Y (162), piericidin A, and piericidin B (163) were discovered by systemic screening for microbial metabolites with insecticidal properties. Destruxins A and B (164) and aspochracin (165) were isolated from a fungus pathogenic to various insects. Tricholomic acid (166) and ibotenic acid (167) were isolated from agaric mushrooms. The antibiotics antimycin A, cycloheximide, hygromycin B, cytovirin, porfiromycin, pactomycin, sparsomycin, novobiocin, oxytetracycline, streptomycin, dihydrostreptomycin, neomycin, bacitracin, and polymyxin were found to have insecticidal activities in conventional tests.

At present, the greatest commercial potential is with preparations of the facultative insect pathogen, *Bacillus thuringiensis,* which infests many caterpillars.

MISCELLANEOUS ACTIVITIES

Microbial decarboxylation of amino acids to amines is a well-known reaction, and fermentation products such as certain cheeses can contain relatively high concentrations of physiologically active amines (168). When the decarboxylation is compounded with hydroxylation of the indole nucleus in tryptophane, psychotropic activity may result. It is not generally recognized that serotonin (5-hydroxytryptamine) has been isolated from microbial sources (169). The hallucinogen bufotenine exists in some species of *Amanita* mushrooms (170). The psychomimetic 4-hydroxytryptamines psilocin (171), psilocybin (171), and baeocystin (172) (Fig. 9) are present in such toxic mushrooms as *Psilocybe, Conocybe,* and *Stropharia.* The central nervous system active isoxazole derivatives ibotenic acid (167), tricholomic acid (166), muscazone (172),

Fig. 9

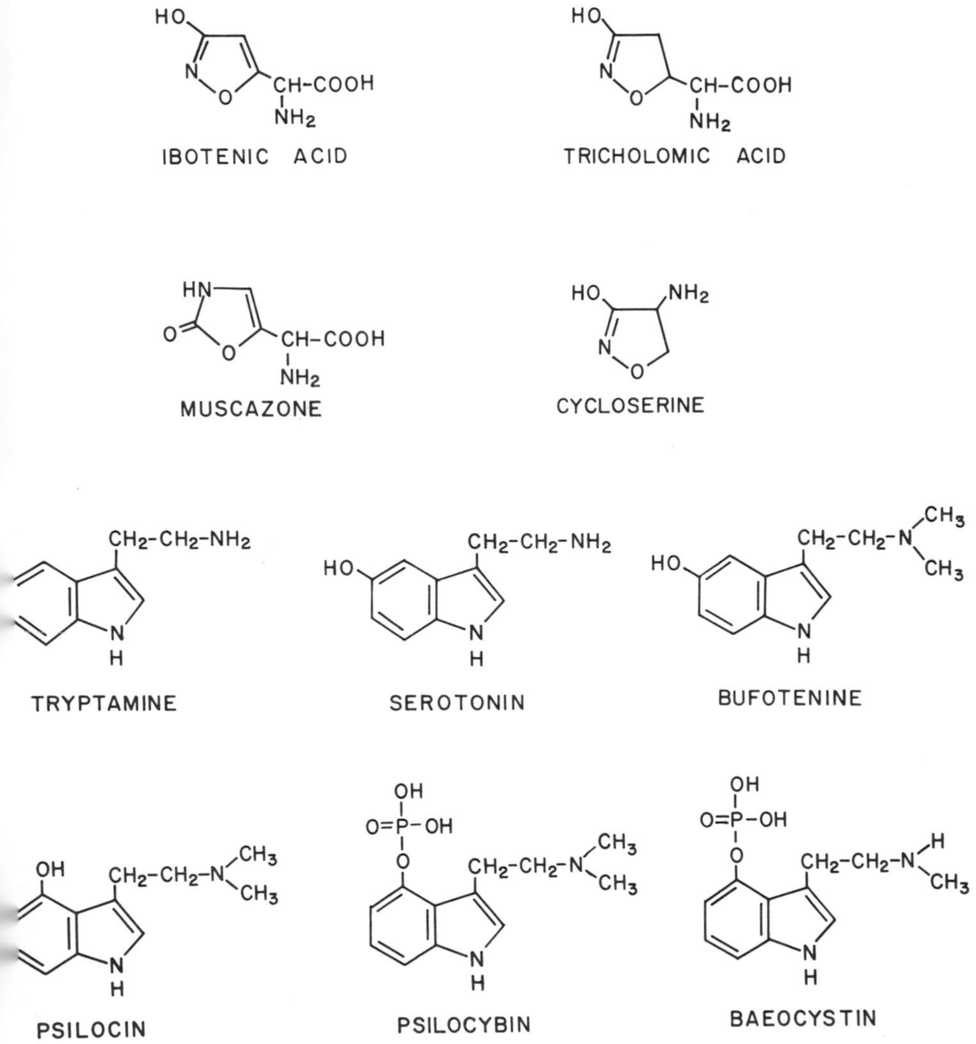

IBOTENIC ACID

TRICHOLOMIC ACID

MUSCAZONE

CYCLOSERINE

TRYPTAMINE

SEROTONIN

BUFOTENINE

PSILOCIN

PSILOCYBIN

BAEOCYSTIN

and pantherine (173) have been isolated from *Amanita* and *Inocybe* mushrooms, and cycloserine (oxamycin) is produced by a *Streptomyces* (174). By contrast with the folklore leads on the above mushroom origin compounds, cycloserine was discovered in a screening program for antitubercular agents, and its central nervous system activity was first observed when it was under clinical study. The development of submerged fermentation processes for the production of ergot or lysergic acid alkaloids has resulted in the discovery of new alkaloids of this series.

Fig. 10

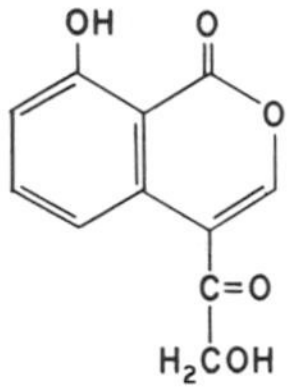

OH
O
O
C=O
H_2COH

OOSPONOL

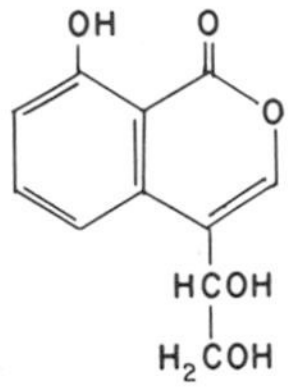

OH
O
O
HCOH
H_2COH

OOSPOGLYCOL

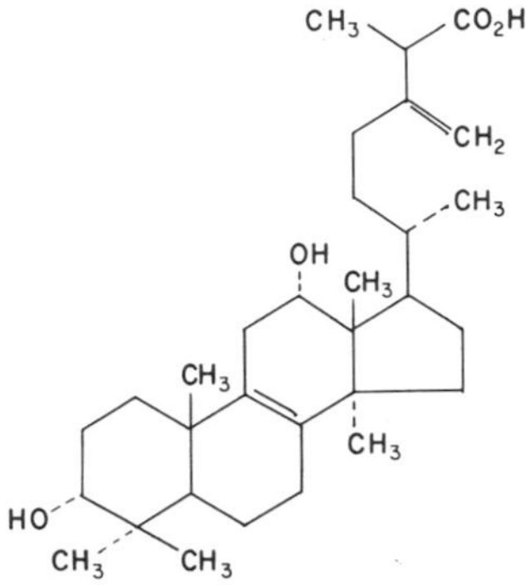

CH_3
CO_2H
CH_2
CH_3
OH
CH_3
CH_3
CH_3
HO
CH_3
CH_3

POLYPORENIC ACID

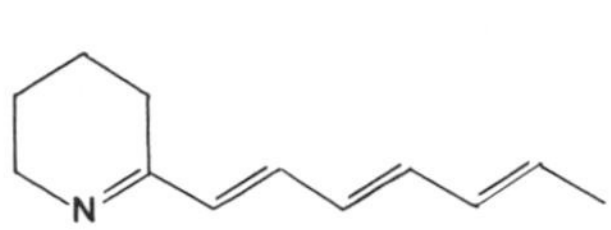

N

NIGRIFACTIN

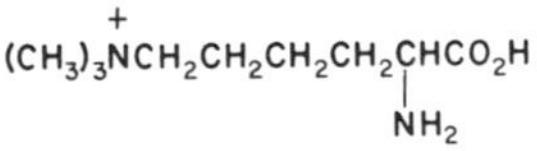

$(CH_3)_3\overset{+}{N}CH_2CH_2CH_2CH_2CHCO_2H$
NH_2

LAMININE

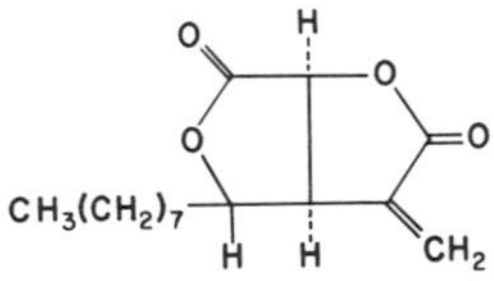

O
H
O
O
O
$CH_3(CH_2)_7$
H
H
CH_2

AVENACIOLIDE

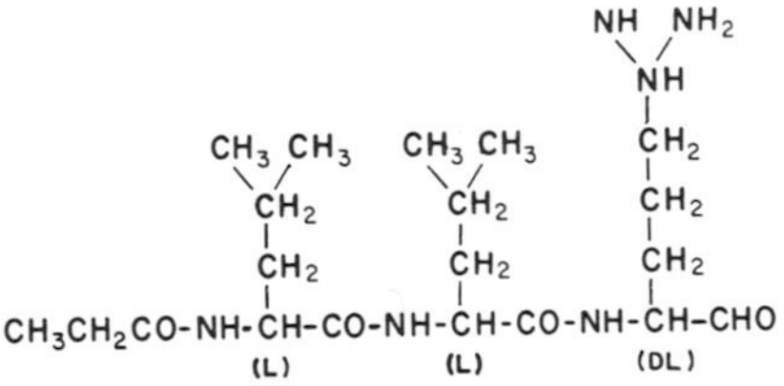

NH NH_2
NH
CH_3 CH_3
CH_3 CH_3
CH_2
CH_2
CH_2
CH_2
CH_2
CH_2
CH_2
CH_3CH_2CO-NH-CH-CO-NH-CH-CO-NH-CH-CHO
(L)
(L)
(DL)

LEUPEPTIN Pr-LL

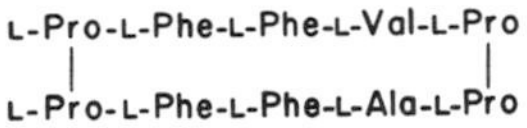

L-Pro-L-Phe-L-Phe-L-Val-L-Pro
L-Pro-L-Phe-L-Phe-L-Ala-L-Pro

ANTAMANID

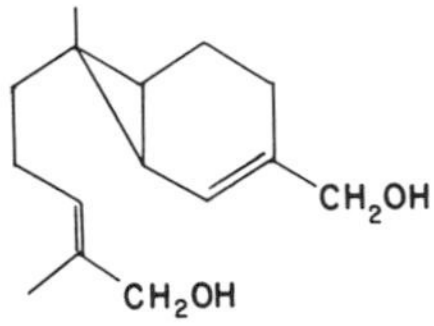

CH_2OH
CH_2OH

SIRENIN

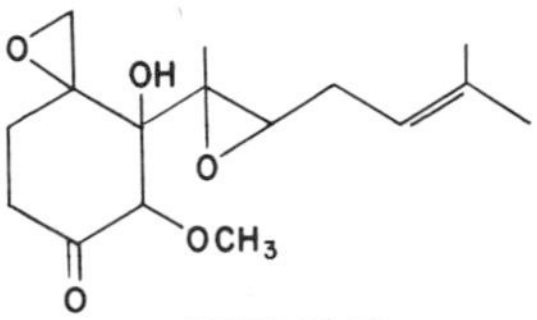

O
OH
O
OCH_3
O

OVALICIN

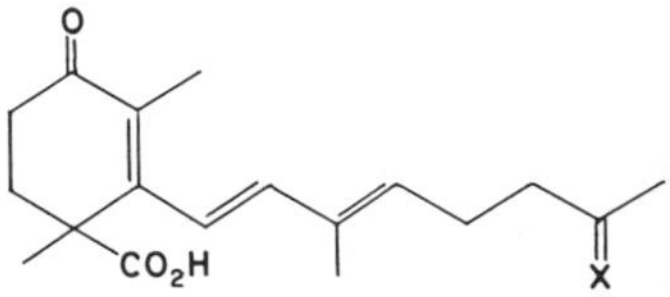

O
CO_2H
X

TRISPORIC ACIDS

a, X = O ; b, X = H, OH.

Oosponol (175) and oospoglycol (176) (Fig. 10) were obtained from an *Oospora* isolated from the air of hospital rooms housing asthmatic patients. Oosponol causes contraction of tracheal muscles of guinea pigs. This activity is not antagonized by atropine, antihistamines, or antiserotonins, but it can be relaxed with epinephrine, papaverine, or theophylline. By contrast, oospoglycol relaxes contraction induced by serotonin, barium chloride, oosponol, and histamine, but not that caused by acetylcholine (177). A completely different metabolite nigrifactin (2-heptatrienyl- Δ'-piperideine) has antihistaminic and hypotensive properties (178).

Polyporenic acid, a bacteriostatic triterpenoid obtained from a *Polyporus,* exhibits a variety of anti-inflammatory activities. In early stages of experimental burns in rats it is more active than cortisone (179). It is somewhat less active than hydrocortisone or deoxycorticosterone acetate in increasing survival time of adrenalectomized rats exposed to low temperatures (180). Pretreatment of rats with polyporenic acid prevented ammonium chloride induced inflammation in lungs, whereas cortisone was inactive under these conditions (179). Eburicoic acid, also a triterpenoid, exhibits selective inhibition of androgenic effects on ventral prostrate glands, seminal vesicles, and coagulating glands (181). It has no effect on total body growth induced by androgens (181). Both polyporenic and eburicoic acids stimulate yeast fermentations and growth of fungi (182).

Avenaciolide, an antifungal lactone (183), inhibits lipolytic hormone and phosphodiesterase inhibitor mediated lipolysis of adipose cells (184). It also abolishes the stimulation of glucose oxidation and lipogenesis caused in adipose cells by insulin or proteolytic enzymes (184).

The bacterial polysaccharide "dextran" and algal "fucoidin" have been used as plasma extenders. Dextran sulfate and fucoidin are heparin-like inducers of clearing factor lipase in blood (185, 186). Vitamin K, a hemagglutinin from *Agaricus campestris* (187), atrometin from *Hydnellum diabolis* (188), and aspergillin O (189) affect blood phenomena. Kojic acid and a number of polyene antibiotics have cardiotonic properties (190, 191). Naematolin exhibits coronary vasodilating action on isolated guinea pig heart (192). Laminine from a marine alga is a hypotensive (193). Surfactin, a crystalline peptide-lipid compound produced by *Bacillus subtilis,* inhibits blood coagulation (194). The leupeptins are protease (trypsin, papain, and kallikrein) inhibitors with antiplasmin activity (195). They inhibit coagulation of rabbit and human blood, inhibit carragenin inflammation, are well adsorbed from the intestinal tract, and have an LD_{50} in mice of 1.5g/kg (oral administration).

Antamanid, a cyclic decapeptide, was isolated from an *Amanita* which produces phalloidine and α-amanitine toxins. When administered to mice before or simultaneously with these toxins it can counteract their lethal effects (196).

The detoxins are a complex of metabolites which decrease the phytotoxicity of blasticidin S without decreasing the effectiveness of this antibiotic in controlling the rice blast fungus *Piricularia oryzae*. The detoxins also decrease the eye irritation caused by blasticidin S (197).

Ovalicin from *Pseudeurotium ovalis* has immunosuppressive and antimitotic properties (198).

Sirenin is a sperm attractant from female gametes of the water mold *Allomyces* (199). The trisporic acids are sexual hormones for *Mucor mucedo* and *Blakeslea trispora* (200).

MICROBIAL TRANSFORMATIONS

A very important characteristic of some microorganisms is their ability to transform compounds to new products, at times to products not obtainable by chemical procedures, or in higher yields or more economically than is possible chemically. The importance of microbial transformations in the U.S. pharmaceutical industry is reflected by the 142 million dollars of manufacturers' shipments of corticoids in 1967. Although some of the corticoids were first made by chemical processing of raw plant materials, it is believed that all of the commercial processes for corticoid manufacture involve at least one fermentation step.

Microbial transformations are generally highly specific with respect to organisms, substrates, and isomerity of products. The most common transformations involve hydroxylation, dehydrogenation, reduction, C-C bond cleavage, aromatization, epoxidation, hydrolysis, esterification, demethylation, deamination, halogenation, N and S oxidations, and isomerization (201, 202). Transformations have been done with steroid hormones, bile acids, sterols, sapogenins, cardenolides, bufadienolides, steroidal alkaloids, indole alkaloids, morphine alkaloids, nicotine, alkanes, fatty acids, naphthalenes, carbohydrates, antibiotics, and so on. Growing cells, resting cells, or spores and cell-free enzyme preparations of bacteria, yeasts, streptomyces, or filamentous fungi have been effectively utilized. Yields have varied up to as much as 90 percent in the 11α-hydroxation of progesterone by *Rhizopus nigricans* (203).

Fig. 11

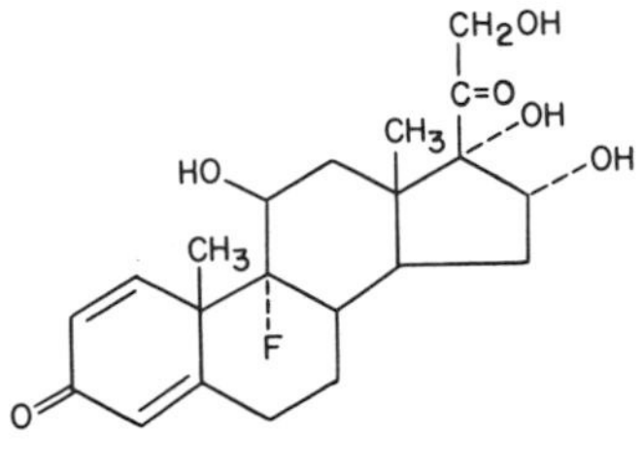

TRIAMCINOLONE

With suitable substrates and organisms, steroid hydroxylation have been reported in the 1α and β, 2β, 5α, 6β, 7α and β, 8α, 9α, 10β, 11α and β, 12α and β, 14α, 15α and β, 16α and β, 17α, 18 and 19 positions. The process for the manufacture of the important glucocorticoid triamcinolone (Fig. 11) requires microbial hydroxylations in the 11 and 16 positions and dehydrogenation at the 1, 2 position. It is doubtful that this compound would be a commercial product without these microbial processes. Selected steroid transformations are indicated in Fig. 12.

An interesting alkaloid transformation (Fig. 13) is the conversion of thebaine to 14-hydroxycodeinone in 40 percent yield by the wood-rotting fungus *Trametes sanguinea* (204). The cinnamic acid ester of this product is 177 times as potent as morphine as an analgesic and more than 500 times as potent as morphine in a test involving pentobarbitone potentiation of grip on a 45 degree plane (205).

In laboratories with collections of transformation organisms, it is becoming increasingly more common to subject sensitive, complex, or rare chemicals to microbial treatment in anticipation of the formation of new biologically active products.

SUMMARY AND CONCLUSIONS

Examples have been selected to illustrate (a) the diversity of chemical structures produced by microorganisms; (b) the many types of activities associated with microbial products; (c) some procedures and circumstances leading to the discovery or development of diodynamic agents; and (d) the characteristics and importance of microbial transformations in the production of pharmacodynamic products.

It is apparent that because they are of biological origin, microbial products have good potential for biological activity. These products are candidates for further chemical or biological transformation, and they are excellent models for synthetic programs. New testing procedures

Fig. 12

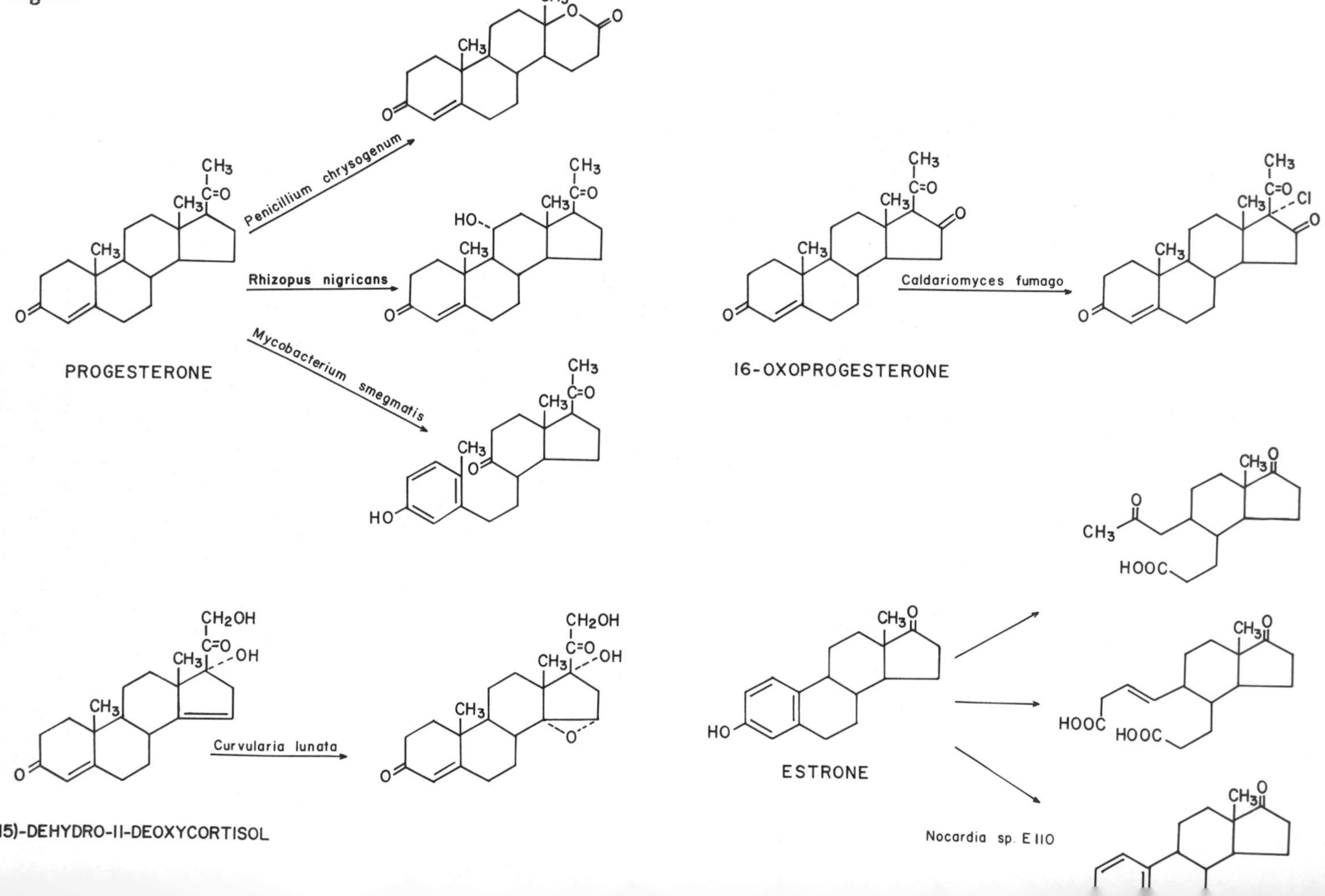

PROGESTERONE

Penicillium chrysogenum

Rhizopus nigricans

Mycobacterium smegmatis

16-OXOPROGESTERONE

Caldariomyces fumago

14(15)-DEHYDRO-11-DEOXYCORTISOL

Curvularia lunata

ESTRONE

Nocardia sp. E 110

Fig. 13

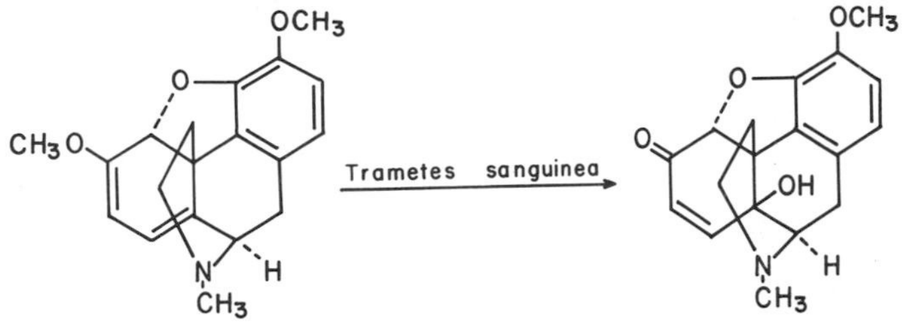

Trametes sanguinea

(-)-THEBAINE (-)-14-HYDROXYCODEINONE

should be adapted to testing microbial preparations. Previously described but inadequately tested compounds and new products should be produced in sufficient quantities for extensive random biological evaluation.

References

1. U.S. Bureau of the Census Current Industrial Reports. Pharmaceutical Preparations, Except Biologicals 1967 Series MA-28G (67-1). (U.S. Government Printing Office, Washington, D.C.), December 4, 1968.

2. H. Umezawa, S. Kondo, K. Maeda, Y. Okami, and K. Takeda, eds. *Index of Antibiotics from Actinomycetes* (University of Tokyo Press, Tokyo and University Park Press, State College, Penn. 1967).

3. T. Korzybski, Z. Kowszyk-Gindifer and W. Kurylowiez, *Antibiotics—Origin, Nature and Properties* (London, Pergamon Press, 1967).

4. S. O. Thomas, V. L. Singleton, J. A. Lowery, R. W. Sharpe, L. M. Pruess, J. N. Porter, J. H. Mowat, and N. Bohonos, *Antibiotics Annual* (1956-57), p. 716.

5. C. W. Waller, J. B. Patrick, W. Fulmor, and W. E. Meyer, *J. Amer. Chem. Soc. 79* (1957), 1101.

6. L. W. Parks, *J. Biol. Chem. 232* (1958), 169.

7. J. A. Mandel and K. Dunham, *J. Biol. Chem. 11* (1912), 85.

8. U. Suzuki, S. Odake, and T. Mori, *Biochem. Z. 154* (1924), 278.

9. F. Schlenk and R. E. DePalma, *J. Biol. Chem. 229* (1957), 1037.

10. S. K. Shapiro and A. N. Mather, *J. Biol. Chem. 233* (1958), 631.

11. F. Schlenk and J. A. Tillotson, *J. Biol. Chem. 206* (1954), 687.

12. K. G. Cunningham, W. Manson, F. S. Spring, and S. A. Hutchinson, *Nature 166* (1950), 949.

13. A. J. Guarino and N. M. Kredich, *Fed. Proc. 23* (1964), 371.

14. N. M. Kredich and A. J. Guarino, *J. Biol. Chem. 236* (1961), 3300.

15. D. Klämbt, G. Thies, and F. Skoog, *Proc. Natl. Acad. Sci.* (U.S.) *56* (1966), 52.

16. M. J. Robins, R. H. Hall and R. Thedford, *Biochemistry 6* (1967), 1837.

17. W. J. Burrows, D. J. Armstrong, F. Skoog, S. M. Hecht, J. T. A. Boyle, N. J. Leonard, and J. Occolowitz, *Science 161* (1968), 691.

18. F. Harada, H. J. Gross, F. Kimiura, S. H. Chang, S. Nishimura, and U. L. RajBhandary, *Biochem. Biophys. Res. Commun. 33* (1968), 299.

19. C. O. Miller, *Science 157* (1967), 1055.
20. R. H. Williams, K. Gerzon, M. Hoehn, M. Gorman, and D. C. DeLong, Abstr. Papers, 158th American Chemical Society, National Meeting, MCT 38 (1969).
21. J. Gripenberg, *Acta chem. scand. 5* (1951), 590.
22. R. Lemberg, *Australian J. Exp. Biol. Med. Sci. 30* (1952), 271.
23. G. W. K. Cavill, B. J. Ralph, J. R. Tetaz, and R. L. Werner, *J. Chem. Soc.* (1953), 525.
24. E. E. Snell and W. H. Peterson, *J. Bact. 39* (1940), 273.
25. B. L. Hutchings, N. Bohonos, D. M. Hegsted, C. A. Elvehjem, and W. H. Peterson, *J. Biol. Chem. 140* (1941), 681.
26. B. L. Hutchings, E. L. R. Stokstad, N. Bohonos, and N. H. Slobodkin, *Science 99* (1944), 371.
27. B. L. Hutchings, E. L. R. Stokstad, J. H. Mowat, J. H. Boothe, C. W. Waller, R. B. Angier, J. Semb, and Y. SubbaRow, *J. Amer. Chem. Soc. 70* (1948), 10.
28. M. A. Pollack, A. Taylor, and R. J. Williams, *Studies on B Vitamins in Human, Rat and Mouse Neoplasms* (The University of Texas Publication no. 4237, 1942), p. 56.
29. C. Leuchtenberger, R. Lewisohn, D. Laszlo and R. Leuchtenberger, *Proc. Soc. Exp. Biol. Med. 55* (1944), 204.
30. R. Leuchtenberger, C. Leuchtenberger, D. Laszlo, and R. Lewisohn, *Science 101* (1945), 46.
31. R. Lewisohn, C. Leuchtenberger, R. Leuchtenberger, and J. C. Keresztesy, *Science 104* (1946), 436.
32. S. Farber, L. K. Diamond, R. D. Mercer, R. F. Sylvester, Jr., and J. A. Wolff, *New Eng. J. Med. 238* (1948), 787.
33. S. Farber, *Blood 4* (1949), 160.
34. C. W. Hesseltine, C. Pidacks, A. R. Whitehill, N. Bohonos, B. L. Hutchings, and J. H. Williams, *J. Amer. Chem. Soc. 74* (1952), 1362.
35. J. B. Nielands, *Structure and Bonding* (New York, Springer Verlag, 1966), I, 59.
36. H. Bickel, G. E. Hall, W. Keller-Schierlein, V. Prelog, E. Vischer, and A. Wettstein, *Helv. chim. Acta 43* (1960), 2129.
37. F. Wöhler, *Med. Klin.* (Munich) *57* (1962), 1370.
38. S. Moeschlin and U. Schnider, *New Eng. J. Med. 269* (1963), 57.
39. K. E. Price, *Advanc. Appl. Microbiol. 11* (1969), 17.
40. R. G. Jones, *Amer. Scientist 58* (1970), 404.
41. J. M. Walshe, *Lancet 1* (1960), 188.
42. A. Goldberg, J. A. Smith, and A. C. Lockhead, *Brit. Med. J. 1* (1963), 1270.
43. J. C. Crawhall, E. F. Scowen, and R. W. E. Watts, *Brit. Med. J. 1* (1963), 588.
44. S. E. Ritzmann, S. I. Coleman, and W. C. Levin, *J. Clin. Invest. 39* (1960), 1320.
45. G. A. Nicolson, A. C. Greiner, W. J. McFarlane, and R. A. Baker, *Lancet 1* (1966), 344.
46. K. Allman and M. S. Tobin, *Proc. Soc. Exp. Biol. Med. 118* (1965), 554.
47. I. A. Jaffe, *Ann. Rheum. Dis. 22* (1963), 71.
48. M. E. Nimini and L. A. Bavetta, *Science 150* (1965), 905.
49. J. H. Boothe, J. Morton II, J. P. Petisi, R. G. Wilkinson, and J. H. Williams, *J. Amer. Chem. Soc. 75* (1953), 4621.
50. L. H. Conover, W. T. Moreland, A. R. English, C. R. Stephens, and F. J. Pilgrim, *J. Amer. Chem. Soc. 75* (1953), 4622.

51. J. H. Martin, A. J. Shay, L. M. Pruess, J. N. Porter, J. H. Mowat and N. Bohonos, *Antibiotics Annual* (1954-1955), p. 853.
52. P. P. Minieri, M. C. Firman, A. G. Mistretta, A. Abbey, C. E. Bricker, N. E. Rigler, and H. Sokol, *Antibiotics Annual* (1954-1955), p. 851.
53. J. R. D. McCormick, N. O. Sjolander, U. Hirsch, E. R. Jensen, and A. P. Doerschuk, *J. Amer. Chem. Soc. 79* (1957), 4561.
54. A. C. Finlay, G. L. Hobby, S. Y. Pan, P. P. Regna, J. B. Routien, D. B. Seeley, G. M. Shull, B. A. Sobin, I. A. Solomons, J. W. Vinson, and J. H. Kane, *Science 111* (1950), 85.
55. R. K. Blackwood, J. J. Beereboom, H. H. Rennhard, M. Schach von Wittenau, and C. R. Stephens, *J. Amer. Chem. 83* (1961), 2772.
56. C. R. Stephens, K. Murai, H. H. Rennhard, L. H. Conover, and K. J. Brunings, *J. Amer. Chem. Soc. 80* (1958), 5324.
57. L. A. Mitscher, J. H. Martin, P. A. Miller, P. Shu, and N. Bohonos, *J. Amer. Chem. Soc. 88* (1966), 3647.
58. G. S. Redin, *Antimicrobial Agents Chemotherap.* (1966), p. 371.
59. P. W. Brian, P. J. Curtis, and H. G. Hemming, *Brit. Mycol. Soc. Trans. 29* (1946), 173.
60. A. E. Oxford, H. Raistrick, and P. Simonart, *Biochem. J. 33* (1939), 240.
61. J. C. Gentles, *Nature 182* (1958), 476.
62. I. M. Lauder and J. G. O'Sullivan, *Vet. Record 70* (1958), 949.
63. D. I. Williams, R. H. Martin, and I. Sarkany, *Lancet 2* (1958), 1212.
64. H. Blank and F. J. Roth, Jr., *Arch. Dermatol. 79* (1959), 259.
65. J. MacMillan, *Chem. and Ind.*, London (1951), 719.
66. J. MacMillan, *J. Chem. Soc.* (1954), 2585.
67. T. Cochrane and A. Tullet, *Brit. Med. J. 2* (1959), 286.
68. J. Logeais, J. Maillard, M. Vincent, P. Delaunay, and Vo-Van-Tri, *Compt. Rend. 262D* (1966), 933.
69. R. Slonin, D. S. Howell, H. E. Brown, Jr., *Arthritis and Rheum. 4* (1961), 124.
70. A. Cohen, J. Goldman, R. Daniels, and W. Kanenson, *J. Amer. Med. Assn. 173* (1960), 542.
71. N. P. Pasquale, J. W. Burke, and G. E. Burch, *J. Amer. Med. Assn. 184* (1963), 421.
72. C. Rimington, P. N. Morgan, K. Nicholls, J. D. Everall, and R. R. Davies, *Lancet 2* (1963), 318.
73. S. I. Cullen and P. M. Catalano, *J. Amer. Med. Assn. 199* (1967), 150.
74. L. L. Barich, T. Nakai, J. Schwarz, and D. J. Barich, *Nature 187* (1960), 335.
75. E. W. Hurst and G. E. Paget, *Brit. J. Dermatol. 75* (1963), 105.
76. H. Raistrick and G. Smith, *Biochem. J. 30* (1963), 1315.
77. J. Balan, L. Ebringer, P. Nemec, S. Kovak, and J. Dobias, *J. Antibiot.* (Tokyo) *A16* (1963), 157.
78. L. Ebringer, J. Balan, G. Catar, K. Horakaua, and J. Ebringerova, *Exp. Parasitol. 16* (1965), 182.
79. K. Maeda, J. Osato, and H. Umezawa, *J. Antibiot.* (Tokyo) *A6* (1953), 182.
80. H. Horie, *J. Antibiot.* (Tokyo) *A9* (1956), 168.
81. R. Despois, S. Pinnert-Sindico, L. Ninet, and J. Preud'homme, *Giorn. Microbiol. 21* (1956), 76.
82. C. Cosar and L. Julou, *Ann. Inst. Pasteur 96* (1959), 238.
83. S. Farber, *J. Amer. Med. Assn. 198* (1966), 826.

84. R. A. Manaker, F. J. Gregory, C. Vining, and S. A. Waksman, *Antibiotics Annual* (1954-1955), p. 853.
85. K. Anzai and S. Suzuki, *J. Antibiot.* (Tokyo) *A14* (1961), 253.
86. L. J. Hanka, J. S. Evans, D. J. Mason, and A. Dietz, *Antimicrob. Ag. Chemother.* (1966), 619; M. E. Bergy and R. R. Herr, *Antimicrob. Ag. Chemother.* (1966), 625.
87. J. Kara, J. Skoda, and F. Sorm, *Coll. Czechoslov. Chem. Commun. 26* (1961), 1836.
88. R. E. Handschumacher and A. D. Welch, *Fed. Proc. 15,* Abstr. no. 871 (1956); J. Skoda, V. F. Hess, and F. Sorm, *Experientia 13* (1957), 150.
89. R. Weindling, *Phytopathology 27* (1937), 1175.
90. R. S. Speirs, *Life Science 4* (1965), 343.
91. N. R. Rose, J. A. Haber, and P. Calabressi, *Proc. Soc. Exp. Biol. Med. 128* (1968), 1121.
92. H. J. Meuwissen and R. A. Good, *Nature 215* (1967), 634.
93. J. B. Flexner, L. B. Flexner, and E. Stellar, *Science 141* (1963), 57.
94. S. H. Barondes and H. D. Cohen, *Brain Res. 4* (1967), 44.
95. A. Ciegler and E. B. Lillehoj, *Advanc. Appl. Microbiol. 10* (1968), 155.
96. R. I. Mateles and G. N. Wogan, eds. *Biochemistry of Some Foodborne Microbial Toxins* (Cambridge, Mass., M.I.T. Press, 1967).
97. C. W. Hesseltine, O. L. Shotwell, J. J. Ellis, and R. D. Stubbelfield, *Bacteriol. Rev. 30* (1966), 795.
98. R. Schoental, *Ann. Rev. Pharmacol.* 7 (1967), 345.
99. L. D. Owens, *Science 165* (1969), 18.
100. P. M. Scott and E. Somers, *J. Agr. Food Chem.* 17 (1969), 430.
101. William J. Kelleher, *Advanc. Appl. Microbiol. 11* (1969), 211.
102. M. A. Stahmann, C. F. Huebner, and K. P. Link, *J. Biol. Chem. 138* (1941), 513.
103. D. M. Bellis, M. S. Spring, and J. R. Stoker, *Biochem. J. 103* (1967), 202.
104. G. N. Wogan, *Bacteriol. Rev. 30* (1966), 460.
105. K. J. van der Merwe, P. S. Steyn, and L. Fourie, *J. Chem. Soc.* (1965), 7083.
106. K. J. van der Merwe, P. S. Steyn, L. Fourie, D. B. Scott, and J. J. Theron, *Nature 205* (1965), 1113.
107. M. J. Pitout and W. Nel, *Biochem. Pharmacol. 18* (1969), 1837.
108. M. Stob, R. S. Baldwin, J. Tuite, F. N. Andrews, and K. G. Gillette, *Nature 196* (1962), 1318.
109. D. Rainey, E. B. Smalley, M. W. Crump, and F. M. Strong, *Nature 205* (1965), 203.
110. S. D. Aust and H. P. Broquist, *Nature 205* (1965), 204.
111. S. D. Aust, *Biochem. Pharmacol. 18* (1969), 929.
112. V. L. Singleton, N. Bohonos, and A. J. Ullstrup, *Nature 181* (1958), 1072.
113. B. Betina, P. Nemec, J. Dobias, and Z. Barath, *Folia Microbiol.* (Prague) 7 (1962), 353.
114. E. Harri, W. Loeffler, H. P. Sigg, H. Stahelin, and Ch. Tamm, *Helv. Chim. Acta 46* (1963), 1235.
115. V. Betina, J. Fuska, A. Kjaer, M. Kutkova, P. Nemec, and R. H. Shapiro, *J. Antibiot.* (Tokyo) *A19* (1966), 115.
116. J. C. Thornton and R. H. Percival, *Nature 182* (1958), 1095.
117. L. D. Scheel, V. B. Perone, R. L. Larkin, and R. E. Larkin, *Biochemistry 2* (1963), 1127.

118. D. J. Birmingham, M. M. Key, G. E. Tabich, and V. B. Perone, *Arch. Dermatol. 83* (1961), 73.
119. T. Wieland, *Science 159* (1968), 946.
120. S. Marumo, *Bull. Agr. Chem. Soc.* (Japan) *23* (1959), 428.
121. K. Uraguchi, T. Tatsuno, F. Sakai, M. Tsukioka, O. Yonemitsu, K. Ito, M. Miyake, M. Saito, M. Enomoto, T. Shikata, and T. Ishiko, *Jap. J. Exp. Med. 31* (1961), 19.
122. T. Kubota and T. Matsuura, *Chem. and Ind.* (1956), 521.
123. W. Rothe, *Deutsch. Med. Woschr. 79* (1954), 1080.
124. A. C. Hetherington and H. Raistrick, *Trans. Roy. Soc.* (London) *B220* (1931), 269.
125. A. C. Hetherington and H. Raistrick, *Trans. Roy. Soc.* (London) *B220* (1931), 209.
126. J. Nagai, M. Hayashi, and K. Mizobe, *Fukuoka Igaku Zasshi 48* (1957), 311 (*Chem. Abstr. 55* (1961), 1914).
127. K. Uraguchi, *Jap. J. Med. Progr. 34* (1947), 155.
128. P. J. Brook and E. P. White, *Ann. Rev. Phytopathol. 4* (1966), 171.
129. B. J. Wilson, C. H. Wilson, and A. W. Hayes, *Nature 220* (1968), 77.
130. A. Ciegler, *Appl. Microbiol. 18* (1969), 128.
131. N. Prentice and A. D. Dickinson, *Biotech. and Bioeng. 10* (1968), 413.
132. J. R. Bamburg, F. M. Strong, and E. B. Smalley, *Agric. Food Chem. 17* (1969), 443.
133. C. J. Mirocha, C. M. Christensen, and G. H. Nelson, *Biotech. and Bioeng. 10* (1968), 469.
134. E. E. Clayton, *J. Agric. Res. 48* (1934), 411.
135. A. C. Braun, *Proc. Nat. Acad. Sci.* (U.S.) *36* (1950), 423.
136. A. C. Braun, *Phytopathology 45* (1955), 659.
137. C. Lamar, S. L. Sinden, Jr., and R. D. Durbin, *Biochem. Pharmacol. 18* (1969), 521.
138. S. L. Sinden, Jr., and R. D. Durbin, *Nature 219* (1968), 379.
139. J. M. Stewart, *J. Amer. Chem. Soc. 83* (1961), 435.
140. L. D. Owens and D. A. Wright, *Plant Physiol. 40* (1965), 927.
141. H. W. Johnson, U. M. Means, and F. E. Clark, *Nature 183* (1959), 308.
142. L. D. Owens, S. Guggenheim, and J. L. Hilton, *Biochim. biophys. Acta 158* (1968), 219.
143. K. Mori, M. Shiozaki, N. Itaya, M. Matsui, and Y. Sumiki, *Tetrahedron 25* (1969), 1293.
144. L. G. Paleg, *Plant Physiol. 35* (1960), 293.
145. D. B. Carlisle, D. J. Osborne, P. E. Ellis, and J. E. Moorhouse, *Nature 200* (1963), 1230.
146. P. E. Ellis, D. B. Carlisle, and D. J. Osborne, *Science 149* (1965), 546.
147. J. L. Nation and F. A. Robinson, *Science 152* (1966), 1765.
148. E. Schwartz, *Naturwissenschaften 54* (1961), 371.
149. E. Schwartz, *Bratisl. lek. Listy 43* (1963), 317.
150. P. deMayo, E. Y. Spencer, and R. W. White, *Can. J. Chem. 39* (1961), 1608.
151. S. Tamura, A. Sakurai, K. Kainuma, and M. Takai, *Agric. Biol. Chem. 27* (1963), 738.
152. D. E. Briggs, *Nature 210* (1966), 418.
153. Y. Satomura and A. Sato, *Agric. Biol. Chem. 29* (1965), 337.

154. T. Sassa, H. Aoki, M. Namiki, and K. Munakata, *Agric. Biol. Chem. 32* (1968), 1432.
155. A. Mahadevan, *Experientia 21* (1965), 433.
156. Y. Koaze, *Bull. Agric. Chem. Soc.* (Japan) *22* (1958), 98.
157. Y. Chen, *Bull. Agric. Chem. Soc.* (Japan) *24* (1960), 372.
158. Y. Koaze, *Bull. Agric. Chem. Soc.* (Japan) *24* (1960), 530.
159. P. Condon, J. Kuć, and H. N. Draudt, *Phytopathology 53* (1963), 1244.
160. I. A. M. Cruickshank and D. R. Perrin, *Life Sciences* 7 (1968), 449.
161. H. T. Huang, in *Fermentation Advances*, D. Perlman, ed. (New York, Academic Press, 1969), p. 591.
162. A. N. Kishaba, D. L. Shankland, R. W. Curtis, and M. C. Wilson, *J. Econ. Entomol. 55* (1962), 211.
163. S. Tamura, N. Takahashi, S. Meyamoto, R. Mori, S. Suzuki, and J. Nagatsu, *Agric. Biol. Chem. 27* (1963), 576.
164. Y. Kodaira, *Agric. Biol. Chem. 26* (1962), 36.
165. R. Myokei, A. Sakurai, C. Chang, Y. Kodaira, N. Takahashi, and S. Tamura, *Agric. Biol. Chem. 33* (1969), 1501.
166. T. Takemoto and T. Nakajima, *Yakugaku Zasshi 84* (1964), 1183.
167. T. Takemoto, T. Yokobe, and T. Nakajima, *Yakugaku Zasshi 84* (1964), 1186.
168. A. M. Asaton, A. J. Levi, and M. D. Milne, *Lancet 2* (1963), 733.
169. V. E. Tyler, Jr., *Science 128* (1958), 718.
170. T. Wieland and W. Motzel, *Ann. Chem. Liebigs 581* (1953), 10.
171. R. Heim, A. Brack, H. Kobel, A. Hofmann, and R. Carlteux, *Comp. Rend. 246* (1958), 1346.
172. C. H. Eugster, C. F. R. Miller, and R. Good, *Tetrahedron L. 23* (1965), 1813.
173. M. Onda, H. Fukushima, and M. Akagawa, *Chem. Pharm. Bull.* (Tokyo) *12* (1964), 751.
174. R. L. Harned, P. H. Hidy, and E. Kropp La Baw, *Antibiot. and Chemother. 5* (1955), 204; F. A. Kuehl Jr., F. J. Wolf, N. R. Trenner, R. L. Peck, R. P. Buhs, I. Putter, R. Ormond, J. E. Lyons, L. Charet, E. Howe, B. D. Hunnewell, G. Downing, E. Newstead, and K. Folkers, *J. Amer. Chem. Soc.* 77 (1955), 2344.
175. J. Yamamoto, *Agric. Biol. Chem. 25* (1961), 400.
176. K. Nitta, Y. Yamamoto, I. Yamamoto, and S. Yamatodani, *Agric. Biol. Chem. 27* (1963), 822.
177. S. Ohashi, M. Yamaguchi, and Y. Kobayashi, *Proc. Japan Acad. 38* (1962), 766.
178. Y. Kaneko, T. Terashima, and Y. Kuroda, *Agric. Biol. Chem. 32* (1968), 783.
179. O. M. Efimenko, T. A. Melnikova, R. N. Zozulya, and N. M. Kostygov, *Antibiotiki 6* (1961), 215.
180. A. R. Ratsimamanga, B. Pasich, P. Bioteau, and M. Nigeon-Dureuil, *Comp. Rend. Soc. Biol. 156* (1962), 1552.
181. Olin Mathieson Corp., British Patent #951,490 (1964).
182. O. M. Efimenko, *Mikrobiologiya 29* (1960), 548.
183. D. Brookes, B. K. Tidd, and W. B. Turner, *J. Chem. Soc.* (1963), 5385.
184. J. F. Kuo, I. K. Dill, C. E. Holmlund, and N. Bohonos, *Biochem. Biopharmacol. 17* (1968), 345.
185. G. T. Stewart, *Brit. J. Exp. Pathol. 39* (1958), 109.
186. W. Shuler and G. F. Springer, *Naturwissenschaften 44* (1957), 265.
187. H. J. Sage and J. J. Vazquez, *J. Biol. Chem. 242* (1967), 120.

188. K. L. Euler, V. E. Tyler, Jr., and L. R. Brady, *Lloydia 28* (1965), 203.
189. M. Karaca, M. Stefanini, F. Soardi, and R. H. Mele, *Proc. Soc. Exp. Biol. Med. 109* (1962), 301.
190. A. Beelik, *Advanc. Carbohydrate Chem. 2* (1956), 145.
191. H. R. K. Arora, *Med. Exp. 13* (1965), 57.
192. Y. Ito, H. Kurita, T. Yamaguchi, M. Sato, and T. Okuda, *Chem. Pharm. Bull. 15* (1967), 2009.
193. T. Takemoto, K. Daigo, and N. Takagi, *Yakugaku Zasshi 84* (1964), 1176.
194. K. Arima, A. Kakinuma, and G. Tamura, *Biochem. Biophys. Res. Commun. 31* (1968), 488.
195. T. Aoyagi, T. Takeuchi, A. Matsuzaki, K. Kawamura, S. Kondo, M. Kondo, M. Hamada, K. Maeda, and H. Umezawa, *J. Antibiot. 22* (1969), 283.
196. Th. Woland, G. Lüben, H. Ottenheym, J. Faesel, J. X. DeVries, A. Prox, and J. Schmid, *Angew. Chem.*, Intern. Ed. 7 (1968), 204.
197. H. Yonehara, H. Seto, S. Aizawa, T. Hidaka, A. Shimazu, and N. Otake, *J. Antibiot. 21* (1968), 369.
198. S. Lazary and H. Stähelin, *Experientia 24* (1968), 1171.
199. L. Machlis, W. H. Nutting, M. W. Williams, and H. Rapoport, *Biochemistry 5* (1966), 2147.
200. D. J. Austin, J. D. Bu'Lock, and G. W. Gooday, *Nature 223* (1969), 1178.
201. H. Iizuka and A. Naito, *Microbial Transformation of Steroids and Alkaloids* (State College, Penn., University Park Press, 1967).
202. W. Charney and H. L. Herzog, *Microbial Transformation of Steroids* (New York, Academic Press, 1967).
203. D. H. Peterson, H. C. Murray, S. H. Eppstein, L. M. Reineke, A. Weintraub, P. D. Meister, and H. M. Leigh, *J. Amer. Chem. Soc. 74* (1952), 5933.
204. K. Iizuka, S. Okuda, K. Aida, T. Asai, K. Tsuda, M. Yamada, and I. Seki, *Chem. Pharm. Bull. 8* (1960), 1056.
205. W. R. Buckett, *J. Pharm. Pharmacol. 17* (1965), 759.

Drugs from Plants of the Sea*

ARA H. DER MARDEROSIAN
Associate Professor of Pharmacognosy
Philadelphia College of Pharmacy and Science,
Philadelphia, Pennsylvania

Oh! call us not weeds, but flowers of the sea,
For lovely, and gay, and bright tinted are we:
Our blush is as deep as the rose of thy bowers,
Then call us not weeds, we are Ocean's gay flowers.

Not nurs'd like the plants of the summer parterre,
Whose gales are but sighs of an evening air:
Our exquisite, fragile, and delicate forms
Are nurs'd by the Ocean, and rock'd by the storms.

Anon.

INTRODUCTION

The ocean has often been considered as the "Mother" of all organisms, providing an environment within which an almost infinite number of genetic permutations has occurred, leading ultimately to the existing array of evolved forms now found in it. At one time, abiogenesis held sway, since a multitude of complex random reactions could have occurred in the "primordial soup." But today the results of these reactions, the organisms in the sea themselves, have taken on the major task of biogenesis in their diverse biochemical habits, assimilating simple organic molecules and building them into systems of enormous complexity (1). A consideration of the biomedical and pharmaceutical aspects of some of these adjuncts is the subject of this chapter.

*Based in part on a review published by the *Journal of Pharmaceutical Sciences*, vol. 58, no. 1 (1969), pp. 1-33, entitled "Marine Pharmaceuticals."

One of the major purposes of this review is to make the botanical and pharmaceutical community aware of the enormous potential which the sea holds as a source of new and different pharmaceuticals of all types. It will become evident that a major difficulty exists in promulgating new discoveries in this field. At least a dozen different disciplines have contributed to unraveling some of the biomedical mysteries of the deep, and the results have been published in hundreds of papers in various journals. Due to the "publication explosion," few people in any one discipline can cope with and keep up with all the new facts as they become available. Furthermore, bits of information gathered here and there give one only a dim awareness of overall developments. However, the recent publication of several reviews aimed at different scientific groups, should make it possible for some of these difficulties to be overcome (2, 3).

By far the most ambitious undertaking is the definitive monograph now in its final stages of completion by Halstead (4). Entitled *Poisonous and Venomous Marine Animals of the World,* it is a three-volume series covering a comprehensive survey of world literature from 3000 B.C. to the present. The material is covered in taxonomic fashion and comprises the history, biology, morphology, toxicology, pharmacology, and chemistry of all known poisonous and venomous animal marine organisms. This work should go a long way toward aiding and promoting investigations in all the medical and paramedical disciplines.

There also exist a number of other reviews and popular articles which have helped set the stage for this chapter. Many of the basic ideas, principles, and background materials are condensed from these widely scattered publications (5-35). Finally, it should be noted that the material on which this paper is mainly based was put together first through the interest generated at a recent conference on the subject of "Drugs from the Sea" held in August 1967 at the University of Rhode Island. This paper is an expansion and revision of a paper entitled "Current Status of Drug Compounds from Marine Sources" presented there by the author (36).

In order to discuss the topic under consideration, it is necessary to define a number of important terms. First, the term "poisonous" is used here in its generic sense, and includes both oral and parenteral poisons, the former being more common. However, the term *biotoxic* is more commonly used when referring to poisonous terrestrial and marine organisms. Two major subdivisions are the plant poisons, or phytotoxins, and the animal poisons, or zootoxins. The route of administration of the poisonous substances gives us further classifications into oral poisons which are toxic on ingestion, and parenteral poisons which are administered externally, usually via some venom apparatus (stings, spines, and so on). Another class of poisons, peculiar to marine organisms, are those found in certain glands without any specific structures to deliver them. These poisons may be absorbed via the pores or by injury to the animal. The oral poisons are generally small molecules, whereas the parenteral venoms are complex mixtures containing enzymes which facilitate penetration, local histamine-releasing agents, pain-producing materials, and large molecular weight peptides, or proteins, which usually are the true toxic principles. The location of poisonous materials in the organism itself gives us several terms: for example, *ichthytoxism* (poisoning from fish), *ichthyosarcotoxic* (toxin in fish flesh), *ichthyohemotoxic* (toxin in fish blood), *ichthyootoxic* (toxin in fish gonads), and *ichthyocrinotoxic* (toxin in glands of fish skin).

Certain terms are used to describe the syndrome, or characteristic pharmacological effects, of undefined poisons. Hence, the term *ciguatera* poisoning, caused by ingestion of a large variety of tropical marine reef or shore fishes and characterized by nervous disorders and gastrointestinal disturbances and capable of causing death; *tetraodon* poisoning, following ingestion of certain puffers and ocean sunfishes characterized by deleterious neuromuscular, respiratory, and central nervous system effects and causing about 60 percent mortality in victims ingesting fish in this group (here tetrodotoxin has been identified as the active toxic principle); *scombroid* poisoning following ingestion of the mackerel-like fishes and characterized by central nervous system effects, burning of the throat, numbness, thirst, and generalized urticaria; *clupeoid* poisoning, following ingestion of certain herring-like fishes of the tropical Pacific and causing symptoms slightly different from those of ciguatera poisoning; *cyclostome* poisoning, following ingestion of the slime and flesh of certain lampreys and hagfishes and characterized by gastrointestinal distress; *elasmobranch* poisoning, due to ingestion of shark musculature or livers and characterized by nausea, vomiting, abdominal pain, headache, diarrhea, oral parasthesia, muscle cramps, and respiratory distress; *paralytic shellfish poisoning,* caused by ingestion of the flesh of molluscs—e.g., mussels and clams which have become toxic by their ingestion of miscoscopic dinoflagellates, and characterized by muscular paralysis; *hallucinatory fish poisoning,* produced on ingestion of certain mullets and goatfish and characterized by nightmares and hallucinations.

Certain toxic dinoflagellates can affect the entire food web by passing along their poisons, which may be concentrated in various organs of different organisms anywhere throughout the food chain. Marine algae have phenomenal ability to concentrate and retain chemical substances from the marine environment. Certain seaweeds still serve as good sources of iodine.

It is estimated that more than 30,000 intoxications occur each year from eating poisonous marine products. Less than 20 percent of the cases are properly diagnosed as to the exact etiological agent (4). Poisoning may be due to toxins passed through the food chain, unknown pathogenic strains of marine bacteria, or other unknown factors. Intensive ecological research on the food web of ichthyotoxic fish is certainly required (1).

CURRENT WORK ON MARINE PHARMACEUTICALS

There are many reasons why the successful development of marine pharmaceuticals continues to be a difficult problem. There is a lack of trained personnel; a lack of a multidisciplinary approach; serious procurement problems, especially for sufficient quantities of material to carry out thorough and complete studies; problems in the culture of marine organisms for study in the laboratory; problems in the screening of crude extracts for biological activity; difficulties inherent in tedious natural product extraction, separation, and characterization; and problems concerned with adequate patent protection for products developed. Even if a drug is found efficacious for a particular disorder, its marketing may take up to seven years. A new drug application to the Food and Drug Administration can cost up to seven million dollars, chiefly for obtaining all the information required for the authorization of full trials. Then the FDA must clear the item for safety and efficacy before it can be marketed.

The answer to some of these difficulties lies in the successful pooling of ideas, talent, and experience of several related overlapping disciplinary areas, as, for example, ecology, venomology, taxonomy, ethnobotany, pharmaceutical chemistry, pharmacognosy, pathology, oceanography, and so forth (18), and concentrating effort on certain problems which are most likely to yield new useful and marketable pharmaceuticals.

Even with all these obstacles, many classic drugs from marine sources continue to be used. These include agar from species of *Gelidium, Gracilaria,* and *Hypnea,* alginate from species of *Fucus,* carrageenan from

Chondrus, cod-liver oil and sodium morrhuate from the cod *Gadus morrhua*, protamine sulfate from the sperm or mature testes of salmon, spermaceti from the sperm whale, and ichthammol from bituminous schists containing fossil fishes, and many others (21).

Today, experimental marine biology has given us a brief hint of many other newer pharmacologically active substances from marine organisms. Some are categorized or considered as toxins or poisons. However, their isolation, characterization, and ultimate pharmaceutical examination should lead to several useful medications. Even if they cannot be used, they may point to the synthesis of analogous compounds.

TAXONOMIC SURVEY OF MARINE PLANTS YIELDING PHARMACEUTICALLY INTERESTING COMPOUNDS

Because of the vast body of information available and the lack of a broad chemical or pharmacological base on which to categorize this information, the marine plant organisms of biomedical interest will be discussed by taxonomic grouping. An attempt to summarize the available taxonomic and pharmacological information for quick reference is given in Table 1. A few examples from each phylum are given in order to demonstrate the broad distribution of substances having potential drug activity in practically all categories of marine plant organisms.

Bacteria. Although disputed (41), the existence of specific marine bacteria is well established and, of the approximate number of living species (ca. 1500), some 12 percent are ubiquitous marine forms. They show enormous range in habitat and are the major sources of mineral nutrients (particularly dissolved carbonates) for plants, besides carrying out the ultimate decomposition of dead marine organisms.

About 95 percent of the marine bacteria are gram-negative rods, and many are active flagellated forms. Some are sedentary and attach themselves tenaciously to solid surfaces via mucilaginous holdfasts. Nearly 70 percent are pigment producers (orange, yellow, brown, pink, green), and many show fluorescence.

The largest populations are found near the shore (50,000-400,000 bacteria per ml. of sea water), while in the open sea the population is low (40 bacteria per ml. of sea water). Up to 160,000 viable bacteria per ml. have been found on the sea-floor muds in the West Indies. Even geological cores taken five meters below the surface of the ocean have yielded living bacteria (42).

Table 1. Types of Pharmacological Activity and Potential Drugs from Marine Organisms

TAXONOMIC GROUP	Type of pharmacological activity observed in various organisms									Nature of toxin and toxicity	Potential pharmacological drug use
	CNS[a]	RS[b]	NMS[c]	ANS[d]	CVS[e]	GI[f]	Local[g]	Antibiotic activity	Other activity		
Schizophyta											
Marine bacteria											
Bacillus spp.								+[h]		Antibacterial	Potential source
Micrococcus spp.								+		principles	of antibiotics
Chromobacterium spp.								+			
Aeromonas spp.								+		Antibacterial	Antibiotics
Pseudomonas spp.								+	Antifungal	antifungal	with control
Vibrio spp.								+	and	and	against yeast and
Flavobacterium spp.								+	Antiyeast	antiyeast	fungal
Alcaligenes spp.								+	activity	principles	pathogens
Flavobacterium piscicida	+							+		Toxin	CNS Drug
Marine Actinomycetes											
Nocardia spp.								+		Antibacterial	Antibiotic
Cyanophyta (Blue-green algae)											
Lyngbya majuscula							+	+	Toxic	Antibacterial	Antibiotic
Microcystis aeruginosa (freshwater)	+								Toxic	Polypeptide Endotoxin	CNS Drug
Anabaena flos-aquae (freshwater)	+								Toxic	Polypeptide	CNS Drug

Table 1 (continued)

Phormidium spp.							+		Stimulate growth of bacteria, plant, animal, algae cultures	Growth stimulant	Wound healing
Nostoc ribulare							+		Carcino-genic	Unknown	Study of cancers
Rhodophyta (Red algae)											
Digenea simplex							+		Anthel-minthic	Kainic acid	Against parasitic intestinal worms (ascaris)
Chondrus crispus									Antiviral	Carrageenan	Antiviral drug
Gelidium cartilagenium									Antiviral	Polysaccharide	Antiviral drug
Phaeophyta (Brown algae)											
Rhodomela larix								+		Brominated phenolic com-pound	Antibiotic
Laminaria spp.									Blood anti-coagulant	Laminarin	Anticoagulant

Table 1 (continued)

TAXONOMIC GROUP	Type of pharmacological activity observed in various organisms									Nature of toxin and toxicity	Potential pharmacological drug use
	CNS[a]	RS[b]	NMS[c]	ANS[d]	CVS[e]	GI[f]	Local[g]	Antibiotic activity	Other activity		
Chlorophyta (Green-algae)											
Chlorella spp.		+[h]	+			+		+	Toxic	Unknown, oxidation products of fatty acids	NMS studies, antibiotics
Chlamydomonas reinhardtii								+		Fatty acids	Antibiotics
Chrysophyta (diatoms)											
Ochromonas spp.	+		+						Ichthyotoxic	Unknown	CNS and Neuromuscular drugs
Prymnesium parvum	+		+						Ichthyotoxic, hemolytic, cytolytic, antispasmodic activities	Prymnesin	CNS and Neuromuscular drugs
Phaeocystis pouchetii						+		+		Acrylic acid	Broad spectrum antibiotic for G.I. Tract

Table 1 (continued)

Pyrrophyta (dinoflagellates)											
Gymnodinium spp.	+	+	+	+	+	+	+			Alkylguanidine compounds	CNS drugs
Gonyaulax spp.	+	+	+	+	+	+	+			Alkylguanidine compounds	CNS drugs

[a]CNS = Central Nervous System (nausea, headache, confusion, visual disturbances, nervousness, drowsiness, etc.)
[b]RS = Respiratory System (depression, distress, syncope, dyspnea, etc.)
[e]NMS = Neuromuscular System (muscle weakness, incoordination, spasms, "curare"-like action, paralysis, etc.)
[d]ANS = Autonomic Nervous System (pupil dilation, anticholinesterase activity, parasympathetic action, etc.)
[e]CVS = Cardiovascular System (cardiac stimulation, bradycardia, congestion, myocardial ischemia, etc.)
[f]GI = Gastrointestinal (vomiting, diarrhea, abdominal pain, etc.)
[g]Local = Pruritis, parasthesias, pain, necrosis, edema, etc.
[h]Plus = presence of activity on the system given; no sign = no activity noted on system given.

Sea water also contains bacteriophages (bacterial viruses) and many free enzymes which are particularly concentrated on the sea bottom and are considered to arise mainly from bacteria and other microorganisms. Some of these enzymes continue to function long after the organisms responsible for their synthesis have disappeared. The enzymes rise to the surface during upswellings, and if seitz-filtered sea water, free of bacteria and other microorganisms, is examined, oxidases, reductases, and other enzymes capable of catalyzing changes in phosphates, oxygen, ammonia, nitrates, and so on, will be found (20).

Reference to Table 1 will indicate that the marine bacteria, like their terrestrial counterparts, are prominent in showing antibiotic activity. Rosenfeld and Zobell (43) recorded the antibacterial activity of almost 60 marine microorganisms and found that six (*Bacillus* and *Micrococcus*) species were effective against several microorganisms. Grein and Meyers (44) found 70 actinomycetes out of some 166 derived from littoral sediments, and materials suspended in sea water were effective against both gram-positive and gram-negative bacteria.

In a study of antibiotic properties of microorganisms isolated from various depths (0-3,500 meters) of the world oceans, Krasilnikova (45) found 124 active microbial isolates out of 326 collected, most (ca. 217) of which were nonspore-forming. Eight yeasts and one actinomycete were likewise obtained. The actinomycete showed a wide-range antibacterial spectrum; antibioses was found against *Staphylococcus aureus, Escherichia coli, Mycobacterium luteum,* and *Saccharomyces cerevisiae.*

In a series of papers by Buck *et al* (46-48) several further marine species having antibacterial and antifungal activity were obtained (seven active out of 132 isolates) in addition to one having anti-yeast activity. The anti-yeast activity was found in an isolate of a marine *Pseudomonas.* The test organisms were a number of terrestrial (including human) and marine yeasts. The possibilities of developing agents for the destruction of yeasts pathogenic to man could come from continued studies of this type.

A fungus, *Cephalosporium acremonium,* claimed as a "healer from the sea" (49), was isolated from a sewage outlet off the coast of Sardinia. This organism is the source of cephalosporin C, from which cephalothin, a semisynthetic derivative, has been prepared. This is an antibiotic with activity similar to that of benzyl penicillin but insensitive to penicillinase and, therefore, active against a number of penicillin-resistant *Staphylococci* and some gram-negative species of bacteria. Cephalothin is marketed as "Keflin" by Eli Lilly and is widely used in medicine to-day (35, 50, 51).

The subject of bacterial toxins is still important, particularly with regard to the ecological balances in the marine environment. Bein (52) has described a new species of bacteria (*Flavobacterium piscicida* Bein) which may have been implicated in the mass mortality of fish on the southwest coast of Florida. An abnormal bloom, or "red tide," occurred here in 1951. Later chemical studies by Meyers in collaboration with Myers, Baslow, et al. (53) indicate the toxic substance to be a fairly stable small volatile molecular species which causes deleterious effects on the nervous system of fish.

The subject of antibiosis of marine microorganisms was considered in a recent symposium (54). All of the leads discussed above point to potential antibiotic, antifungal, anti-yeast, and central nervous system-active agents.

In the Cyanophyta (blue-green algae) of the moneran kingdom, several marine and fresh-water genera show interesting pharmacological activity. *Lyngbya majuscula* has been implicated in outbreaks of dermatitis among swimmers (55-57), and toxicity in fish and mice (58). However, *Lyngbya majuscula* has also shown antimicrobial, antiviral, fungicidal, and other types of growth-inhibitory properties in preliminary pharmacological studies (4).

Toxic cyclic polypeptides capable of producing quick death in laboratory animals have been isolated from *Microcystis aeruginosa* and *Anabaena flos-aqua* (59). The LD_{50} for mice, of the purified peptide from *M. aeruginosa* is 0.5 mg/kg. The MLD for mice of a 95 percent ethanol extract of dried *Anabaena flos-aqua* is 40-320 mg/kg. The toxin of the former organism affects the liver and the central nervous system while the toxin of the latter affects only the nervous system (60). These two organisms, in addition to *Aphanizomenon,* seem to be the primary causative agents in many dramatic and serious poisonings in livestock. Gorham (59) reports that some animals, even those as large as a mature cow, have died in less than 30 minutes after ingesting these algae.

Schwimmer and Schwimmer (24) give 235 references to the toxic properties of various marine and fresh-water algae. Gastrointestinal, hepatic, neuromuscular, respiratory, and cardiovascular effects of poisoning by several genera on a variety of animals and man are given. Human mycoses and tumor formations associated with algae are also discussed. In the latter, the use of *Nostoc rivulare* to produce tumors experimentally may be helpful in elucidating the mechanism of cancer induction.

Antibacterial substances produced by marine algae are described by Sieburth (61), and other important properties are reported by Lewin (62) and Jackson (63).

The growth-stimulating properties of algal extracts on bacteria, plant and animal tissue cultures, and other algae had been noted early by Feller (64). In a continued and much later research effort concerning growth-stimulating substances produced by certain algae, Lefevre (65) concluded that fresh-water algae (*Phormidium* spp.), can be used as a source of therapeutically active compounds. These had only growth-stimulant and no antibiotic properties. Over 40 clinical assays on humans and animals showed positive results, characterized as "spectacular," in the treatment of infected wounds, dermal ulcers, and in scar healing.

As an example of how an ecological study can turn up interesting biomedically useful facts, Sieburth (20, 78), in studying why Antarctic penguins had a sterile intestinal tract, found that they fed on the ubiquitous Krill (crustacean) which, in turn, fed on the blue-green algae, *Phaeocystis pouchetti,* which elaborates acrylic acid. Acrylic acid is a strongly active and effective antibiotic against a number of pathogenic organisms, including bacteria and yeasts. It is obvious that these results call out for further research both in phycology and phycotherapy.

Algae, Diatoms, and Dinoflagellates. In the red, brown, and green algae, the diatoms and dinoflagellates, there are many members possessing a wide variety of pharmacological activity. Generally, certain genera of the red, brown, and green algae contain useful phycocolloids, are good sources of thiamine, niacin, riboflavin, folic acid, α-tocopherol, vitamin A, ascorbic acid, and ergosterol and have a relatively high mineral content (up to 38.9 percent): for example, halides, sulfates, phosphates, and oxides of calcium, magnesium, potassium, and sodium, and trace elements (18).

The phycocolloids continue to be used extensively in the food, drug, textile, and cosmetic industries because of the gelling, emulsifying, thickening, suspending, and sizing properties which they possess.

The siliaceous diatoms have traditionally been widely used as filtering aids, scouring powders, and absorbents, while the dinoflagellates continue to serve as important nutritional sources in the food chain of the sea. However, certain species of dinoflagellates are toxic.

Because phytoplanktonic organisms and higher algae are the primary producers in the marine environment, they give rise to the majority of materials found in it. Of interest here, besides those which have been mentioned previously, are those substances which have a broad spectrum antibiotic activity. The types of compounds thus far characterized as antibacterials, are fatty acids of varying molecular weight, derivatives of certain terpenes and of chlorophyll, brominated phenolic compounds, acrylic acid and certain polysaccharide sulfates (2, 8, 24, 61, 64, 66-89).

Most of these studies have been carried out by either one or two researchers and have been limited to the detection of antibiotic activity using the standard agar culture technique with test organisms. Relatively few investigations have been carried to the point where the antibiotic principles have been identified, due probably to limited amounts of materials available. Thus, the time is long overdue for large-scale collection of those species that show activity and for research efforts directed toward growing these organisms on a pilot plant scale in order to facilitate the isolation of new potential antibiotics. By use of the fermentation techniques commonly employed in industry, this might be achieved and would possibly yield new antibiotics that would prove useful in combating resistant strains of common pathogenic organisms and, perhaps, even destroying those which have thus far been resistant to all known chemotherapeutic agents. If any such research has been carried out so far by any of the pharmaceutical firms, little about it has appeared in the scientific literature.

In addition to the references to antibiotic substances elaborated by algae and given above, there are many others which refer to additional pharmacological activity of other algae diatoms and dinoflagellates (90-109). Among those of interest are a bacterial toxin type of phospholipase (lecithinase C) in a marine phytoplanktonic chrysomonad (90); *Gonyaulax catenella* (dinoflagellate) toxin (91); *Caulerpa* (green algae) toxin (70); several marine algae toxins (93); vitamins in marine algae (97); algae growth-inhibition by extracts from *Fucus vesiculosis* (brown algae) (98); toxicity of *Prymnesium parvum* (dinoflagellate) (99); induced shellfish poisoning in chicks with *Gymnodinium breve* (101); electrophoretic separation and analysis of serum proteins, hemoglobin, lipoproteins, and isoenzymes using agarose from seaweed (102); fish toxins in *Ochromonas* (diatom) (103); hypocholesterolemic agents derived from sterols of marine algae (104); anti-coliform activity of sea water associated with the termination of *Skeletonema costatum* blooms (105); antibacterial and antiviral activities of algae extracts (107); toxicity of algae (108); and antiviral kelp (brown seaweed) extracts (109).

In the phylum *Pyrrophyta,* most of the toxicity reported appears to be due to members of the dinoflagellates. The dinoflagellates contain nearly 1,000 species, many being components of plankton. At least 22 species are implicated in poisoning (4). During certain times, weather disturbances and other factors cause the "overgrowth" of these organisms to the point where there is considerable discoloration of the water to produce what is known as "red tide." The excessive accumulation of the dinoflagellates ("blooms") often produces a mass mortality of many fish and other organisms in the surrounding environment.

While many of the factors responsible for such toxicity may be physical in nature (oxygen depletion in water, physical asphyxiation, etc.), there is a possibility that certain toxins may also be implicated. It is known, for example, that toxic dinoflagellates are the cause of paralytic shellfish poisoning. Chemical and pharmacological studies have revealed that this toxin (also referred to as *Gonyaulax* toxin, saxitoxin, mussel poison, mytilotoxin) appears to be a single chemical entity or, at least, a series of closely related compounds, and is among the most toxic materials known to man. The equivalent of 1 mg of purified toxin on ingestion has caused death in man. Nigrelli (2) and Russell (3) have recently summarized the current knowledge of this toxin. Essentially, an alkylguanidine group is common in some of the suggested formulas (4) and a molecular formula of $C_{10}H_{.17}N_7O_4.2\ HCl$ with a molecular weight of 372 has been assigned to it (3, 4, 23, 110-117). Pharmacological studies have revealed that this compound is a potent neurotoxin having dramatic central and peripheral effects (113, 118-125). The central effects include strong action on the cardiovascular and respiratory centers, and the peripheral effects include action on the neuromuscular junction and sensory nerve endings. Experiments show that the toxin has direct effect on the heart, and this is believed to be the cause of the cardiovascular crisis and subsequent precipitous deliterious changes in the systemic arterial blood pressure. Recent studies show further that, by specifically preventing an increase in ionic permeability usually associated with sodium influx, the toxin blocks action potentials in nerves and muscles. It does this without changing chloride or potassium conductances (126, 127). To summarize, the toxin produced by *Gonyaulax catanella* is one of the most potent and pharmacologically active substances yet isolated from marine organisms. It is active in nanomolar concentrations and is about 100,000 times more active than the known conventional local anesthetics (procaine, cocaine) (126). The molecular structure of the toxin, thus far thought to be a substituted purine compound with certain unidentified side chains, certainly offers a model on which the synthesis of new and potent cardioactive, central nervous system active or local anesthetics may be based.

Another interesting toxic factor from the fresh-water chrysomonad (dinoflagellate), *Prymnesium parvum*, is "prymnesin." Although still uncharacterized and believed to be a mixture, it has shown hemolytic, ichthyotoxic, antispasmodic, and cytolytic effects on various animals (99, 128-132).

The red alga, *Digenea simplex*, long used in Japanese folk medicine as an anthelmintic, has yielded kainic acid (2-carboxy-3-carboxymethyl-4-isopropenyl-pyrrolidine; digenic acid). Kainic acid is the active principle in the alga and is widely used in Japan for its vermifuge or anthelmintic properties against the parasitic round worm, *Ascaris lumbricoides*, the whip worm *Trichuris trichura*, and the tapeworm *Taenia* spp. Kainic acid has few side effects and has produced no pathological changes in the digestive tract when administered to mice. It acts apparently by dissolution and separation of the intestinal epithelium of the round worm (*Ascaris lumbricoides*), in addition to causing mucoid degeneration of its epithelial cells. It also causes motor paralysis and inhibition of the action of dehydrogenase in the muscles, so that tissue respiration is depressed. Kainic acid, in combination with santonin, is available in powder and tablet form. "Digesan" syrup used in the treatment of ascariasis, trichuriasis, and oxyuriasis, is also available and contains these same ingredients in combination with piperazine adipate; all of these products are manufactured by the Takeda Pharmaceutical Industries, Japan. The products are, however, not generally available in the United States because of certain FDA regulations (21, 133-142).

Domoic acid (2-carboxy-4-(1-methyl-5-carboxy-*trans-trans-S-trans*-1, 3-hexadienyl)-3-pyrrolidinacetic acid), obtained from *Chondria armata* (red alga), is currently being investigated for its potent exterminating effects on *Oxyuris* as well as *Ascaris* worms (139, 143-149). These are good examples of instances in which folklore has led to the isolation of useful drugs.

It was observed quite early that agar, when used as a substrate for certain types of virus-infected tissue (e.g. EMC, or encephalomyocardites virus), possessed the ability to inhibit the development of the virus (2). Subsequent investigations revealed that the active principle was a polysaccharide sulfate. *Gelidium cartilagenium*, a common source of agar, yielded an active component which was found to be a linear polysaccharide of D-galactose linked 1 to 3 with sulfate groups in low number through the polymer.

In the case of *Chondrus crispus,* the active compound was identified as carrageenan, which is also a linear polysaccharide consisting of D-galactose units linked 1 to 3. Here, some L-galactose, a ketose, and ester sulfate moieties were also found.

The antiviral properties of both polysaccharides have been attributed to their galactan units, since other similar polysaccharides which lack such units have no antiviral activity. The specific antiviral activity was shown against influenza B and mumps virus in embryonated chicken eggs, even after 24-hour inoculation. Protection on the order of 70 percent was also afforded mice given intranasal viral infections (PVM) (150).

Takemoto (151) tested a number of natural and synthetic polyanionic substances and the sulfated polysaccharides mentioned above and found that they could adversely affect the growth of animal viruses in tissue cultures. These included herpes viruses, picornaviruses, arboviruses, and myxoviruses (2, 151).

Another seaweed, *Laminaria coloustoni,* and other related species, yield the polysaccharide, laminarin, which consists essentially of *D*-glucose residues joined mainly through 1-3-type linkages. The highly sulfated derivatives of laminarin have anticoagulant properties comparable to heparin. Laminarins with few sulfate groups have antilipemic properties, without anticoagulant activity. This latter property allows the use of these lower-sulfated laminarin derivatives as effective antilipemic agents without the hazards of concomitant anticoagulant action. It would appear that further assessment of the effectiveness of these agents to lower the amount of fatty substances in the blood, against atherosclerosis, would be useful.

Much biological testing, both *in vivo* and *in vitro,* has shown that the highly sulfated laminarin derivatives are about one-third as effective as heparin in delaying blood coagulation in dogs and guinea pigs. Single large doses appear to be nontoxic, whereas continued dosage may have deleterious effects on the gastrointestinal tract (2, 152-160).

Carrageenan has recently been found to be useful for its anti-peptic or anti-ulcer properties as well as a potentially useful anticoagulant and antithrombic substance. Anderson, using the guinea pig, (161-166) has studied the effects of this algal polyanion on the inhibition of peptic activity, on protection against histamine ulceration, and on acidity and volume of histamine-stimulated gastric secretion.

Bianchi has reported on the antipeptic and anti-ulcerogenic properties of a synthetic sulfated polysaccharide (167), and Hawkins on the antithrombic activity of carrageenan in human blood (168). Heineken (169) has discussed carrageenan in the management of peptic ulcer, and Houck (170, 171) gives anticoagulant, lipemia clearing and other effects of anionic polysaccharides extracted from seaweed. Studies on the antilipemic properties of the algae *Sargassum vulgare* and *Polysiphonia subulifera* are reported by Atkin (172).

Another line of investigation yielded through research based on folk medicine led to the isolation of laminine dioxalate (5-amino-5-carboxypentyl-trimethyl ammonium dioxalate), a purported hypotensive agent. Certain species of *Laminaria* have been used in folk therapy in the northeastern part of Japan, particularly for prevention and treatment of hypertension. Kameda and Osato (173) reported that *Laminaria* extracts were effective in controlling experimentally induced arteriosclerosis and hypertension in rabbits fed high doses of cholestrol. In addition, patients with high blood pressure and hypertension who were fed these extracts obtained relief from their symptoms.

Takemoto et al. (174-177), following these observations, isolated laminine dioxalate and determined the concentration of this principle in 19 different species in the Laminariaceae. They found that *Laminaria angustata* contained the highest concentration of laminine, while others showed a slightly lower concentration—for example, *L. yezoensis, L. cichoriods,* and *Ecklonia cava. Heterochordaria abietina*, which belongs in the *Chordariaceae,* also contained large concentrations of laminine. Pharmacological and clinical studies are continuing on the hypotensive agents of these marine algae.

It should be noted that kainic acid, domoic acid, and laminine dioxalate are all new and novel biologically active amino acids obtained from marine algae.

Alginic acid (a long chain of uronic acid groups, joined by 1:4 glycosidic linkages) and derivatives obtained from the brown seaweeds (kelp) —species of *Fucus* and *Macrocystis,* for example—continue to be useful as sizing materials (textiles), in adhesive formulations, as stabilizers and emulsifiers in food products, as cosmetic and pharmaceutical ointment bases, as suspending agents, emulsifying agents, and so forth. Some of the relatively new uses of the alginates as pharmaceuticals include the use of alginic acid as a tablet disintegrating agent (3-10 percent alginic acid in tablets disintegrates them much faster than those containing 15 percent starch); blood anticoagulants (the sulfuric acid ester of low viscosity alginic acid requires slightly higher doses over heparin, but effects last twice as long); absorbable hemostatic material for control of surface bleeding (mixed sodium-calcium alginate in the form of fine wool or powder has been successfully used in clinical trials and has the advantage of ease of sterilization); in the preparation of sustained release formulations (propylene glycol alginate and alginate gel slow rate of absorption of drugs administered by intramuscular injection); and in formulations successful as dental impression materials (178-183).

One unique use of sodium alginate is based upon its ability to inhibit uptake of radioactive strontium from the human gastrointestinal tract. In human tests, 0.36 microcurie of ^{85}Sr was administered orally 20 minutes after an oral administration of 10 grams of sodium alginate. Twenty-six days later, 0.48 microcurie of ^{85}Sr was given orally. In both phases of the experiment, samples of excreta and blood were collected, and the body retention of the radioactive strontium was measured by means of the Windscale whole body counter. Based on blood plasma, urine, and body retention measurements, sodium alginate reduced the uptake of radioactive strontium from the gastrointestinal tract by a factor of about 9. This discovery is of great importance, since ^{90}Sr is probably the most hazardous of all the long-lived fission products occurring in nuclear weapon fallout. The use of sodium alginate appears to be able to remove this contaminant from the body without seriously affecting the availability of Ca, Na, or K to the body (184-185). Sodium alginate is available as "Kelgin" (Kelco Co., Clark, New Jersey) and "Manucol" and "Manutex" (Alginate Industries Ltd., London, England).

Table 2 shows the number of plants in the different plant phyla and divisions which are found in the marine environment (42). It can be seen that examples cited in this paper are only a small fraction of the total number of between 10,000 and 14,000 species of marine plants which yield useful pharmaceuticals and biodynamic principles. As man progresses ever closer to the sea for the necessities of life, he will find newer biodynamic agents which will ease his pain and allay his suffering from disease and old age.

Table 2. Number of Plants In Different Plant Phyla Found in Marine[a] Environment

Phylum	Approx. no. of living species	Proportion (marine (%)	No. of species marine
Schizophyta (Bacterial)	1,500	12	180
Cyanophyta (Blue-green algae)	7,500 described taxa; probably 200 autonomous species	±75	150 (75% of 200)
Rhodophyta (Red algae)	4,000	98	3,920
Phaeophyta (Brown algae)	1,500	99.7	1,495
Chlorophyta (Green algae)	7,000	13	910
Pyrrophyta (Dinoflagellates)	1,100+	93	1,023
Charophyta (Stonewarts)	76	13	10
Euglenophyta (Euglenoids)	400	3	12
Chrysophyta			
Golden-grown algae	650	±20	130
Coccolithophorids	200	96	192
Diatoms	6,000-10,000	30-50	1,800 up to 5,000
Xanthophyta			
Vaucheria	60	15	9
Mycophyta			
Fungi	75,000	0.4	300
Lichens	16,000	0.1	16
Bryophyta			
Liverworts and mosses	25,000	0	0
Tracheophyta			
Psilopsida, club mosses, horsetails, ferns, cyads, conifers	10,000	0	0
Flowering plants	250,000	0.018 (Sea grasses)	45
			10,192-13,392 Total species (marine)

[a]Adapted from Dawson (42).

References

1. D. L. Fox, *Ann. N.Y. Acad. Sci. 90* (1960), 617.

2. R. F. Nigrelli, M. F. Stempien, G. D. Ruggieri, V. R. Liguori, and J. T. Cecil, *Fed. Proc. 26* (1967), 1197.

3. F. E. Russell, *Fed. Proc. 26* (1967), 2106.

4. B. W. Halstead, "Poisonous and Venomous Marine Animals of the World," (Washington, D.C., Government Printing Office, 1965-1967), vols. I-III.

5. G. A. Emerson and C. H. Taft, *Texas Rep. Biol. Med. 3* (1945), 302.

6. V. J. Chapman, *Seaweeds and Their Uses* (London, Methuen, 1950), p. 287.

7. D. Tressler and J. Lemon, "Marine Products of Commerce," (New York, Reinhold, 1951), p. 782.

8. M. Schwimmer and D. Schwimmer, *The Role of Algae and Plankton in Medicine* (New York, Greene and Stratton, 1955), 85 pp.

9. T. Baarud and V. Sorensen, *Second International Seaweed Symposium* (New York, Pergamon Press, 1956), p. 220.

10. E. Buckley, ed. *Venoms,* Amer. Assn. Advanc. Sci. Publication no. 44 (1956), p. 467.

11. B. W. Halstead, *Med. Arts Sci. 11* (1957), 72.

12. R. F. Nigrelli, *Trans. N.Y. Acad. Sci. 20* (1958), 248.

13. B. W. Halstead, *Dangerous Marine Animals* (Cambridge, Md., Cornell Maritime Press, 1959), p. 146.

14. A. Osol and G. Farrar, "The Dispensatory of the United States of America," 25th ed. (1960), p. 2139.

15. J. Conniff, *Today's Health,* May 1960, pp. 52-63.

16. R. F. Nigrelli, ed., *Ann. N.Y. Acad. Sci. 90* (1960), 615-950.

17. F. Crescitelli and T. A. Geissman, *Ann. Rev. Pharmacol. 2* (1962), 143-192.

18. B. W. Halstead, *J. Amer. Pharm. Assn. N53* (1963), 129.

19. H. L. Keegan, and W. V. Macfarlane, *Venomous and Poisonous Animals and Noxious Plants of the Pacific Region* (London, Pergamon, 1963), p. 456.

20. R. F. Nigrelli, *Metabolites of the Sea,* AIBS, BSCS Pamphlet No. 7 (Boston, Mass., D. C. Heath Co., 1963), p. 35.

21. P. Burkholder, *Armed Forces Chem. J., 27* (1963), 1-8.

22. M. De Clercq, *Ann. Biol. 3* (1964), 429.

23. P. J. Scheuer, "The Chemistry of Toxins Isolated from Some Marine Organisms," in *Progress in Chem. of Organic Natural Products,* L. Zechmeister, ed. (New York, Vienna, Springer-Verlag, 1964), pp. 265-278.

24. D. Schwimmer and M. Schwimmer, "Algae and Medicine," in *Algae and Man* (New York, Plenum Press, 1964), 368-412.

25. J. Welch, *Ann. Rev. Pharmacol. 4* (1964), 293.

26. F. E. Russell, *Advanc. Mar. Biol. 3* (1965), 255.

27. L. Goodman and A. Gilman, *The Pharmacological Basis of Therapeutics,* 3rd ed. (New York, Macmillan, 1965), p. 1785.

28. R. Endean, *Sci. J.,* September 1966, p. 57.

29. D. F. Hornig (chairman), "Effective Use of the Sea." A Report of the Panel

on Oceanography, President's Science Advisory Committee (Washington, D.C., Government Printing Office, 1966), p. 114.
30. J. C. Devlin, *Today's Health,* April 1966, 29-33.
31. H. Bull, ed., *Physician's Desk Reference to Pharmaceutical Specialties and Biologicals,* 20th ed. (Medical Economics Inc., 1966), p. 1131.
32. E. Martin, ed., *Remington's Pharmaceutical Sciences* (Easton, Pa., Mack Publishing Co., 1965), p. 1954.
33. F. E. Russell and P. R. Saunders, *Animal Toxins* (New York, Pergamon Press, 1966), p. 428.
34. R. Hillman, *Oceanology,* September/October, 1967, p. 33.
35. A. Osol, et al., *The United States Dispensatory and Physicians' Pharmacology,* 26th ed. (New York, J. B. Lippincott, 1967), p. 1277.
36. A. H. Der Marderosian, "Current status of drug compounds from marine sources," in Transactions of Symposium, "Drugs from the Sea," Marine Technology Society, Washington, D.C., 1968.
37. B. E. Read, *Peiping Nat. Hist. Bull.* (1939), 136.
38. R. A. Gosselin, *Lloydia 25* (1962), 241.
39. N. R. Farnsworth, *J. Pharm. Sci. 55* (1966), 225.
40. W. E. Yasso, *Oceanography: A Study of Inner Space* (New York, Holt, Rinehart and Winston, 1965), p. 176.
41. R. MacLeod, *Bacteriol Rev. 29* (1965), 9.
42. Y. E. Dawson, *Marine Botany* (New York, Holt, Rinehart and Winston, 1966), p. 371.
43. W. D. Rosenfeld and C. E. Zobell, *J. Bact. 54* (1947), 393.
44. A. Grein and S. P. Mcyers, *J. Bact. 76* (1958), 457.
45. E. N. Krasil'nikova, *Microbiology 30* (1962), 545.
46. J. D. Buck, S. P. Meyers and K. M. Kamp, *Science 138* (1962), 1339.
47. J. D. Buck, D. C. Ahearn, F. J. Roth, Jr., and S. P. Meyers, *J. Bact. 85* (1963), 1132.
48. J. D. Buck and S. P. Meyers, *Limnol. Oceanogr. 10* (1965), 385.
49. Anonymous, *Tile and Till, 51* (1965), 42.
50. Brontzu, G., *Lav. Inst. Ig.,* University of Cagliari (1948).
51. E. Abraham, *Pharmacol. Rev. 14* (1962), 473.
52. S. J. Bein, *Bull. Marine Sci.* (Gulf and Caribbean) *4* (1954), 110.
53. S. P. Meyers, M. H. Baslow, S. J. Bein, and C. E. Marks, *J. Bact. 78* (1959), 225.
54. C. H. Oppenheimer, ed., *Symposium on Marine Microbiology* (Springfield, Ill., Thomas, 1963), p. 769.
55. A. H. Banner, *Hawaii Med. J. 19* (1959), 35.
56. F. H. Grauer, *Hawaii Med. J. 19* (1959), 32.
57. F. H. Grauer and H. L. Arnold, *Arch. Dermat. 84* (1962-62), 720.
58. A. H. Banner, P. J. Scheuer, S. Sasaki, Helfrich P., and C. B. Alender, *Ann. N.Y. Acad. Sci. 90* (1960), 770.
59. P. Gorham, "Toxic Algae," in *Algae and Man* (New York, Plenum Press, 1964), 307-336.
60. P. G. Lauw, *S. Afr. Indust. Chemist. 4* (1950), 62.
61. J. McN. Sieburth, *Develop. Indust. Microbiol. 5* (1964), 124.
62. R. A. Lewin, ed., *Physiology and Biochemistry of Algae* (New York, Academic Press, 1962).
63. D. F. Jackson, ed., *Algae and Man* (New York, Plenum Press, 1964), p. 434.

64. B. Feller, "Contribution á l'étude des plaies traitées par un antibiotic derive des algues." These veterinaire (Paris, Alfort, 1948).
65. M. Lefevre, "Extracellular products of algae," in *Algae and Man* (New York, Plenum Press, 1964), pp. 337-367.
66. K. Kamimoto, *Nippon Saikingaku Zasshi 10* (1955), 897.
67. P. R. Burkholder, L. M. Burkholder, and L. R. Almodovar, *Botanica Marina 2* (1960), 149.
68. C. Chesters and J. Stott, "The Production of Antibiotic Substances by Seaweeds," *Second International Seaweed Symposium* (New York, Pergamon Press, 1956), pp. 49-53.
69. K. Kamimoto, *Nippon Saikingaku Zasshi 11* (1956), 307.
70. M. Doty and G. Aguilar, *Nature 211* (1966), 984.
71. D. Duff and D. Bruce, *Can. J. Microbiol. 12* (1966), 877.
72. T. Katayama, "Volatile Constituents," in *Physiology and Biochemistry of Algae,* ed. R. A. Lewin (New York, Academic Press, 1962), pp. 467-473.
73. G. Fassina, *Arch. Ital. Sci. Farmacol. 12* (1962), 238.
74. T. Katayama, *Bull. Japan Soc. Sci. Fisheries 26* (1960), 29.
75. G. E. Fogg, "Extracellular Products," in *Physiology and Biochemistry of Algae,* ed. R. A. Lewin (New York, Academic Press, 1962).
76. K. Saito and Y. Makamura, *J. Chem. Soc. Japan* (Pure Chem. Sect.) *72* (1951), 992.
77. K. Saito and J. Sameshima, *J. Agric. Chem. Soc. Japan 29* (1955), 427.
78. J. McN. Sieburth, *Limnol. Oceanogr. 4* (1959), 419.
79. J. McN. Sieburth, *Science 132* (1960), 676.
80. H. C. Mautner, G. M. Gardner, and R. Pratt, *J. Amer. Pharm. Assn. 42* (1953), 294.
81. R. Nigrelli, *Trans. N.Y. Acad. Sci. 24* (1962), 496.
82. E. Neilsen, *Deep-Sea Research 3* (1955), 281.
83. P. E. Olesen, *Botanica Marina 6* (1964), 224.
84. R. Pratt, H. Mautner, G. Gardner, Hsien Sha, and J. Dufrenoy, *J. Amer. Pharm. Assn. 40* (1951), 575.
85. V. W. Proctor, *Limnol. Oceanogr. 2* (1957), 125.
86. J. Sieburth, *J. Bacteriol. 77* (1959), 521.
87. J. Sieburth, *J. Bacteriol. 82* (1961), 72.
88. D. Vacca and R. Walsh, *J. Amer. Pharm. Assn. 43* (1954), 24.
89. B. Wolters, *Planta Medica 12* (1964), 85.
90. N. Antia and E. Bilinski, *J. Fish. Res. Bd.* (Canada) *24* (1967), 201.
91. J. M. Burke, J. Marchisotto, J. A. McLaughlin, and L. Provasoli, *Ann. N.Y. Acad. Sci. 90* (1960), 837.
92. J. H. Fraser and A. Lyell, *Lancet,* Jan. 5, 1963, p. 6.
93. R. Habekost, I. Fraser, and B. Halstead, *J. Wash. Acad. Sci. 45* (1955), 101.
94. E. Jorgensen, *Physiol. Plantarum 15* (1962), 530.
95. R. A. Lwein, "Vitamin-bezonoj de algoi," in *Sciencaj Studoj,* ed. P. Neergaard, (Copenhagen, Modersmaalet, Haderslve, 1958), pp. 187-192.
96. P. G. Lauw, *S. Afr. Indust. Chemist 4* (1950), 6.
97. H. Lundin and L. Ericson, "On the Occurrence of Vitamins in Marine Algae," in *Second International Seaweed Symposium* (New York, Pergamon Press, 1956), p. 39.
98. J. McLaughlan and J. Craige, *Canad. J. Bot. 42* (1964), 288.

99. I. Parnas, *Israel J. Zool. 12* (1963), 15.
100. L. Provasoli, J. McLaughlan, M. Droop, *Archiv. mikrobiol. 25* (1957), 392.
101. S. M. Ray and D. V. Aldrich, *Science 148* (1965), 1748.
102. W. M. Rees, *Focus 38* (1967), 4.
103. K. Reich and M. Spiegelstein, *Israel J. Zool. 13* (1964), 141.
104. E. Reiner, *Canad. J. Biochem. Physiol. 40* (1962), 1401.
105. J. Sieburth and D. Pratt, *Trans. N.Y. Acad. Sci. 24* (1962), 498.
106. H. A. Spoehr, J. Smith, H. Strain, H. Milner, and G. T. Hardin, *Carnegie Inst. Wash. Publ. No. 586* (1949), pp. 1-67.
107. J. Starr, *Texas Rep. Biol. Med. 20* (1962), 271.
108. R. E. Wheeler, J. Lackey, L. Schott, *Public Health Reports 57* (1942), 1695.
109. R. H. Kathan, *Ann. N.Y. Acad. Sci. 130* (1965), 390.
110. J. D. Mold, W. L. Howard, J. P. Bowden, and E. J. Schantz, Chem. Corps. Res. Dev. Command. Biol. Warfare Lab., Allied Sci. Div. Special Rept. 250 (1956).
111. E. J. Schantz, J. D. Mold, D. W. Stanger, J. Shanel, F. J. Riel, J. P. Bowden, J. M. Lynch, R. W. Wyler, B. Riegel, and H. Sommer, *J. Amer. Chem. Soc. 79* (1957), 5230.
112. E. J. Schantz, E. F. McFarren, M. L. Schafer, and K. H. Lewis, *J. Assn. Offic. Agric. Chemists 41* (1958), 160.
113. E. J. Schantz, *Ann. N.Y. Acad. Sci. 90* (1960), 843.
114. E. J. Schantz, "Studies on the paralytic poisons found in mussels and clams along the North American Pacific Coast," in *Venomous and Poisonous Animals and Noxious Plants of the Pacific Region,* ed. H. L. Keegan and W. V. Macfarlane (Oxford, Pergamon Press, 1963), 75-82.
115. E. J. Schantz, J. M. Lynch, G. Vayvada, K. Matsumoto, and H. Rapoport, *Biochemistry 5* (1966), 1191.
116. E. J. Schantz, "Biochemical studies on purified *Gonyaulax catenella* poison," in *Animal Toxins,* F. E. Russell and P. R. Saunders, eds. (Oxford, Pergamon Press, 1967), p. 91.
117. H. Rapoport, M. S. Brown, R. Ossterlin, and Schuett, W., *147th National Meetings of the American Chemical Society,* Phila., Pa. (1964).
118. M. Prinzmetal, H. Sommer, and C. D. Leake, *J. Pharmacol. Exp. Ther. 46* (1932), 63.
119. C. H. Kellaway, *Australian J. Exp. Biol. Med. Sci. 13* (1935), 79.
120. M. Fingerman, R. H. Forester, and J. H. Stover, Jr., *Proc. Soc. Exp. Biol. Med. 84* (1953), 643.
121. B. L. Bolton, A. D. Bergner, J. J. O'Neill, and P. F. Wagley, *Bull. Johns Hopkins Hosp. 105* (1959), 233.
122. W. J. Pepler, *J. Formosan Med. Assn. 59* (1960), 1073.
123. E. F. Murtha, *Ann. N.Y. Acad. Sci. 90* (1960), 820.
124. W. D. Dettborn, H. Higman, P. Rosenberg, and D. Nachmansohn, *Science 132* (1960), 300.
125. C. Y. Kao, *Pharmacol. Rev. 18* (1966), 997.
126. C. Y. Kao, "Comparison of the Biological Actions of Tetrodoxin and Saxitoxin," in *Animal Toxins,* F. E. Russell and P. R. Saunders, eds. (Oxford, Pergamon Press, 1967), p. 109.
127. M. H. Evans, "Block of sensory nerve conduction in the cat by mussel poison and tetrodotoxin," in *Animal Toxins,* F. E. Russell and P. R. Saunders, eds. (Oxford, Pergamon Press, 1967), p. 97.

128. E. Reich and A. Aschner, *Palestine J. Bot. 4* (1947), 14.
129. F. Bergmann, I. Parnas, and K. Reich, *Toxicol. Appl. Pharm. 5* (1963), 637.
130. F. Bergmann, I. Parnas, and K. Reich, *Brit. J. Pharm. Chemother. 22* (1964), 47.
131. K. Reich, F. Bergmann, and M. Kidron, *Toxicon 3* (1965), 33.
132. S. Ulitzur and M. Shilo, *J. Protozool. 13* (1966), 332.
133. Anon., "Digesan, combined vermifuge of kainic acid and santonin," *Technical Bulletin* (n.d.), Takeda Pharmaceutical Industries, Ltd., Osaka, Japan, p. 1-4.
134. M. Miyasaki, *Yakugaku Zasshi 75* (1955), 692.
135. H. Morimoto et al., *Proc. Japan Acad. 32* (1956), 41.
136. H. Morimoto et al., *Yakugaku Zasshi 76* (1956), 294.
137. S. Murakami et al., *Yakugaku Zasshi 73* (1953), 1026.
138. S. Murakami et al., *Yakugaku Zasshi 74* (1954), 540.
139. T. Takemoto *Japan. Med. Gazette 20* (1966), 1-3.
140. K. Tanaka et al., *Proc. Japan. Acad. 33* (1957), 53.
141. J. Ueyanagi et al., *Yakugaku Zasshi* 77 (1957), 618.
142. *Merck Index of Chemicals and Drugs. Helminal: A Dry Brown Extract Prepared from a Sea Alga, a Species of* Digenea (Rhodomelaceae), 6th ed. (Rahway, N.J., Merck & Co., Inc., 1952), p. 48.
143. T. Takemoto, T. Nakajima, K. Daigo, *Jap. J. Pharm. Chem. 34* (1959), 404.
144. T. Takamoto and D. Daigo, *Arch. Pharm. 293/65 6* (1960), 627.
145. T. Takemoto, K. Daigo, Y. Kondo, and K. Kondo, *Yakugaku Zasshi, 86* (1966), 874.
146. T. Takemoto, *Jap. J. Pharm. Chem. 32* (1960), 645.
147. T. Takemoto and T. Sai, *Yakugaku Zasshi 85* (1965), 33.
148. T. Takemoto, K. Daigo, and T. Sai, *Yakugaku Zasshi 85* (1965), 83.
149. K. Tsunematsu et al., *Yakugaku Zasshi 86* (1966), 874.
150. P. Garber, J. D. Dutcher, E. V. Adams, and J. H. Sherman, *Proc. Soc. Exp. Biol. Med. 99* (1958), 590.
151. K. Takemoto and S. Spicer, *Ann. N.Y. Acad. Sci. 130* (1965), 365.
152. E. Besterman and J. Evans, *British Med. J.*, Feb. 9, 1957, p. 310.
153. J. Connell, E. Hirst, and E. Percival, *J. Chem. Soc.*, pt. 4 (1950), 3494.
154. E. Dewar, "Sodium Laminarin Sulphate as a Blood Anticoagulant," in *Second International Seaweed Symposium* (New York, Pergamon Press, 1956), pp. 55-61.
155. W. Hawkins and V. Leonard, *Canad. J. Biochem. Physiol. 36* (1958), 161.
156. W. Hawkins and H. O'Neill, *Canad. J. Biochem. Physiol. 33* (1955), 545.
157. S. Mookerjea and W. Hawkins, *Canadian J. Biochem. Physiol. 36* (1958), 261.
158. S. Peat, W. Whelan, and H. Lawley, *J. Chem. Soc.*, pt. I (1958), 724.
159. E. Percival and A. Ross, *J. Chem. Soc.*, pt. I (1951), 720.
160. J. Evans, *British Med. J.*, Feb. 9, 1957, p. 310.
161. W. Anderson and J. Watt, *J. Pharm. Pharmacol. 11* (1959), 318.
162. W. Anderson, *J. Pharm. Pharmacol. 147* (1959), 52.
163. W. Anderson, *J. Pharm. Pharmacol. 13* (1960), 139.
164. W. Anderson, *J. Pharm. Pharmacol. 14* (1962), 119.
165. W. Anderson, *Nature 199* (1963), 389.
166. W. Anderson, *Nature 101* (1965).
167. Bianchi, *Gastroenterology 47* (1964), 409.
168. Wand Hawkins and V. Leonard, *Canad. J. Biochem. Physiol. 41* (1963), 1325.

169. T. Heineken, *Amer. J. Gastroenterology 35* (1961), 619.
170. J. Houck, R. Morris, and E. Lazaro, *Proc. Soc. Exp. Biol. 96* (1957), 528.
171. J. Hauck, J. Bhayana, and T. Lee, *Gastroenterology 39* (1960), 196.
172. E. Atkin, *Qual. Plant. Mater. Veg. 12* (1965), 210.
173. T. Takemoto, *Jap. Med. Gaz.*, May 20, 1966, p. 1.
174. T. Takemoto, K. Daigo, and N. Takagi, *Yakugaku Zasshi 84* (1964), 1176.
175. T. Takemoto, K. Daigo, and N. Takagi, *Yakugaku Zasshi 84* (1964), 1180.
176. T. Takemoto, K. Daigo, and N. Takagi, *Yakugaku Zasshi 85* (1965), 37.
177. T. Takemoto, N. Takagi, and K. Daigo, *Yakugaku Zasshi 85* (1965), 843.
178. Anon.: "Alginates in Pharmaceuticals and Cosmetics," Technical Bulletin (Alginate Industries, Ltd., London, W.C. 2) (1966), pp. 1-12.
179. Anon.: "Algin for Impression Materials, Dental, Facial, and Technical," Technical Bulletin PH no. 5 (Kelco Co., Clark, N.J., 1961).
180. R. H. McDowell, "Properties of Alginates" (Alginate Industries, Ltd., London, W.C. 2), 1st reprint, 2nd ed. (1963), pp. 1-61.
181. A. Myers, *Canad. Pharm. J. 98* (1965), 28.
182. G. Richardson, *Pharm. J., 192* (1964), 527.
183. A. Steiner and W. McNeely, "Algin in review," in *Advanc. in Chem. 11* (1954), 68.
184. R. Hesp and B. Ramsbottom, *Nature 208* (1965), 1341.
185. R. Hesp and B. Ramsbottom, "The effect of sodium alginate in inhibiting uptake of radiostrontium from the human gastrointestinal tract." United Kingdom Atomic Energy Authority (1965), Production Group Report 686 (W), pp. 1-9.

Ergot–A Rich Source of Pharmacologically Active Substances

ALBERT HOFMANN

INTRODUCTION

Since the discovery of penicillin, research on the chemistry of fungi has intensified to an unforeseen extent. Ergot, the subject of this chapter, is a fungus with no antibiotic activity, but it has engaged the interest of doctors, pharmacologists, and chemists for centuries, long before the antibiotic era.

What is commonly known as ergot (*Secale cornutum*) is the sclerotium of the fungus *Claviceps purpurea* (Fries) Tulasne which commonly grows on rye. The grains which are infected with the fungus develop a purple-brown curved body as can be seen in Fig. 1. Cereals other than rye, as well as wild grasses, can be infected by *C. purpurea* and other species of *Claviceps,* as will be described later.

HISTORY OF ERGOT

Ergot has a fascinating history. Over the centuries its role and significance have undergone a complete metamorphosis. Once a dreaded poisonous contaminant, it has come to be regarded as a rich treasure house of drugs (1). Some of the most important dates are listed in Table 1.

Ergot began its history as a poisonous contaminant of edible grain. As early as 600 B.C., an Assyrian tablet alluded to a "noxious pustule in the ear of grain." In the Middle Ages, bizarre epidemics occurred in Europe which cost tens of thousands of people their lives, caused by bread made from rye contaminated with ergot. According to ancient records, 40,000 people died in the south of France during a severe epidemic in 994, and 12,000 died in the Cambrai region in 1129. This

Fig. 1 Ergot growing on rye.

Table 1. History of Ergot

	A poisonous contamination of edible grain:
600 B.C.	Assyrian tablet "noxious pustule in the ear of grain."
Middle Ages	Epidemics of Ergotismus Convulsivus and Ergotismus Gangraenosus described as "ignis sacer," "holy fire," "St. Antony's fire."
	A remedy for quickening childbirth:
1582	Adam Lonitzer ". . . a proved means of producing pains in the womb."
1808	John Stearns in the Medical Repository of New York "Account of the Pulvis Parturiens."
	A remedy to control post-partum hemorrhage:
1824	D. Hosack in "Observations on Ergot" recommended ergot to be used only to control post-partum hemorrhage.
	A source of pharmacologically useful alkaloids:
1906	G. Barger and F. H. Carr: Isolation of ergotoxine and discovery of its adrenolytic activity.
1918	A. Stoll: Isolation of ergotamine, the first pure pharmacologically active ergot alkaloid.
1935	H. W. Dudley and C. Moir and other groups: Isolation of ergonovine (ergometrine, ergobasine, ergotocine), the oxytocic principle of ergot.
1935 onward	Extensive investigations on the chemistry of ergot alkaloids by W. A. Jacobs and L. C. Craig (U.S.A.); S. Smith and G. M. Timmis (England); A. Stoll, A. Hofmann et al. (Switzerland), and other groups; extensive pharmacological and clinical investigations by E. Rothlin, A. Cerletti et al. (Switzerland), and other groups.

Fig. 2 St. Anthony, patron saint of the Order, caring for sufferers from ergotism.

scourge, in which gangrenous manifestations leading to mummification of the extremities were a prominent feature, was known as "ignis sacer," "holy fire," or "mal des ardents."

In 1093, a religious Order was founded in southern France for the purpose of caring for those afflicted by ergotism. The new Order chose St. Anthony as its patron saint (Fig. 2). Figure 2 shows St. Anthony surrounded by patients stricken with ergotism. From this time, "mal des ardents," or the "holy fire," came to be called also "St. Anthony's fire."

The cause of the epidemics was recognized in the seventeenth century, and since then there have been only sporadic outbreaks of ergot poisoning.

Ergot was first mentioned as a remedy used by midwives for quickening labor by the German physician Lonitzer in 1582. The first scientific report on the use of ergot as an oxytocic agent, "Account of the Pulvis Parturiens," was given by the American physician Stearns in 1808. But in 1824, Hosack, recognizing the dangers of using ergot for accelerating child-birth, recommended that the drug be used only to control post-partum hemorrhage. Since that time ergot has been used in obstetrics mainly for this purpose.

The last and most important chapter in the history of ergot, and one which is still not completed, concerns ergot as a source of pharmacologically useful alkaloids. It started with the isolation of ergotoxine in 1906 by Barger and Carr and the discovery of its adrenolytic activity. In 1918, Stoll isolated ergotamine, the first ergot alkaloid to find widespread therapeutic use in obstetrics and internal medicine. Another important step was the discovery in 1935 of the specific oxytocic principle of ergot by Dudley and Moir, which resulted in the isolation of the alkaloid ergonovine simultaneously in four separate laboratories.

Since 1935, extensive investigations on the chemistry of ergot alkaloids have been carried out mainly by Jacobs and Craig in the United States, Smith and Timmis in England, and Stoll, Hofmann et al., paralleled by pharmacological and clinical investigations by Rothlin, Cerletti et al., in Switzerland (2).

THE CHEMISTRY OF THE ERGOT ALKALOIDS

The ergot alkaloids belong to the large and important class of indole alkaloids. All ergot alkaloids contain a tetracyclic ring system, which has been named ergoline (I).

The most important ergot alkaloids are derivatives of lysergic acid, which is 6-methyl-ergolene-($\Delta^{9,10}$)-8β-carboxylic acid (II).

Lysergic acid (II) and its derivatives, the ergot alkaloids, epimerize very easily at position 8, giving rise to isolysergic acid (III) and its derivatives, respectively. The latter compounds are usually much less physiologically active than the genuine alkaloids, and great care must be taken to avoid such isomerization during extraction.

The structure of lysergic acid (II) was confirmed by total synthesis by Kornfeld and co-workers in the Lilly Laboratories in 1954 (3). This

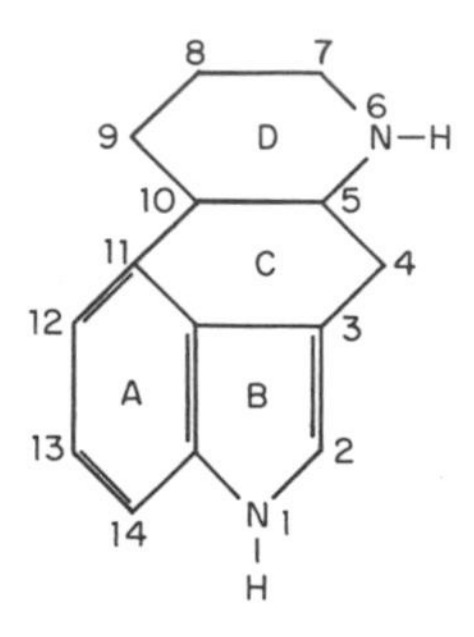

I. Ergoline

synthesis has not found industrial application because of the poor yield, and because, as will be described later, lysergic acid can be produced today by fermentation processes on an industrial scale.

Up to the present time, some two dozen alkaloids have been isolated from various species of ergot and their structures and stereochemistry have been completely elucidated (Fig. 3). They can be divided into two main groups, A and B. Group A comprises amides of lysergic acid: subgroup A I contains simple amides and subgroup A II, amides of the peptide type. In the first subgroup, the amide radical may be simply NH_2, as in ergine, or it may be an amino-alcohol such as 2-amino-propanol in ergometrine, the oxytocic principle of ergot. The therapeutically most important ergot alkaloids belong to subgroup A II, in which lysergic acid is combined with various cyclic tripeptide (Fig. 3). These tripeptides all contain a proline residue, joined to a second amino acid, which can be either L-phenylalanine, L-leucine, or L-valine, and an α-hydroxy-α-amino-acid, either α-hydroxy-alanine, α-hydroxy-valine, or α-hydroxy-α-amino-butyric acid.

In 1961, the first total synthesis of a peptide-type ergot alkaloid, namely of ergotamine, was accomplished in my laboratory (4). This synthesis, which will be discussed later, afforded confirmation of the special, cyclol structure of the peptide moiety.

The ergot alkaloids of the second main group (clavine type, Fig. 3) differ from those of Group A in that the carboxyl group of lysergic acid is replaced by a hydroxymethyl or a methyl group. The alkaloids of the clavine type occur mainly in ergot growing on wild grasses and are of no value in medicine.

Fig. 3 Ergot alkaloids.

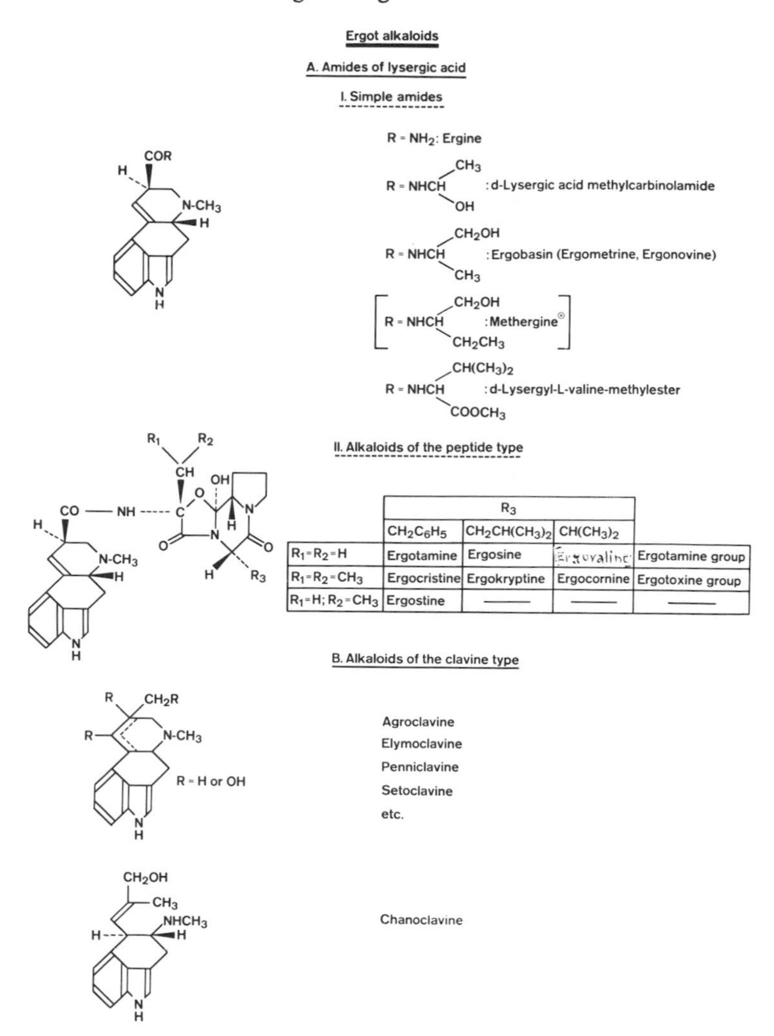

	R_3			
	$CH_2C_6H_5$	$CH_2CH(CH_3)_2$	$CH(CH_3)_2$	
$R_1 = R_2 = H$	Ergotamine	Ergosine	Ergovaline	Ergotamine group
$R_1 = R_2 = CH_3$	Ergocristine	Ergokryptine	Ergocornine	Ergotoxine group
$R_1 = H; R_2 = CH_3$	Ergostine	——	——	——

PRODUCTION OF ERGOT ALKALOIDS BY FERMENTATION

When the medicinal importance of the ergot alkaloids was realized, many attempts were made to produce them by growing the *Claviceps* fungus *in vitro*, instead of on rye.

Abe, in Japan, discovered in 1951 a species of *Claviceps* growing on wild grasses which produced ergot alkaloids in good yield in submerged cultures (5). However, these alkaloids were of the clavine type, which, as I have already pointed out, are of no use in medicine. In 1953, Stoll Brack, Hofmann, and Kobel first succeeded in producing ergotamine by *in vitro* cultivation of *C. purpurea* (6). However, the yields were low, and the surface culture method they used was not suitable for industrial production. Recently, Tonolo (7), and somewhat later, Amici et al. (8) reported that they had isolated a strain of *C. purpurea* which produced ergotamine in appreciable yield in submerged culture. Unfortunately, this strain too cannot be used for the industrial production of ergotamine as the yield is too low.

A discovery of industrial importance was the observation in 1960 by Chain and his co-workers, at the Istituto Superiore di Sanità in Rome, that an Italian strain of *C. paspali* was able to produce lysergic acid amide and simple derivatives of lysergic acid amides in high yield in submerged culture (9). These lysergic acid amides can be readily hydrolyzed to lysergic acid, which can be used as starting material for the synthetic production of therapeutically useful pharmaceutical preparations.

More recently, after investigation of many hundreds of ergot samples from all over the world, Kobel, Schreier, and Rutschmann of the Sandoz Laboratories succeeded in isolating, from ergot found in Portugal on *Paspalum dilatatum*, a *Claviceps* strain capable of producing excellent yields of a mixture of free lysergic acid isomers in submerged cultures (10). This mixture consists of some 30 percent of lysergic acid (II) with a small amount of isolysergic acid (III) and of some 70 percent of a new isomer of lysergic acid. We have named this new acid from *Paspalum* ergot *paspalic acid* (IV). The structure and stereochemistry of paspalic acid and its relationship to lysergic and isolysergic acid is illustrated in Fig. 4.

The structure and configuration of the new acid were apparent from the fact that $LiAlH_4$-reduction of its methyl ester yielded a mixture of elymoclavine and lysergol, two ergot alkaloids of known structure. Paspalic acid is thus 6-methyl-($\Delta^{8,9}$) ergolene-8-carboxylic acid. The isolated double bond migrates very easily from the 8,9-position into the 9,10-position under alkaline conditions, producing a mixture of lysergic and isolysergic acid. Paspalic acid, which can be produced on an industrial scale, is therefore a very suitable starting material for the synthesis of lysergic acid which, in turn, can be used for the synthetic production of useful ergot alkaloids or their derivatives.

Fig. 4 Paspalic acid and its products of transformation.

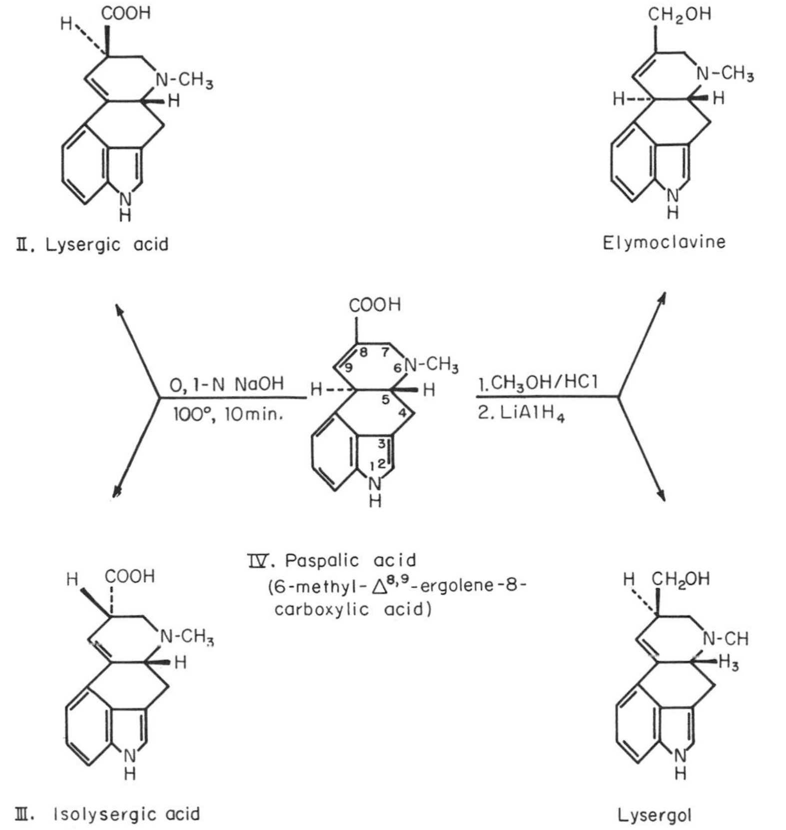

II. Lysergic acid

Elymoclavine

0,1-N NaOH 100°, 10min.

1. CH_3OH/HCl 2. $LiAlH_4$

IV. Paspalic acid (6-methyl-$\Delta^{8,9}$-ergolene-8-carboxylic acid)

III. Isolysergic acid

Lysergol

SYNTHESIS OF AMIDES OF LYSERGIC ACID

Today, all the naturally occurring ergot alkaloids can be synthesized. Within the scope of this review, the discussion will be limited to the synthesis of those ergot alkaloids which are of therapeutic interest, namely those of the lysergic acid amide type. As already mentioned, none of the alkaloids of the clavine type (Fig. 3) has been found to be of value in therapy.

Ergometrine and Other Simple Amides. The synthesis of simple amides of lysergic acid (Fig. 3, subgroup A I) was accomplished a number of decades ago. Lysergic acid being a labile, very sensitive compound, the problem was to prepare a suitable activated derivative of lysergic acid which can be reacted with the appropriate amine. The first successful method of amidation was that using the Curtius reaction via lysergic acid hydrazide and azide (11). Later, other methods were used for the preparation of lysergic acid amides (12-15). By the first procedure, ergometrine (V, ergobasine, ergonovine = d-lysergic acid L-isopropanolamide), the oxytocic principle of ergot, and a large number of other lysergic acid amides were synthesized (11).

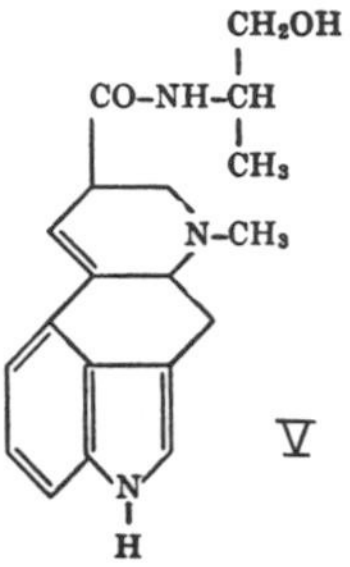

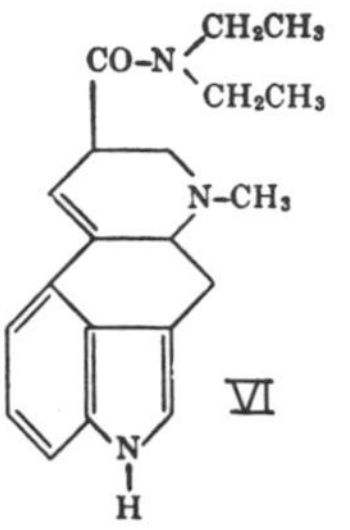

It was during these investigations that I prepared the diethylamide of lysergic acid (VI) or LSD-25. The idea which urged me to synthesize this compound was a certain structural similarity with coramine (VII, nikethamide), a proven analeptic. Pharmacological analysis showed lysergic acid diethylamide to be a strong oxytocic agent with about 70 percent the activity of ergometrine.

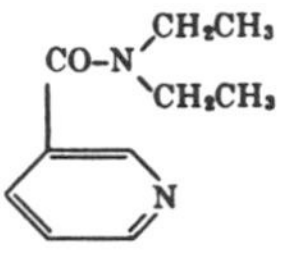

Some years later, I prepared LSD a second time in order to provide our pharmacologists with substance for more profound pharmacological investigation. When I was purifying lysergic acid diethylamide (VI) in the form of its tartrate, I experienced a strange, dream-like state which wore off after some hours. The nature and course of this extraordinary disturbance aroused my suspicions that some exogenic intoxication might be involved and that the substance with which I had been working, lysergic acid diethylamide tartrate, could be responsible. In order to ascertain whether or not this was so, I decided to test the compound in question on myself. Being by nature a cautious man, I started my experiment with the lowest dose which presumably could have any effect, taking 0.25 mg LSD tartrate. This first planned experiment with LSD took a dramatic turn and led to the discovery of the extraordinarily high psychotomimetic activity of this compound. The fascinating psychic effects of the diethylamidemide of lysergic acid prompted the synthesis of a large number of analogues, homologues, and derivatives. None of them proved to be more active in its hallucinogenic psychotomimetic properties. But these investigations were successful in other

Fig. 5 Synthesis of peptide-type ergot alkaloids.

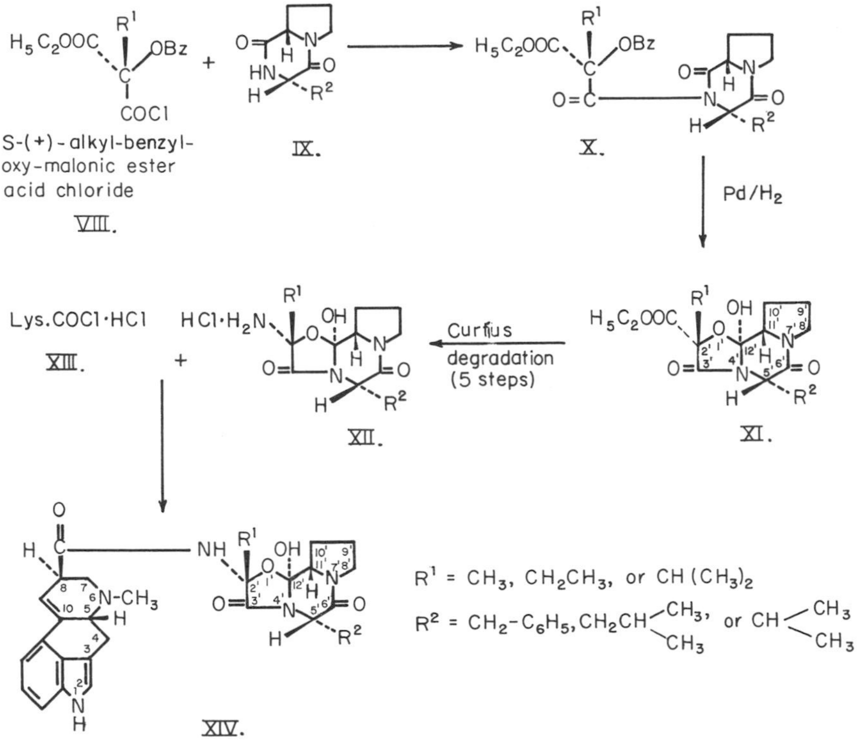

S-(+)-alkyl-benzyloxy-malonic ester acid chloride VIII. + IX. → X.

X. → (Pd/H_2) → XI.

XI. → (Curtius degradation (5 steps)) → XII.

Lys.COCl·HCl XIII. + XII. → XIV.

$R^1 = CH_3$, CH_2CH_3, or $CH(CH_3)_2$

$R^2 = CH_2{-}C_6H_5$, $CH_2CH(CH_3)_2$, or $CH(CH_3)_2$

respects. Compounds with other valuable pharmacological properties were found among these derivatives, as will be described later.

Ergot Alkaloids of the Peptide Type. The most important recent developments in ergot research have been in the synthesis of new alkaloids of the peptide type (Fig. 3, subgroup A II). Since the first synthesis of a peptide-type ergot alkaloid, ergotamine, in our laboratory in 1961 (16), we have improved the various steps of the synthesis, achieving an appreciable increase in overall yield. In the meantime, the same scheme and sequence of reactions that we used for the synthesis of ergotamine has been applied for the synthesis of the other naturally occurring ergot alkaloids of the ergotamine and ergotoxine group. The general scheme used for these syntheses is shown in Fig. 5.

A suitable substituted alkyl-benzyloxy-malonic ester acid chloride is reacted with a dioxopiperazine (IX) consisting of the radical of L-proline and of a variable amino-acid, either L-phenylalanine, L-leucine, or L-valine. The preparation of the malonic acid derivatives (VIII), where R^1

stands for either methyl, ethyl, or isopropyl, and their separation into the optically active forms is almost a synthesis in its own right. The S-form possesses the stereochemistry corresponding to the configuration at the asymmetric center 2′ in the natural alkaloids. The acylated dioxopiperazine (X) (Fig. 5) is treated with hydrogen and Pd-catalyst in order to remove the benzyl group. The resulting compound with the free hydroxyl undergoes spontaneous cyclolization in a stereo-specific manner to yield the cyclol ester XI. The ethoxycarbonyl group of (XI) is replaced by an amino group following the steps of a Curtius degradation (i.e. via free acid, acid chloride, azide, carbobenzoxyamide) and removal of the benzyl group by catalytic hydrogenation in acidic solution to give the hydrochloride of the aminocyclol (XII). The free aminocyclol (XII), which represents the complete peptide part of the corresponding natural alkaloids, is an extremely labile compound, but it can be isolated in the form of relatively stable crystalline salts, e.g. in the form of the hydrochloride. Using special conditions, it is possible to acylate (XII) with lysergic acid chloride hydrochloride (XIII) to obtain the corresponding natural peptide alkaloid (XIV) in good yield. Starting with malonic ester and the free amino acids and with lysergic acid, this synthesis of a peptide-type ergot alkaloid comprises 22 steps. Using this procedure, the following naturally occurring alkaloids have already been synthesized: ergotamine, ergosine, and the missing link in the natural system of peptide ergot alkaloids with valine as the variable amino acid, which we have named ergovaline (Fig. 3) (17). Ergostine, an alkaloid which occurs only in traces in ergot, and which is characterized by an ethyl substituent at the position 2′ (18), and quite recently, the three alkaloids of the ergotoxine group, which are characterized by an isopropyl substituent at position 2′, i.e. ergocristine, ergokryptine, and ergocornine, have also been synthesized.

With this method developed for the synthesis of the natural ergot alkaloids, we are now in the position to synthesize a great variety of lysergic acid peptides with amino acids other than those occurring naturally, and to study the structure-activity relationship in pharmacological agents of this type. As an example of this line of our current research, suffice it to mention the synthetic analogue of ergotamine in which the L-phenylalanine residue is replaced by an α-methyl-alanine residue. This compound, the activity spectrum of which shows true differences from that of ergotamine, possesses valuable pharmacological properties, as is shown in the section on pharmacology.

CHEMICAL MODIFICATIONS OF THE LYSERGIC ACID MOIETY

Much work has been done and is still in progress to modify or replace the lysergic acid part in the natural ergot alkaloids and in synthetic lysergic acid derivatives. Only a few of the more important modifications which have led to derivatives with interesting pharmacological properties will be discussed.

Hydrogenation of the 9, 10 Double Bond. Catalytic hydrogenation of lysergic acid or of its derivatives yields the corresponding 9,10-dihydro derivatives, a new asymmetric center being formed at C 10. Whereas lysergic acid and its derivatives give only one of the two theoretically possible stereoisomers (XV, XVII) namely, the epimer (XV) with the hydrogen atom in α-position at C 10, isolysergic acid and its derivatives yield both epimers (XVI, XVIII) (19, 20). The stereochemistry of these dihydro derivatives is depicted on Fig. 6. The dihydro derivatives of the ergot alkaloids differ fundamentally in pharmacological activity from the natural compounds, as will be discussed later.

Saturation of the double bond in the 9,10 position can be achieved also by addition of the elements of water. This occurs when an acidic solution of the alkaloids is irradiated with u.v. light (21). These so-called lumi-derivatives are of no pharmacological interest.

Substitution at Positions 1 and 2. Of the various substitutions which have been carried out at position 1 of lysergic acid and dihydrolysergic acid derivatives (Fig. 7) alkylation, principally methylation (22, 23) cause an interesting shift in pharmacological activity. Halogenation, especially bromination, in position 2 (24) (Fig. 7) can also produce fundamental changes in pharmacological and clinical properties, for example, of LSD.

PHARMACOLOGY AND THERAPEUTIC USE OF ERGOT PREPARATIONS

Before we deal with the changes in the pharmacological properties resulting from chemical modifications of the natural ergot alkaloids, let us first take a look at the main pharmacological activities of the natural alkaloids themselves. The ergot alkaloids have an astonishingly wide spectrum of action, a multiplicity of different pharmacological activities such as is rarely found in any other group of natural products.

The pharmacological effects of ergot alkaloids fall into six categories as listed in Fig. 8. They can be divided into three groups, depending on the

Fig. 6 Stereochemistry of the dihydro derivatives of d-lysergic and d-isolysergic acids.

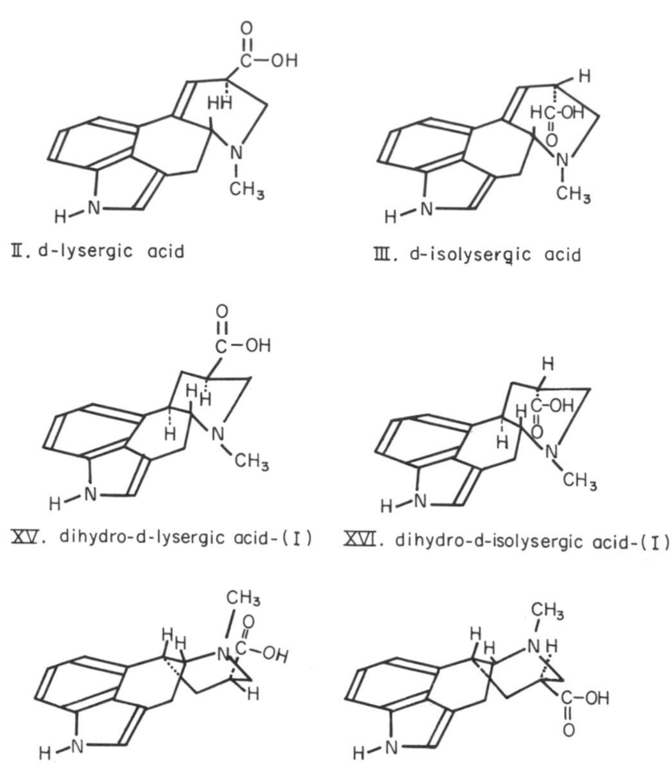

Fig. 7 Substitutions at positions 1 and 2 of lysergic acid and dihydrolysergic acid derivatives.

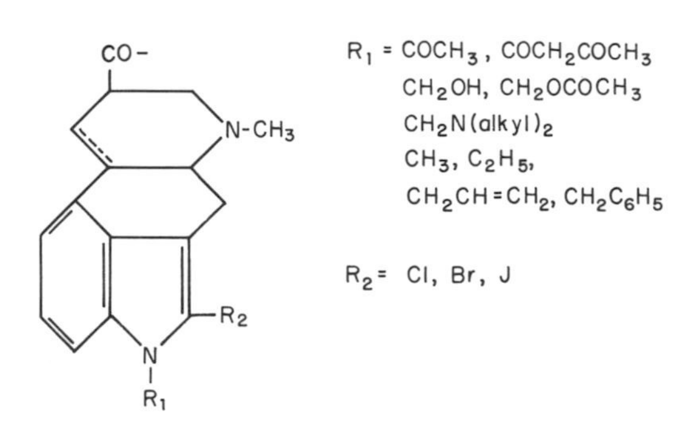

Fig. 8 The main pharmacological effects of ergot alkaloids.

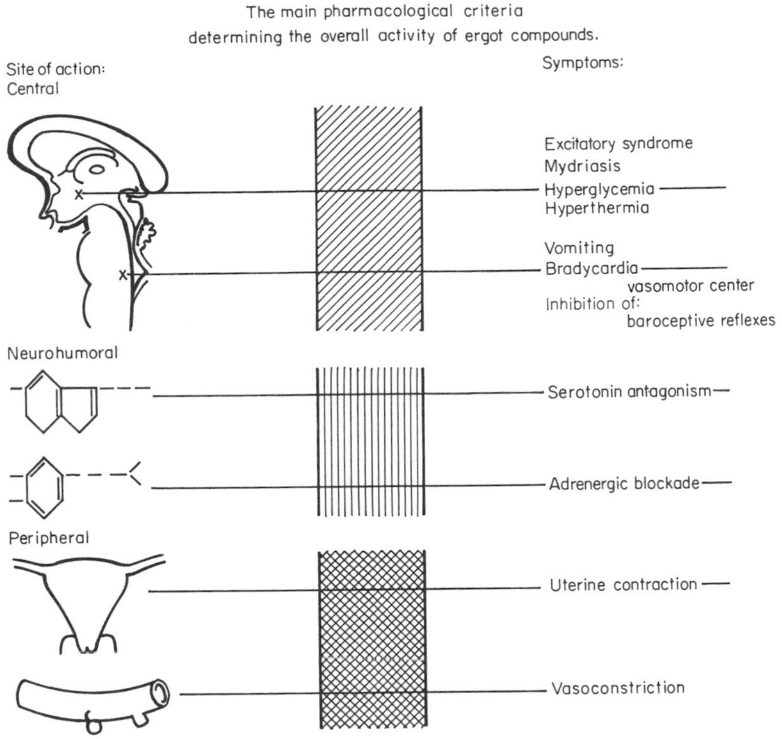

The main pharmacological criteria determining the overall activity of ergot compounds.

Site of action:
Central

Symptoms:

Excitatory syndrome
Mydriasis
Hyperglycemia
Hyperthermia

Vomiting
Bradycardia
Inhibition of: vasomotor center, baroceptive reflexes

Neurohumoral

Serotonin antagonism

Adrenergic blockade

Peripheral

Uterine contraction

Vasoconstriction

site of action, a distinction being made between central, neurohumoral and peripheral effects. The peripheral action on smooth muscle is manifest as vasoconstriction and uterine contraction. The classical indication for ergot alkaloids, i.e., their use in obstetrics to arrest hemorrhage and promote uterine contractions, is based on this effect.

Neurohumoral effects are antagonism to adrenaline and to serotonin. Antagonism to adrenaline and to the effects of postganglionic sympathetic nerve stimulation, i.e. adrenolytic and sympathicolytic activities, account for many of the uses of the ergot preparations in internal medicine. The other neurohumoral effect, antagonism to serotonin, has been selectively developed in certain ergot derivatives, as will be shown later.

Central effects occupy an important place in the activity spectrum of the ergot alkaloids. The site of action is in the medulla oblongata and in the midbrain. The ergot alkaloids reduce the activity of the vasomotor center in the medulla oblongata, and this reduced activity is responsible for the vasodilator, hypotensive and bradycardic effects of certain ergot alkaloids. Many ergot alkaloids stimulate the vomiting center in the medulla oblongata. Sympathetic structures in the midbrain, particularly in the hypothalamus, are stimulated. This leads to a comprehensive excitation syndrome, with such signs as mydriasis, hyperglycaemia, hyperthermia, tachycardia, and so on. This excitation syndrome is connected with the psychotomimetic and hallucinogenic actions of certain ergot derivatives such as LSD.

The various structural types of natural ergot alkaloids and their derivatives differ in biological activity in that the relative predominance of these six main effects varies from compound to compound. One or more of these activity components may be almost completely absent; other effects may remain unaltered or may even be enhanced. The goal of chemical modification is to arrive at compounds with a narrower range of activity, but with more selective specific effects.

Cerletti has delineated the activity spectra of the various types of ergot alkaloids with reference to the six main effects described (25). He depicted his findings graphically by selecting a relative scale for each compound, beginning with the smallest effective dose (top) and finishing with the 100 percent lethal dose (bottom). The maximum value indicates which of the six main effects is particularly prominent.

It will be seen that in the case of ergotamine (Fig. 9), the ratio between the various effects is rather well balanced. Thus, this alkaloid exerts the full effects of ergot, in that it causes the uterus to contract, reduces the activity of the adrenergic system, and elicits central effects by inhibiting the vasomotor center. This spectrum of actions accounts for the use of ergotamine in obstetrics as a hemostatic and in internal medicine and neurology as an agent blocking the sympathetic nervous system and as a cranial vasoconstrictor in migraine and related headache syndromes.

Saturation of the double bond at position 9,10 in the lysergic acid moiety has furnished a number of pharmacologically interesting derivatives. As can be seen from the spectrum of dihydroergotamine (Fig. 9), hydrogenation results in a fundamental change in pharmacological actions. The dihydro derivatives of the other peptide alkaloids possess similar spectra of activity. The vasoconstrictor and uterotonic actions, the classical effects of ergot, and the stimulation of central sympathetic structures are greatly attenuated, so that they are barely present within the therapeutic dose range. Instead, the dihydro derivatives of the peptide alkaloids exert a marked sympathicolytic-adrenolytic effect and reduce the activity of the vasomotor center. These effects of dihydroergotamine are exploited

Fig. 9 Activity spectra for various ergot alkaloids and synthetic analogs.

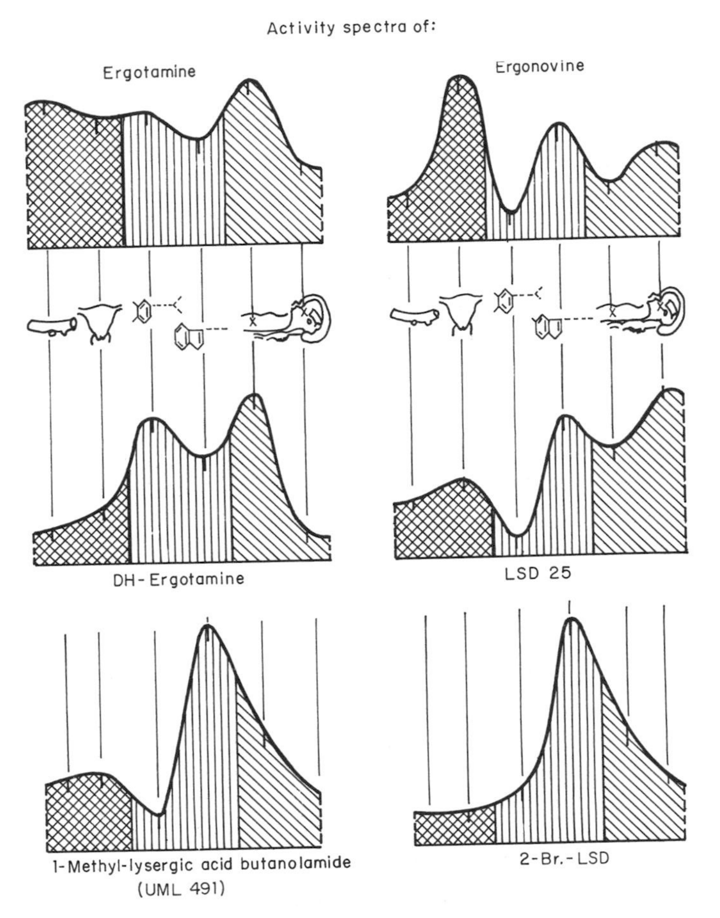

therapeutically, e.g. in the pharmaceutical preparation "Dihydergot" ®. The adrenolytic, vasodilator, and hypotensive effects are even more pronounced in the case of the dihydro derivatives of the alkaloids of the ergotoxin group. A combination of equal amounts of dihydroergocristine, dihydroergokryptine, and dihydroergocornine is used under the brand name "Hydergine" ® for the treatment of vascular disease in order to improve the peripheral and cerebral circulation, especially in geriatric patients.

As can be seen from Fig. 8, the activity spectrum of ergonovine is quite different from that of ergotamine and the other peptide alkaloids. It exerts practically no adrenolytic action. It retains considerable antiserotonin activity. The predominant effect is that on the uterus—the hemostatic and oxytocic action. For this reason, ergonovine is used almost exclusively in obstetrics. Methylergometrine, a synthetic compound in which the L-2-aminopropanol side chain of ergonovine is replaced by a L-2-aminobutanol residue, possesses the same activity spectrum as ergonovine. This derivative, which is also known under the trade name "Methergine" ®, has a somewhat stronger and longer-lasting vasoconstrictor effect on the uterus than the natural alkaloid.

LSD, which differs from the natural alkaloid ergonovine only in having a diethylamide instead of an isopropanolamide side chain, shows an activity spectrum which is quite different from that of ergonovine. LSD exhibits marked antagonism to serotonin (Fig. 9). However, its curve attains its maximum for stimulation of central nervous structures in the hypothalamus. The syndrome of central excitation is elicited even by extremely minute doses and is characterized by mydriasis, hyperthermia, hyperglycaemia, etc. This central excitation syndrome seems to be related to the impressive psychic effects of LSD, which have rendered this substance of considerable importance in experimental psychiatry, neurophysiology, and many other areas. The LSD problem, of course, could be the subject of a chapter of its own.

As examples of how chemical modifications in the lysergic acid radical can change pharmacological activity, let us now briefly discuss the effect of bromination at position 2, and methylation at position 1 (Fig. 7).

Bromination produces fundamental changes in the pharmacological effects of LSD, as can be seen from the activity spectrum of 2-bromo-LSD in Fig. 9. The high psychotomimetic activity of LSD has disappeared. The outstanding pharmacological property of 2-bromo-LSD is its specific antiserotonin activity.

A similar profile of activity, namely, predominance of the antiserotonin component, results when the hydrogen atom on the indole nitrogen is replaced by a methyl group. 1-Methyl ergot derivatives include the most potent serotonin antagonists yet discovered. Today, serotonin antagonists are playing an important role in pharmacological research, because it is with their aid that we are able to study the biological functions of serotonin which is an important neurohumoral factor with manifold effects on major structures and functions of the organism. Only a few of these effects can be mentioned here. The brain stem and the hypothalamus have a particularly high serotonin content, which suggests that the compound is important to the function of these structures. Serotonin increases permeability and elicits pain; it has been postulated that it may be the humoral pain factor in migraine. This will suffice to show that substances exhibiting a specific antagonism to serotonin are not merely of academic interest; they may also be of great importance from the point of view of therapy.

From the large number of 1-methyl ergot derivatives which have been studied in our laboratories as serotonin antagonists, one compound with especially favorable pharmacological properties was selected for introduction into therapy, namely, 1-methyl lysergic acid L-butanolamide (XIX), which is also known as UML-491, and marketed under the brand name "Sansert" ®. Its activity spectrum is depicted in Fig. 9. It has found widespread use in the prophylactic treatment of migraine and other vascular headaches between attacks.

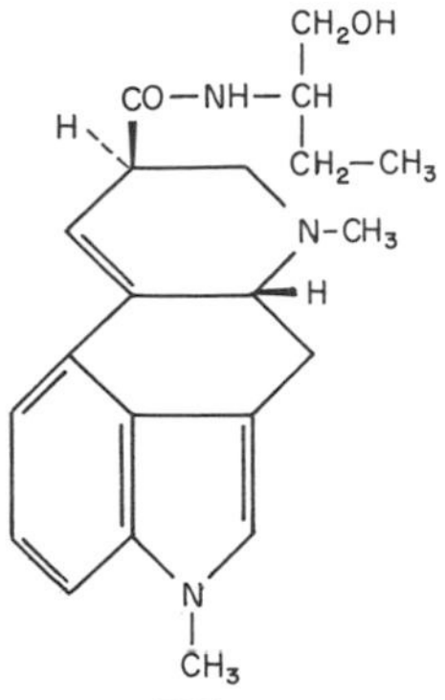

XIX

As a last example of how the pharmacological properties of the natural ergot alkaloids are changed by chemical modifications, I should like to say something about our recent investigations in the field of synthetic peptide-type ergot alkaloids. From the many synthetic analogues which we have lately prepared, only one modification will be mentioned here, i.e. the compound in which we have replaced the L-phenylalanine residue of ergotamine (XX) by an α-methylalanine residue and which was named 5′-methylergoalanine (XXI) (26). As may be seen from Table 2, the con-

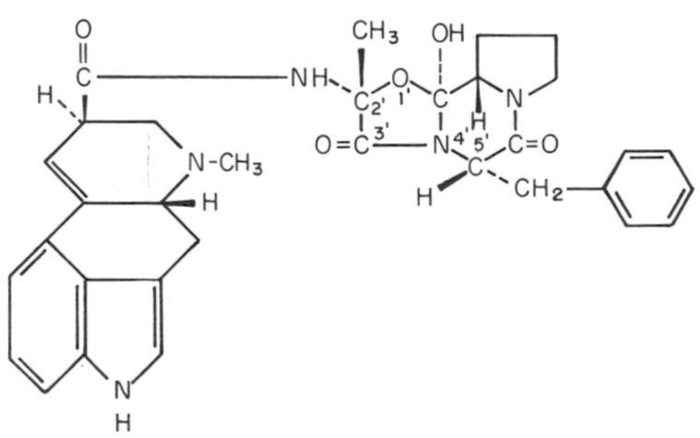

XX Ergotamine

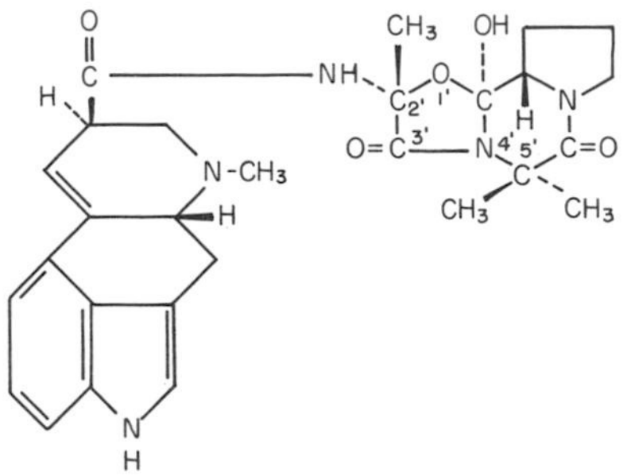

XXI 5′-Methylergoalanine

Table 2. Pharmacology of 5′-Methyl-ergoalanine

Substance	Vasoconstrictor effect: blood pressure increase in spinal cats i.v.(%) (A)	Uterotonic effect in non-pregnant oestrous rabbits i.v.(%) (B)	Emetic effect in conscious dogs i.v.(%) (C)	$\frac{A}{B}$	$\frac{A}{C}$
Ergotamine (XX)	100	100	100	1	1
Ergostine (Fig. 3)	100 ±15	42 ±11	41	2.4	2.4
5′-Methyl-ergoalanine (XXI)	155 ±29	5 ±1	32	31	4.8

tractile effect of ergotamine on smooth muscle, which is manifest on vascular smooth muscle and on extravascular smooth muscle, notably on the uterus, is modified in 5′-methylergoalanine. Whereas the vasoconstrictor effect is increased by 55 percent, the uterotonic effect is decreased by 95 percent (27). It is generally accepted that the therapeutic effect of ergotamine in the migraine attack is due mainly to its vasoconstrictor activity (28). Its uterotonic effect is not desired in the treatment of migraine attack. It can be concluded, therefore, that 5′-methylergoalanine, which is a more specific vasoconstrictor than ergotamine, might have been an improved medicament for the therapy of migraine attack. Furthermore, the emetic effect of 5′-methylergoalanine is less pronounced than that of ergotamine, which was another favorable feature in the activity profile of this chemical modification of the natural alkaloid. Recent clinical investigations, however, showed that this drug elicits certain undesirable side effects which has prevented its therapeutic use.

These few examples may suffice to illustrate how chemical modification of the natural ergot alkaloids, which are themselves useful therapeutic agents, can lead to a variety of compounds with more interesting pharmacological profiles. It will also have shown that ergot is indeed a treasure house of pharmacological active principles.

At this point, my thesis on ergot as a rich source of pharmacological constituents could be brought to a close. But there is another fascinating aspect of research in this field which I should like to report briefly.

OCCURRENCE OF ERGOT ALKALOIDS IN "OLOLIUQUI" AND OTHER CONVOLVULACEAE

The discovery of LSD in the course of our investigations on the alkaloids of ergot awoke our interest in psychotropic agents and in psychopharmacological research in general. This led us to examine the "magic" mushrooms of Mexico, which were reported to elicit psychic effects similar to those of LSD.

The survival of the ancient Indian mushroom cult in the remote mountains of southern Mexico was discovered by Schultes and Wasson. As far back as 1939, Schultes was the first to offer an identification of "teonanacatl," the Aztec hallucinogen, as a mushroom (29). And on several expeditions between 1953 and 1956, Gordon Wasson and his wife studied and described in a masterful manner the ancient and present-day ceremonial use of the hallucinogenic teonanacatl mushrooms. Through the help of Professor Roger Heim, who continued the work of Schultes in identifying and cultivating the "magic" mushrooms, my laboratory obtained samples, and we were able to isolate the active principles, psilocybin and psilocin, elucidate their structure, and synthesize them (30).

After the mushroom problem had been resolved, we decided to tackle the chemical investigation of another enigmatic magic plant of Mexico, namely "ololiuqui." Here again, I was able to rely on the basic research of Schultes, who had published an excellent review on the historical, botanical, and ethnological aspects of ololiuqui in 1941, entitled "A contribution to our knowledge of *Rivea corymbosa*, the narcotic ololiuqui of the Aztecs" (31). And Gordon Wasson again participated in the project. He provided me with original ololiuqui seeds collected in the Mexican state of Oaxaca, where the Indians of several tribes still use these seeds for divinatory purposes in their medical-mystical practices.

Ololiuqui is the Aztec name for the seeds of *Rivea corymbosa* (L.) Hall. f, which is shown in flower in Fig. 10; the seeds are shown in Fig. 12.

The Zapotec Indians use for the same purposes the seeds of another morning glory, *Ipomoea violacea* L., which has been spread all over the world as an ornamental plant. It is the Morning Glory of our gardens (Fig. 11 and 12).

We were surprised to find in ololiuqui seeds active principles familiar to us for a long time, that is, the ergot alkaloids (32, 33). From the phytochemical point of view this finding was quite unexpected and of particular chemotaxonomic interest, for lysergic acid alkaloids which had hither-

Fig. 10 *Rivea corymbosa* in flower.

Fig. 11 *Ipomoea violacea* in flower.

Fig. 12 Seeds of Ololiuqui. Left: *Rivea corymbosa.* Right: *Ipomoea tricolor.*

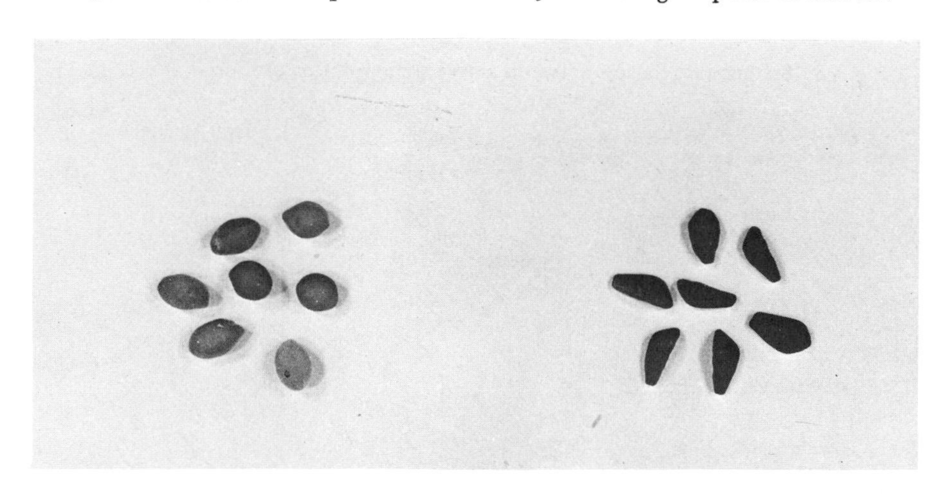

Table 3. Plants of the Family Convolvulaceae Containing Ergot Alkaloids

Plants	Alkaloids	Authors
Rivea corymbosa (L.) HALL.f. *Ipomoea violacea* L. [Ololiuqui]	ergine, isoergine d-lysergic acid methyl-carbinolamide, chanoclavine, elymoclavine, lysergol, ergometrine	A. Hofmann u. H. Tscherter (1960) (32). A. Hofmann (1961) (33). A. Hofmann and A. Cerletti (1961) (34).
Ipomoea rubro-caerulea HOOK *Ipomoea coccinea* L.	same as above	D. Gröger (1963) (35).
Ipomoea and *Convolvulus* spec. (ornamental varieties)	same as above and penniclavine	W. A. Taber, L. C. Vining and R. A. Heacock (1963) (36).
Argyreia nervosa	same as above	J. W. Hylin and D. P. Watson (1965) (37).
Ipomoea argyrophylla VATKE	ergosine, ergosinine, agroclavine	D. Stauffacher, H. Tscherter and A. Hofmann (1965) (38).
Ipomoea hildebrandtii VATKE	Cycloclavine	D. Stauffacher, H. Tscherter and A. Hofmann (1966) (39).

Fig. 13 Structural relation between active principles of ololiuqui and LSD-25.

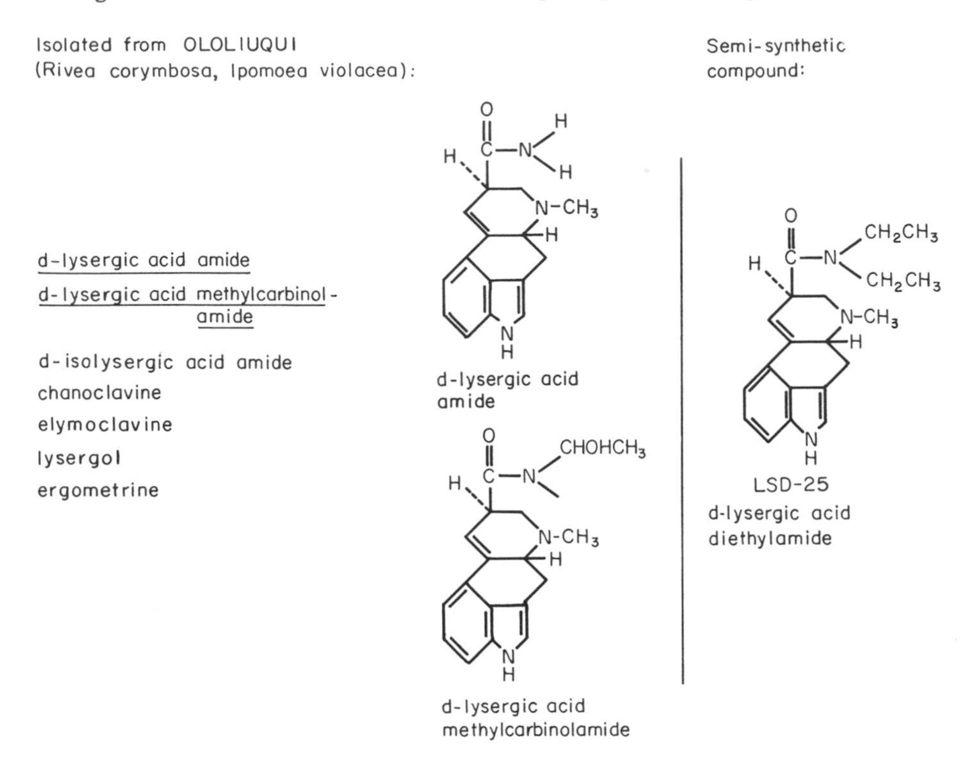

to been found only in lower fungi of the genus *Claviceps,* were now shown for the first time to be present in higher plants also (34).

The alkaloids isolated from ololiuqui are listed in Table 3. The main active principle of ololiuqui is d-lysergic acid amide, also named ergine, and lysergic acid methylcarbinolamide. Present as minor constituents were chanoclavine, elymoclavine, lysergol, and ergometrine. The occurrence of ergot alkaloids in *Ipomoea* species was later confirmed in other laboratories (35, 36, 37). In African species, we also found ergot alkaloids of the peptide type (38) and a new ergoline alkaloid, cycloclavine, which has so far not been discovered in ergot (39).

The main active principles of ololiuqui, i.e. d-lysergic acid amide and d-lysergic acid methylcarbinolamide (=d-lysergic acid α-hydroxy-ethylamide) are closely related to LSD, as can be seen from Fig. 13.

Nevertheless, this slight difference in structure is responsible for a pronounced qualitative and quantitative difference in activity between LSD and the ololiuqui alkaloids. The latter are about 20 times less active than LSD, and their action is more narcotic than hallucinogenic.

CONCLUSIONS

With the isolation of lysergic acid amides from ololiuqui, a series of researches in my laboratory had gone the full circle. The series began with the synthesis of lysergic acid diethylamide and the discovery of its hallucinogenic properties, proceeded via the investigation on the sacred Mexican mushroom teonanacatl, and then led us to investigate another magic Mexican plant, ololiuqui, which was found to contain lysergic acid amides closely related to lysergic acid diethylamide, LSD, the starting point in the series of researches.

This discovery of relatives of LSD in the magic plants of the New World in Mexico may have completed the fascinating picture presented by the ergot alkaloids, alkaloids which are playing an important role in the development of modern medicine.

References

1. G. Barger "Ergot and Ergotism," (London, Gurney and Jackson, 1931).
2. Last review on the chemistry of ergot alkaloids, including pharmacology and botany is given by A. Hofmann, in his monograph *Die Mutterkornalkaloide* (Stuttgart, F. Enke Verlag, 1964).
3. E. C. Kornfeld, E. J. Fornefeld, G. B. Kline, M. J. Mann, R. G. Jones, and R. B. Woodward, *J. Amer. Chem. Soc. 76* (1954), 5256.
4. A. Hofmann, A. J. Frey, and H. Ott, *Experientia* (Basel) *17* (1961), 206.
5. M. Abe, *Ann. Rep. Takeda Research* Lab. *10* (1951), 73, 129.
6. A. Stoll, A. Brack, A. Hofmann, and H. Kobel (Sandoz Ltd., Basel), Swiss Pat. No. 321.323, 10.4.1953.
7. A. Tonolo, *Nature 209* (1966), 1134.
8. A. M. Amici, A. Minghetti, T. Scotti, C. Spalla, and L. Tognoli, *Experientia* (Basel) *22* (1966), 415.
9. F. Arcamone, C. Bonino, E. B. Chain, A. Ferretti, P. Pennella, A. Tonolo, and L. Vero, *Nature 187* (1960), 238.
10. H. Kobel, E. Schreier, and J. Rutschmann, *Helv. Chim. Acta 47* (1964), 1052.
11. A. Stoll and A. Hofmann, *Helv. Chim. Acta 26* (1943), 944.
12. W. L. Garbrecht, *J. Org. Chem. 24* (1959), 368.
13. R. P. Pioch, U.S. Pat. 2.736.728 (1956).
14. Franz. Pat. No. 1.308.758, Sandoz A.G., Basel.
15. R. Paul and G. W. Anderson, *J. Amer. Chem. Soc. 82* (1960), 4596.
16. A. Hofmann, H. Ott, R. Griot, P. A. Stadler, and A. J. Frey, *Helv. Chim. Acta 46* (1963), 2306.
17. P. A. Stadler, A. J. Frey, H. Ott, and A. Hofmann, *Helv. Chim. Acta 47* (1964), 1911.
18. W. Schlientz, R. Brunner, P. A. Stadler, A. J. Frey, H. Ott, and A. Hofmann, *Helv. Chim. Acta 47* (1964), 1921.
19. A. Stoll and A. Hofmann, *Helv. Chim. Acta 26* (1943), 2070.

20. A. Stoll, A. Hofmann, and Th. Petrzilka, *Helv. Chim. Acta 29* (1946), 635.
21. A. Stoll and W. Schlientz, *Helv. Chim. Acta 38* (1955), 585.
22. F. Troxler and A. Hofmann, *Helv. Chim. Acta 40* (1957), 1706.
23. F. Troxler and A. Hofmann, *Helv. Chim. Acta 40* (1957), 1721.
24. F. Troxler and A. Hofmann, *Helv. Chim. Acta 40* (1957), 2160.
25. A. Cerletti, "Proceedings of the 1st International Congress of Neuro-Pharmacology, Rome (1958)," in *Neuro-Psychopharmacology,* ed. P. B. Bradley, P. Deniker, and C. Radouco-Thomas (Amsterdam-London-New York-Princeton, Elsevier Publ. Co., 1959), p. 117.
26. P. Stadler, A. Hofmann, and F. Troxler, Swiss Patent Application No. 5236/67 (1967).
27. A. Cerletti and B. Berde, "New Approaches in the Development of Compounds from Ergot with Potential Therapeutic Use in Migraine. Second Migraine Symposium, London, November 1967.
28. H. G. Wolff, *Headache and other head pain.* 2nd ed. (New York, Oxford University Press, 1963).
29. R. E. Schultes, "The identification of teonanacatl." Botanical Museum Leaflets, Harvard University 7, no. 3 (1939).
30. A. Hofmann, R. Heim, A. Brack, H. Kobel, A. Frey, H. Ott, Th. Petrzilka, and F. Troxler, *Helv. Chim. Acta 42* (1959), 1557.
31. R. E. Schultes, "A contribution to our knowledge of Rivea Corymbosa. The narcotic ololiuqui of the Aztecs." Botanical Museum of Harvard University, Cambridge (Mass.), 1941.
32. A. Hofmann, and H. Tscherter, *Experientia 16* (1960), 414.
33. A. Hofmann, *Planta Medica* (Stuttgart) *9* (1961), 354.
34. A. Hofmann and A. Cerletti, *Deutsche Med. Wochschr. 86* (1961), 885.
35. D. Gröger, *Flora 153* (1963), 373.
36. W. A. Taber, L. C. Vining, and R. A. Heacock, *Phytochemistry 2* (1963), 65.
37. J. W. Hylin and D. P. Watson, *Science 148* (1965), 499.
38. D. Stauffacher, H. Tscherter, and A. Hofmann, *Helv. Chim. Acta 48* (1965), 1379.
39. D. Stauffacher, H. Tscherter, and A. Hofmann. Paper read at the 4th International Symposium on the Chemistry of Natural Products of IUPAC, Stockholm, 1966.

Recent Advances in the Chemistry of Tumor Inhibitors of Plant Origin*

S. MORRIS KUPCHAN

Department of Chemistry,
University of Virginia, Charlottesville, Virginia

INTRODUCTION

The past two decades have witnessed the synthesis of many hundreds of chemical variants of known classes of cancer chemotherapeutic agents. Synthesis of modifications of presently known drugs does and should continue. However, some pessimism is evident among workers in the field because of the relatively small improvements over the prototype drugs that have resulted from the extensive synthetic efforts to date. There exists a need for new prototypes, or templates, for the synthetic organic chemist to use in the design of potential chemotherapeutic agents. Recent studies in the isolation and structural elucidation of tumor inhibitors of plant origin are yielding a fascinating array of novel types of growth-inhibitory compounds. There appears to be reason for confidence that this approach may point the way to useful templates for new synthetic approaches to cancer chemotherapy.

Studies of plant-derived tumor inhibitors are proceeding in many laboratories of wide geographic distribution. However, I will review only recent contributions from my own laboratory.

*Part LV in the series entitled "Tumor Inhibitors." Part LIV is: R. L. Hanson, H. A. Lardy, and S. M. Kupchan, *Science 168* (1970), 378. The work described herein was supported by grants from the National Cancer Institute (CA-04500) and the American Cancer Society (T-275), and a contract from the Cancer Chemotherapy National Service Center (CCNSC), National Cancer Institute, National Institutes of Health (PH 43-64-551).

The program started modestly, in 1959, with a screening study of crude extracts of a limited number of accessible plants for inhibitory activity against animal-tumor systems. Some plants were procured by summer collections in Wisconsin, others by cooperative arrangements with botanists in India, Costa Rica, and other countries. The results of testing of the first plant extracts prepared in our laboratory and elsewhere revealed that a small but significant number of the extracts showed reproducible tumor-inhibitory activity. Encouraged by these results, the Cancer Chemotherapy National Service Center (CCSNC) of the National Institutes of Health arranged with the U.S. Department of Agriculture to procure several thousand plant samples per year for evaluation. Shortly thereafter, the CCNSC arranged a contract with the Wisconsin Alumni Research Foundation, in Madison, to execute the initial extraction and screening studies. From that point onward, our program concentrated on the isolation and structural elucidation of new tumor inhibitors. To date, the active principles of more than fifty active plants have been isolated in our program, and the chemical studies of the most interesting compounds constitute the focus of this review.

One aspect of the approach of our program differs significantly from the classical, and most widely practiced, approach to the biological study of plant constituents. In the classical phytochemical approach, those compounds are studied which are most easily separated from a plant extract and most easily crystallized. In our program, however, the fractionation and isolation studies are guided at every stage by biological assays. The systematic fractionation, guided by biological assays, has made possible the isolation of important minor constituents which would have been missed in the classical approach.

STEROIDAL DERIVATIVES

Solanum dulcamara L., collected near Madison, Wisconsin, was one of the first plants found active in the program. Figure 1 summarizes the fractionation procedure which led to isolation of the tumor-inhibitory principle, the steroid alkaloid glycoside β-solamarine (I). It is noteworthy that *S. dulcamara* L. has been used to treat cancers, tumors, and warts from the time of Galen (ca. A.D. 180), and references to its use have appeared in the literature of many countries (1).

The leaves of *Acnistus arborescens* (L.) Schlecht have been used for many years to treat cancerous growths, and an extract of the leaves was forwarded to us by Professor J. A. Saenz Renauld, of the University of

Fig. 1 Fractionation of tumor-inhibitory extract from *Solanum dulcamara* L.

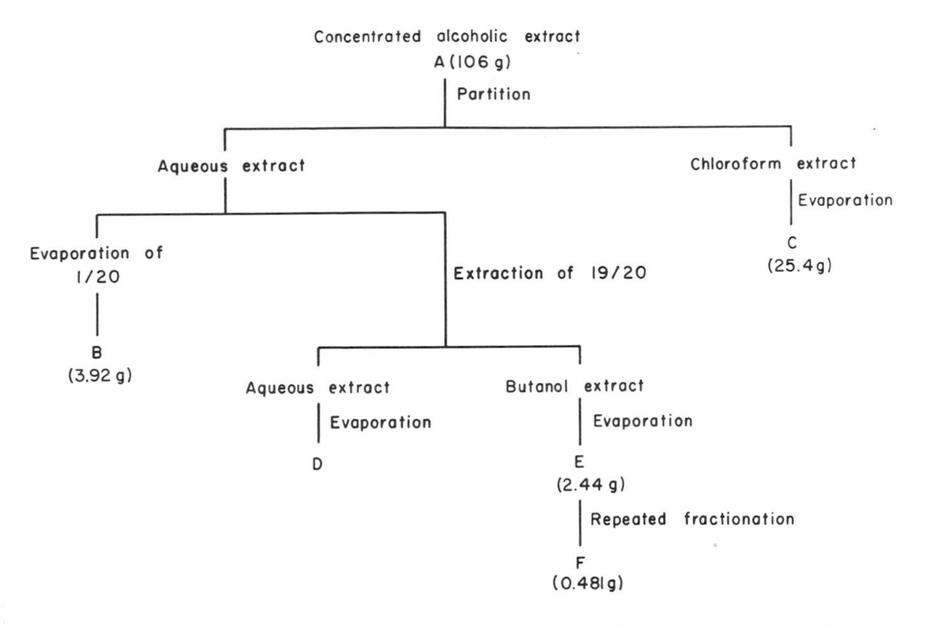

Costa Rica. Fractionation by the procedure outlined in Fig. 2 led to isolation of the tumor-inhibitory principle, withaferin A. A combination of degradative, spectral, and X-ray crystallographic studies resulted in elucidation of the structure II for withaferin A (2). Withaferin A was the prototype of a novel class of polyfunctional steroid lactones. Further chemical and biological studies are under way.

SA-ACTIVE PRINCIPLE OF *Solanum dulcamara* L.

I. β-Solamarine

Fig. 2 Fractionation of tumor-inhibitory extract from *Acnistus arborescens.*

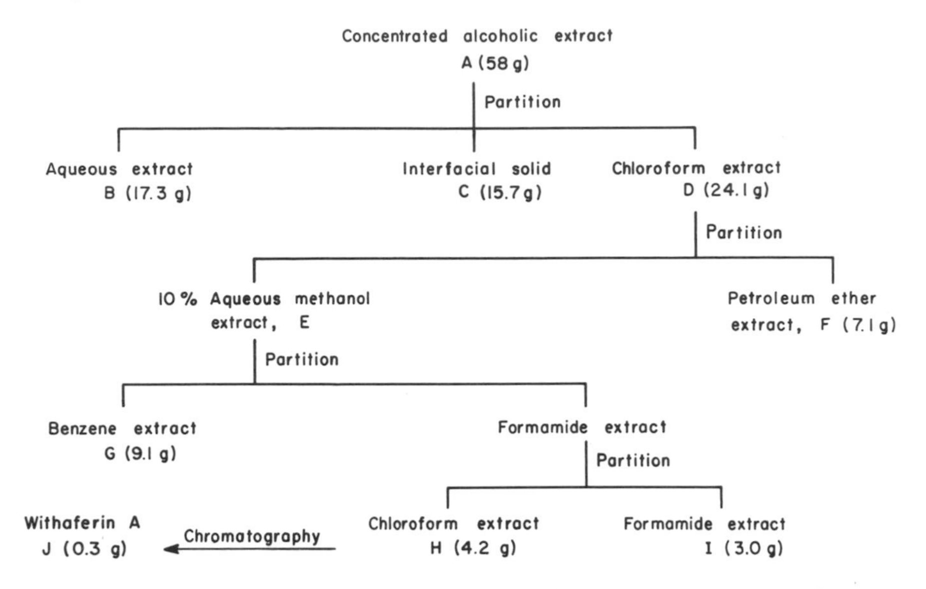

Systematic fractionation of an exceedingly cytotoxic extract from *Marah oreganus,* from California, led to isolation of the four previously known cucurbitacins (III-VI, Fig. 3) (3).

Several of the alkaloids isolated from *Buxus sempervirens* L. showed *in vitro* cytotoxic and *in vivo* tumor-inhibitory properties in animals.

SA-ACTIVE PRINCIPLE OF Acnistus arborescens

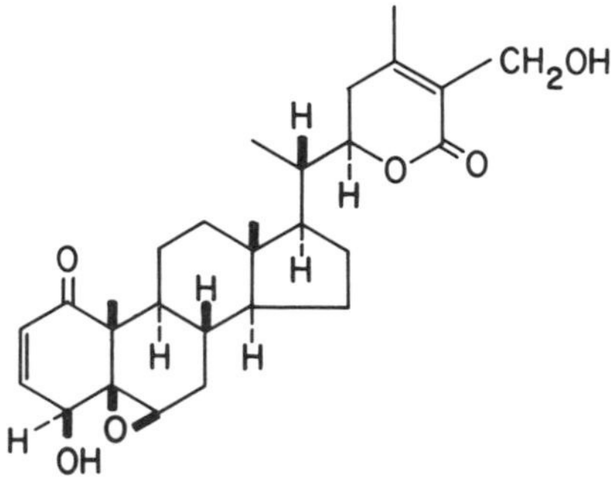

CH_2OH

II. Withaferin-A

Fig. 3 KB-active principles of *Marah oreganus*.

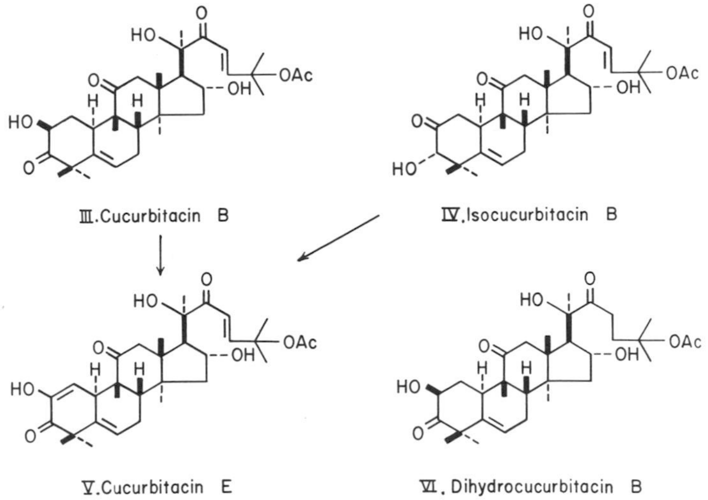

III. Cucurbitacin B

IV. Isocucurbitacin B

V. Cucurbitacin E

VI. Dihydrocucurbitacin B

The 3,20-diamines, exemplified by cycloprotobuxine-C (VII) showed both types of activity, whereas the monoamines exemplified by cyclobuxoxine (VIII) (4) showed only cytotoxic activity.

ISOQUINOLINE AND OTHER ALKALOIDS

The alkaloid thalicarpine (IX) (5, 6), was isolated from *Thalictrum dasycarpum,* the purple meadow rue, collected in Wisconsin. Thalicarpine shows significant inhibitory activity against the Walker intramuscular carcinosarcoma 256 in rats over a wide dosage range. The National Cancer Institute has procured a substantial supply of thalicarpine for advanced biological testing and clinical trial.

WM-ACTIVE PRINCIPLES OF Buxus sempervirens L.

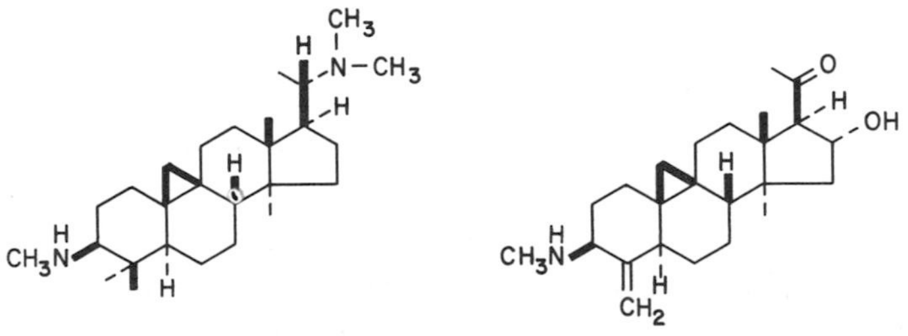

VII. Cycloprotobuxine – C

VIII. Cyclobuxoxine

WM-ACTIVE PRINCIPLE OF *Thalictrum dasycarpum*

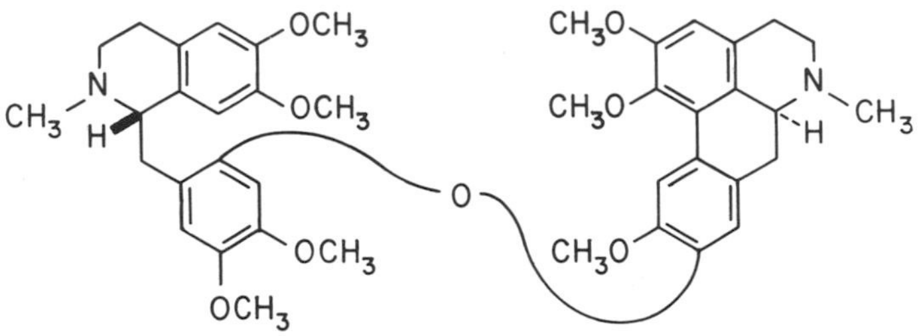

IX. Thalicarpine

WM-ACTIVE PRINCIPLE OF *Cyclea peltata*

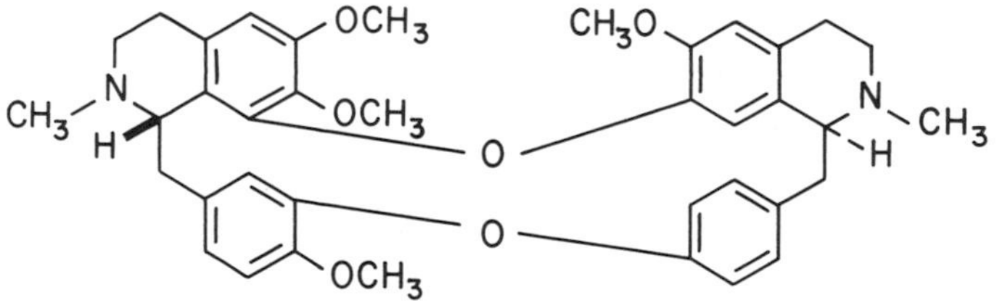

X. Tetrandrine

The alkaloid tetrandrine (X) was isolated from *Cyclea peltata* (7). The promising preliminary results with this alkaloid, in tests with the Walker 256 tumor system, have led the National Cancer Institute to arrange for procurement of a large supply for advanced biological studies.

The alkaloid thalidasine (XI) is a novel and active bisbenzylisoquinoline alkaloid isolated from *Thalictrum dasycarpum*. The degradations which led to structural elucidation of thalidasine are summarized in Fig. 4 (8).

Systematic studies of a tumor-inhibitory extract of *Solanum tripartitum* Dunal from Bolivia yielded the novel liquid alkaloids solapalmitine (XII) and solapalmitenine (XIII). Figures 5-8 summarize the studies on the structural elucidation and synthesis of the alkaloids (9).

THALIDASINE, A WM-ACTIVE PRINCIPLE OF *Thalictrum dasycarpum*

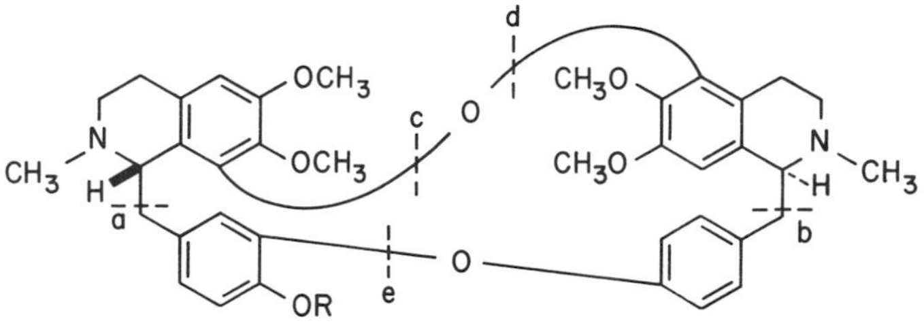

d
OCH3 CH3O
c O
N N
CH3 OCH3 CH3O CH3
H H
a b
O
OR e

XI. Thalidasine : R = CH_3

Thalfoetidine : R = H

Fig. 4 Sodium-liquid ammonia reduction products of thalidasine (XI).

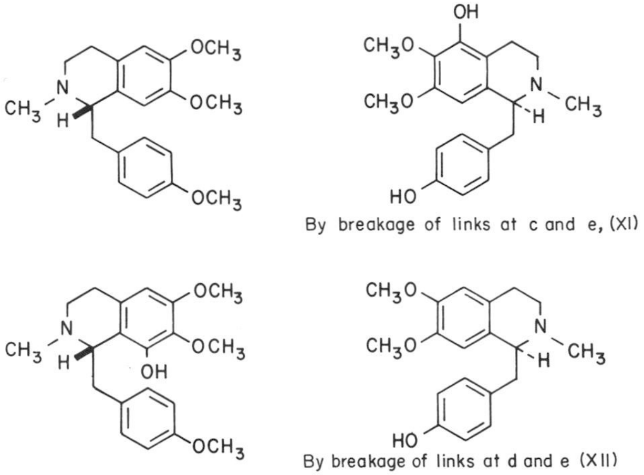

OH
OCH3 CH3O
N N
CH3 OCH3 CH3O CH3
H H
OCH3 HO

By breakage of links at c and e, (XI)

OCH3 CH3O
N N
CH3 OCH3 CH3O CH3
H OH H
OCH3 HO

By breakage of links at d and e (XII)

Fig. 5 WM-active principle of *Solanum tripartitum*.

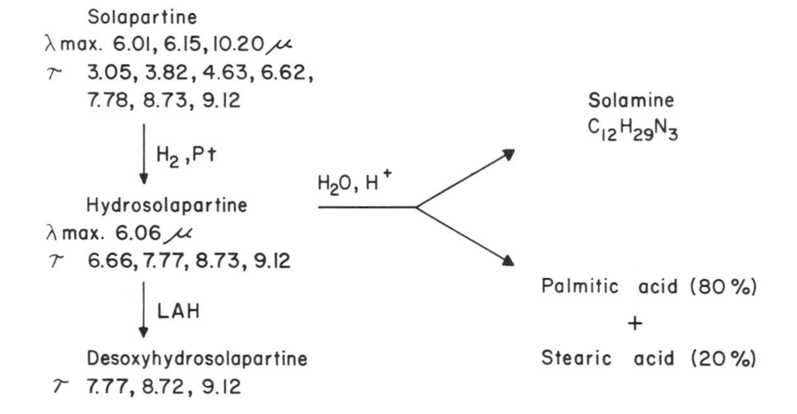

Fig. 6 Structural elucidation of solamine.

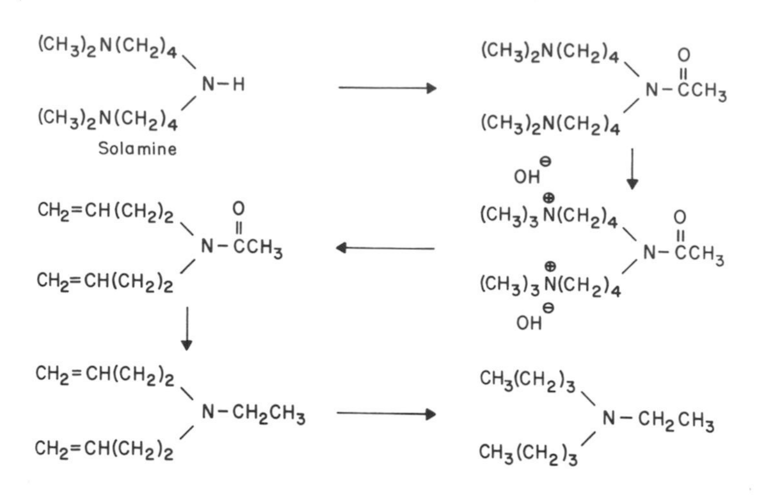

Fig. 7 Isolation of solapalmitine (XII) and solapalmitenine (XIII).

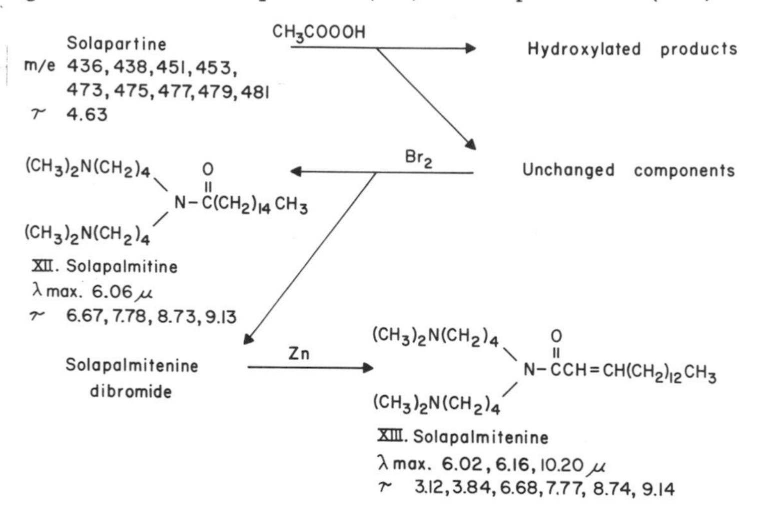

Fig. 8 Synthesis of solapalmitine (XII) and solapalmitenine (XIII).

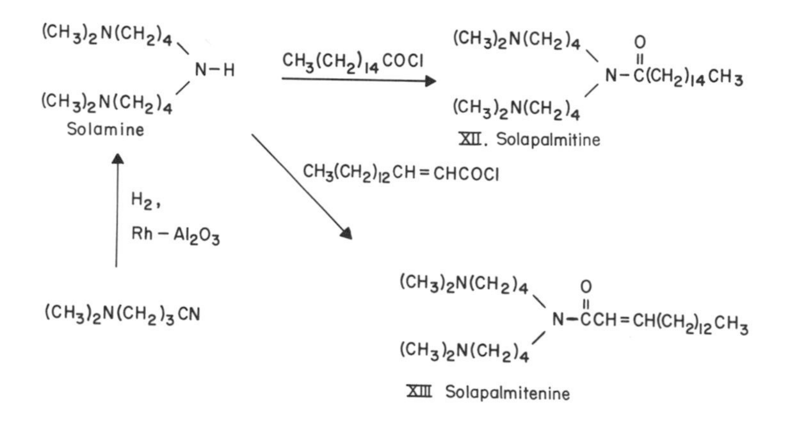

Fig. 9 Crotepoxide (XIV), A WM-active principle of *Croton macrostachys.*

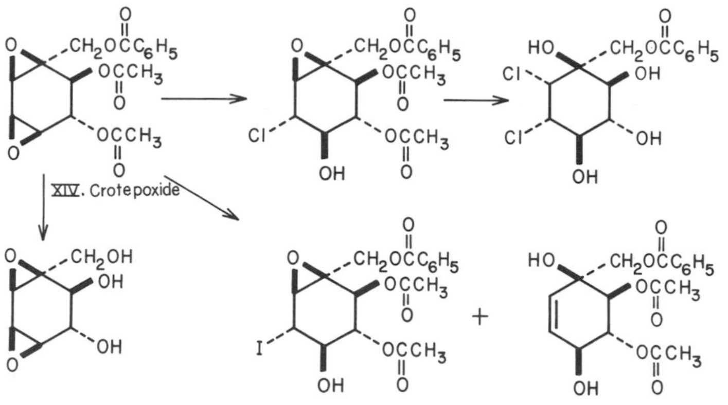

XIV. Crotepoxide

SESQUITERPENOID LACTONES AND OTHER COMPOUNDS

A new tumor-inhibitory principle from *Croton macrostachys,* crotepoxide (XIV), has been shown to possess the novel cyclohexane diepoxide structure (Fig. 9) (10).

A systematic study of the cytotoxic principles of *Elephantopus elatus* led to isolation of the tumor-inhibitory principles elephantin and elephantopin. These compounds were shown to possess the very novel sesquiterpene dilactone structures XV and XVI respectively (11).

Eupatorium rotundifolium L. initially yielded the new and novel sesquiterpenoid tumor inhibitors euparotin and euparotin acetate, XVII and XVIII (12). Further study of the cytotoxic principles led to isolation

WM-ACTIVE PRINCIPLES OF Elephantopus elatus

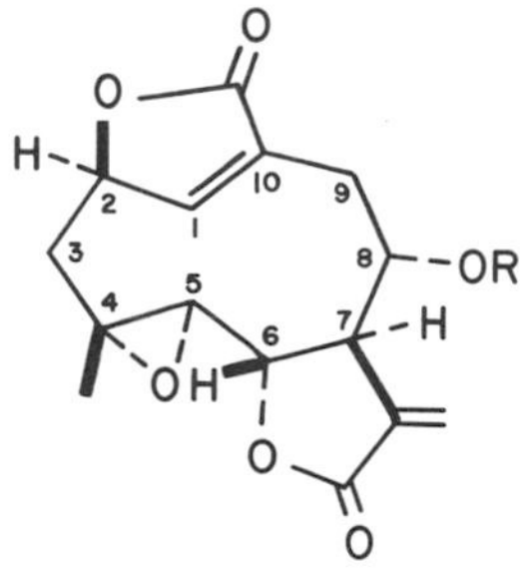

XV. Elephantin : R : $(CH_3)_2C=CHCO-$

XVI. Elephantopin : R : $CH_2=C(CH_3)CO-$

Fig. 10 Cytotoxic principles from *Eupatorium rotundifolium* L.

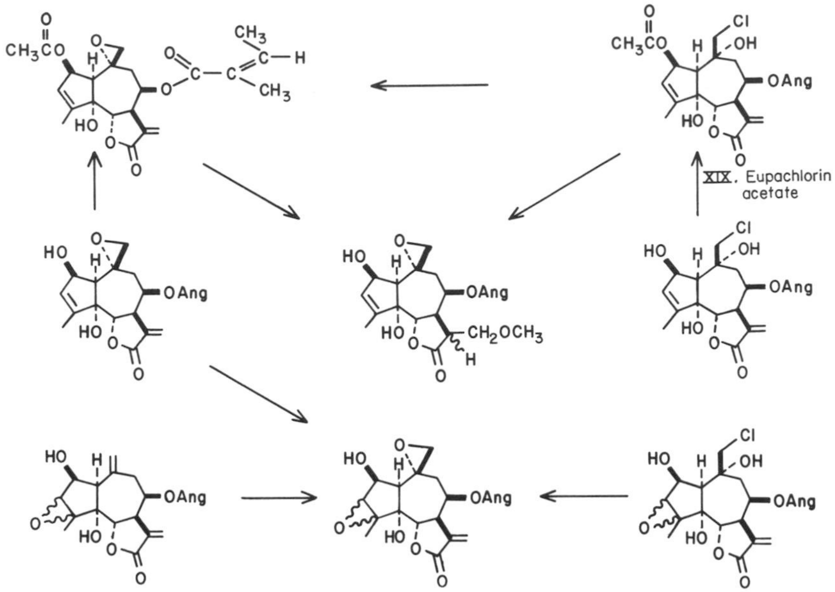

XIX. Eupachlorin acetate

and structural elucidation of the novel eupachlorin acetate (XIX) and five other cytotoxic sesquiterpene lactones, shown in Fig. 10 (13).

Recent studies of *Vernonia hymenolepis* led to isolation and structural elucidation of vernolepin (XX), a novel sesquiterpene dilactone tumor inhibitor whose structure is shown in Fig. 11 (14). Subsequent study of

STRUCTURES OF CYTOTOXIC PRINCIPLES

FROM Eupatorium rotundifolium L.

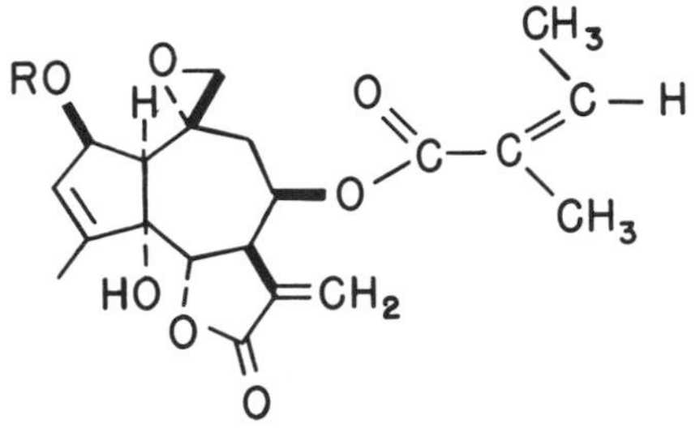

XVII. Euparotin : R = H

XVIII. Euparotin acetate : R = CH_3CO

Fig. 11 Structures of vernolepin (XX) and vernomenin (XXI).

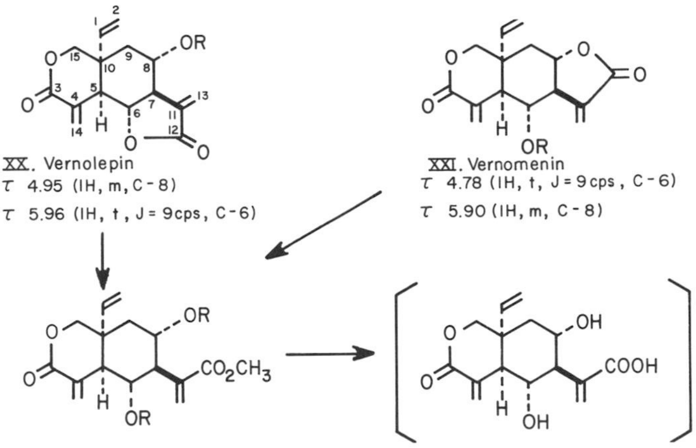

XX. Vernolepin
τ 4.95 (IH, m, C-8)
τ 5.96 (IH, t, J = 9cps, C-6)

XXI. Vernomenin
τ 4.78 (IH, t, J = 9cps, C-6)
τ 5.90 (IH, m, C-8)

Fig. 12 Inhibition of growth of wheat coleoptiles by vernolepin.

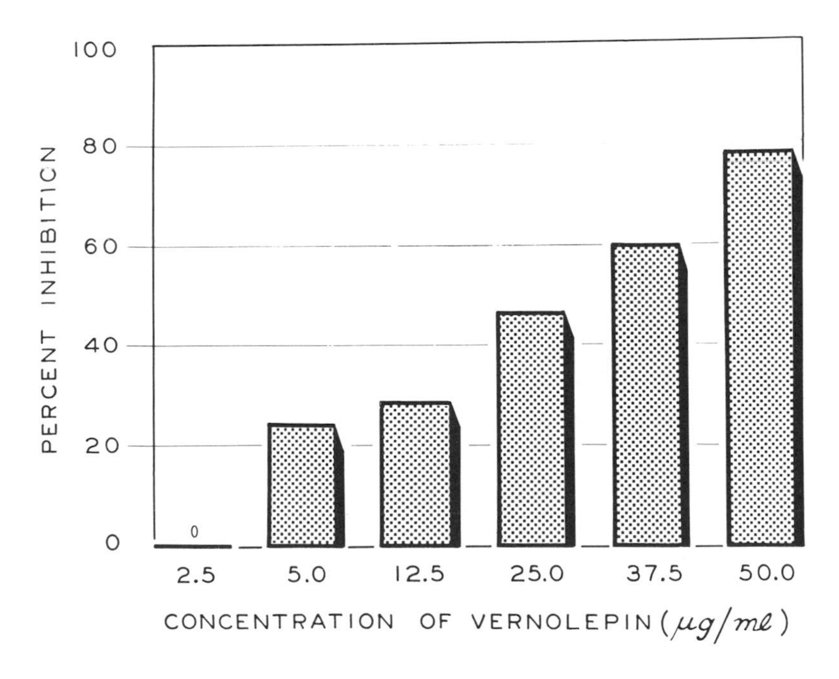

vernolepin for possible effects on the growth of wheat coleoptile tissue showed that the compound possesses very marked inhibitory activity (Figs. 12 and 13) (15). The fact that the plant growth inhibitory activity is reversible indicates that vernolepin may have a natural function in the regulation of plant growth.

Fig. 13 Interaction of vernolepin and IAA on growth of wheat coleoptiles.

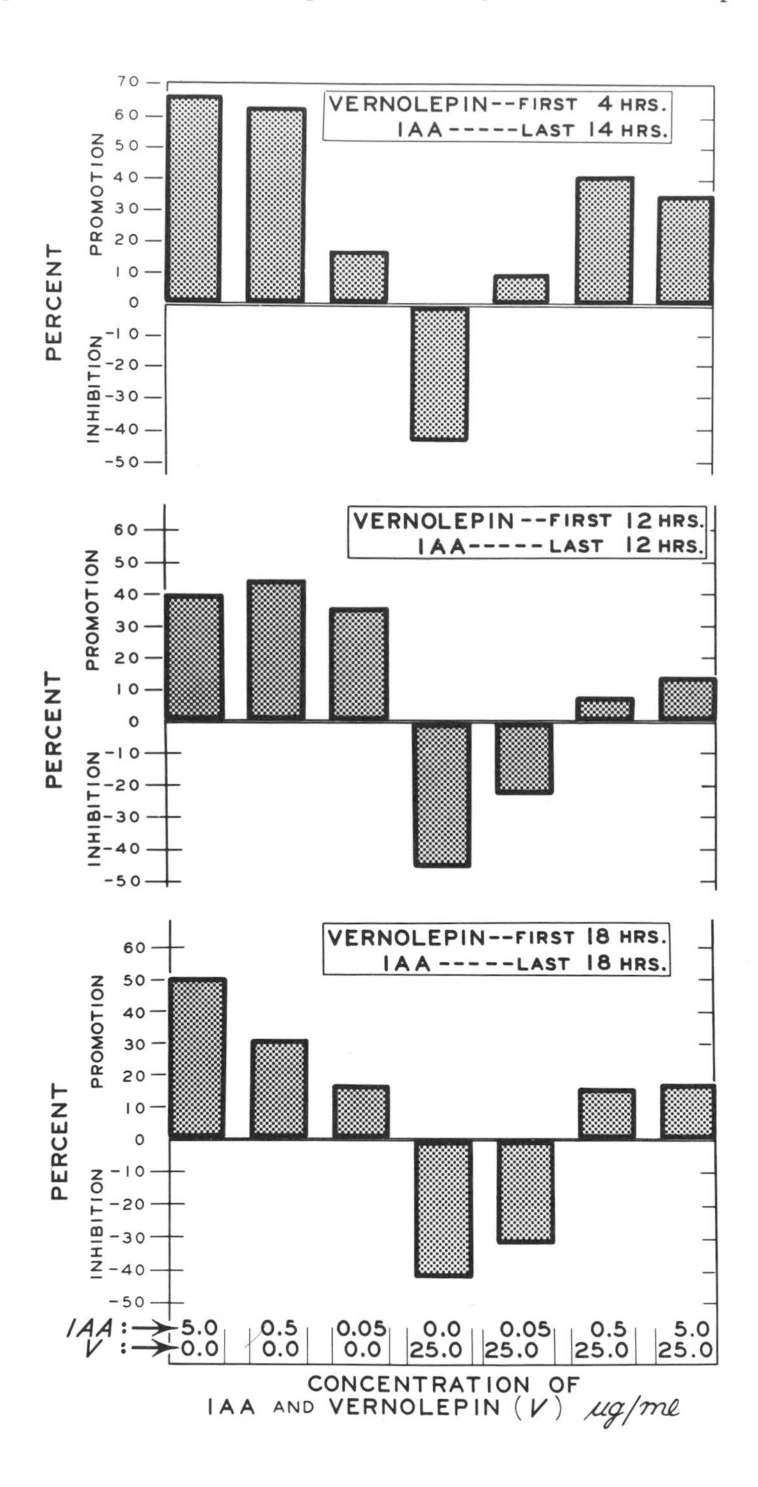

KB-ACTIVE PRINCIPLE OF *Asclepias curassavica* L.

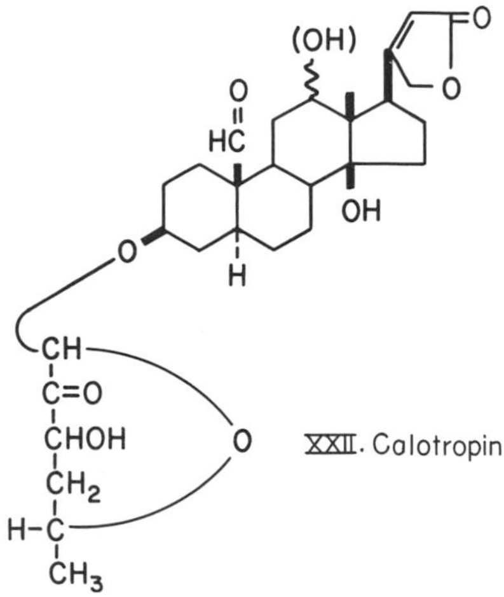

XXII. Calotropin

KB-ACTIVE PRINCIPLE OF *Apocynum cannabinum* L.

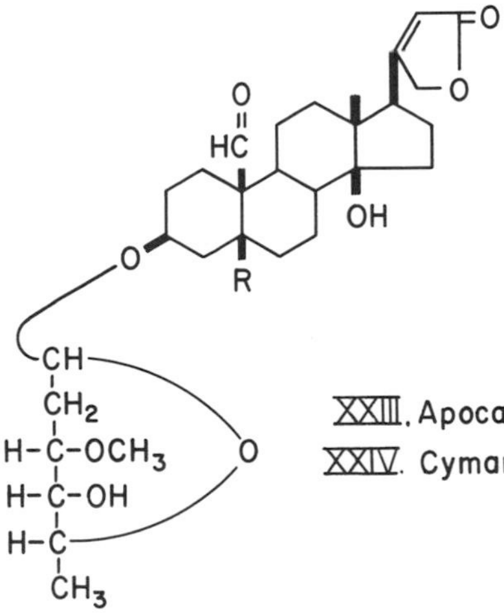

XXIII. Apocannoside : R = H
XXIV. Cymarin : R = OH

CARDENOLIDES

A search for the cytotoxic principle of *Asclepias currassavica* L. led to isolation and characterization of calotropin (XXII) (16). In a parallel study, the cardenolide glycosides apocannoside and cymarin (XXIII and XXIV) were identified as the cytotoxic principles of *Apocynum cannabinum* L. (17).

Fig. 14 Strophanthidin and derivatives.

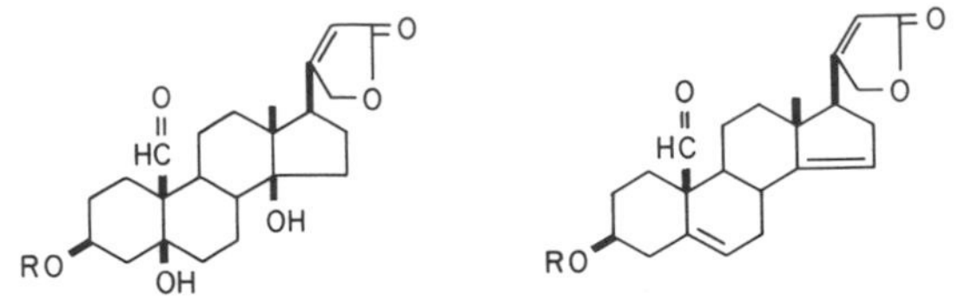

XXV. Strophanthidin
STR: R = H
SA: R = CH_3CO
SIA: R = ICH_2CO
SBA: R = $BrCH_2CO$

DASTR: R = H
DASA: R = CH_3CO
DASIA: R = ICH_2CO

Fig. 15 Effect of various strophanthidin derivatives on transport ATPase activity. ▨ Before washing. ☐ After washing.

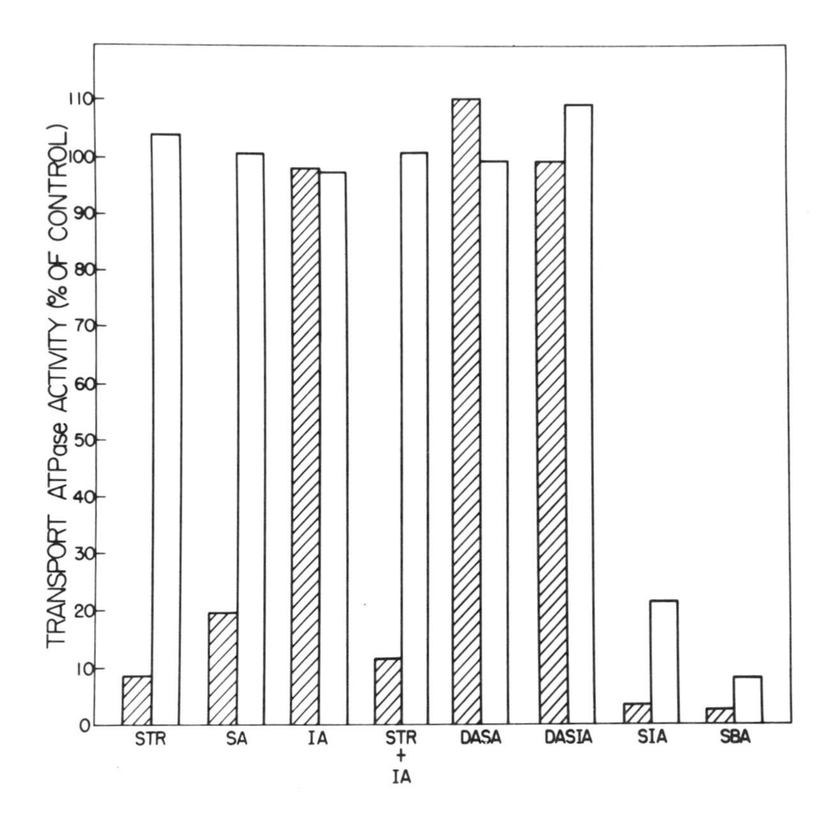

A study of the chemistry *vs.* the biological activity of cardenolides as cytotoxic agents, cardiotonic agents, and ATPase inhibitors revealed that only those compounds modified solely in Ring A retained the major

Fig. 16 Effect of repeated washes on the activity of transport ATPase inhibited by strophanthin iodoacetate (□) or ouabain (●).

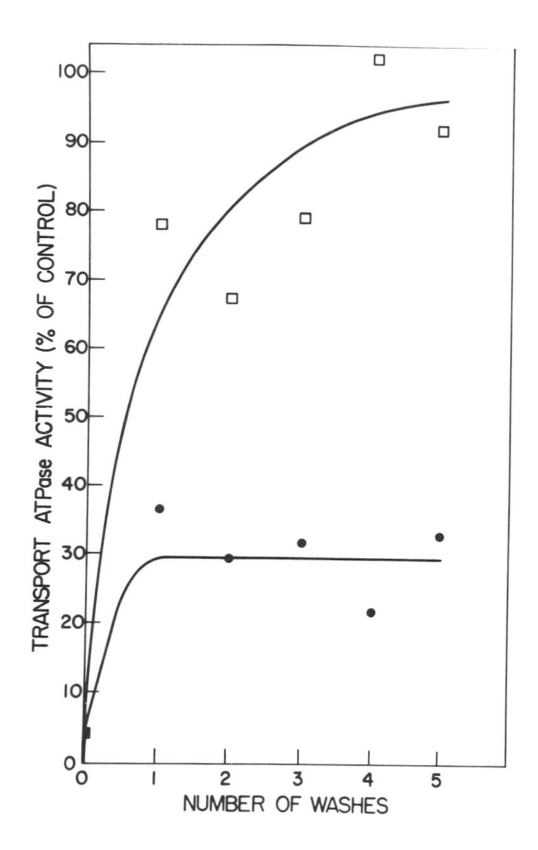

proportion of biological activity (18). In a search for a selective irreversible inhibitor of transport ATPase, in collaboration with Professor L. E. Hokin, it was found that strophanthidin 3-haloacetates (Fig. 14) were effective alkylators of the cardiotonic steroid site (19). Figure 15 indicates level of enzyme activity before and after washing; Fig. 16 shows effects of repeated washes upon SIA and ouabain-inhibited enzyme.

Recent studies of *Bersama abyssinica* from Ethiopia led to isolation and characterization of hellebrigenin 3-acetate and hellebrigenin 3,5-diacetate (XXVI and XXVII) as the tumor inhibitory principles (21). When it was found that hellebrigenin has 30 times the affinity of strophanthidin for transport ATPase, a series of hellebrigenin 3-haloacetates were synthesized. Figure 17 shows that the 3-iodoacetate is 100 times more potent as an irreversible inhibitor of the enzyme than SBA (Fig. 14) is, and 20 times more potent than the 3-bromoacetate (21).

WM-ACTIVE PRINCIPLES OF *Bersama abyssinica*

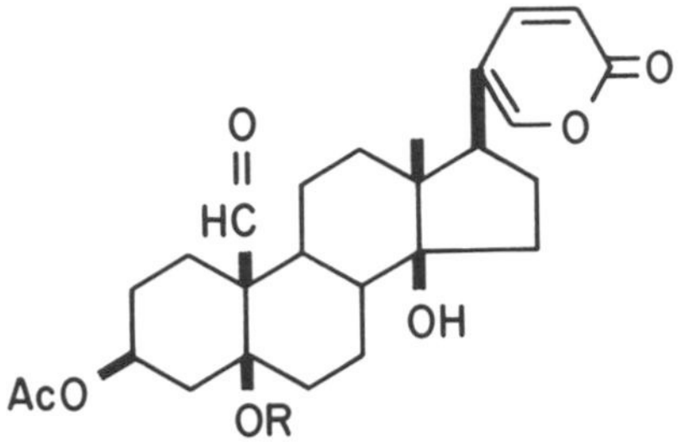

XXVI. Hellebrigenin Acetate : R = H
XXVII. Hellebrigenin Diacetate : R = Ac

CONCLUSION

The program to date has demonstrated that several new types of compounds show significant growth-inhibitory activity against standard tumor systems in the National Cancer Institute's screen. Thus, for instance, bisbenzylisoquinoline alkaloids, cardiotonic steroid lactones, and sesquiterpene lactones represent chemical types not recognized previously as growth inhibitors.

We are optimistic about the future of our approach, from several points of view. First, we hope that some of the new natural products will show sufficient promise in the advanced preclinical animal studies now in progress to become candidates for clinical trial. Secondly, we are encouraged by the fact that several of the new and remarkably cytotoxic compounds are showing usefulness as tools for studying biochemical phenomena. Finally, from a long-range point of view, we are optimistic that some of the unusual types of compounds may serve significant roles as novel chemical templates for new synthetic approaches to cancer chemotherapy.

Fig. 17 Inhibition of transport ATPase by hellebrigenin-3-iodocetate (■), 3-bromo-acetate (●) and strophanthidin-3-bromoacetate (○).

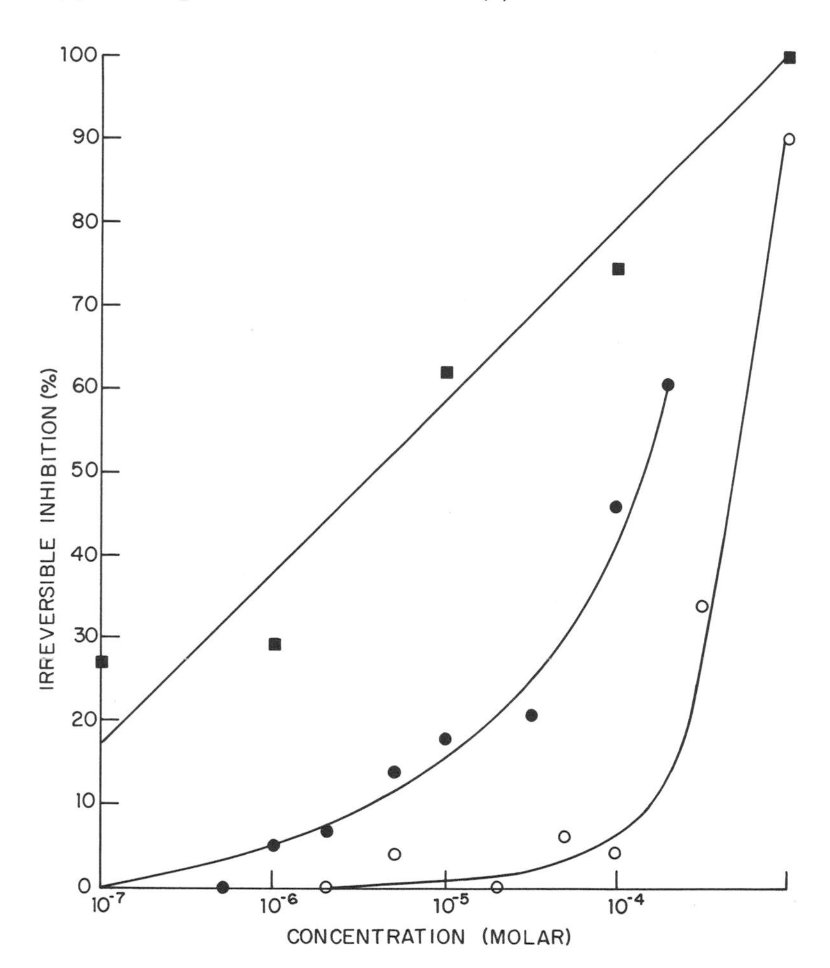

References

1. S. M. Kupchan, S. J. Barboutis, J. R. Knox, and C. A. Lau Cam, *Science 150* (1965), 1827.

2. S. M. Kupchan, R. W. Doskotch, P. Bollinger, A. T. McPhail, G. A. Sim, and J. A. Saenz Renauld, *J. Amer. Chem. Soc. 87* (1965), 5805; S. M. Kupchan, W. K. Anderson, P. Bollinger, R. W. Doskotch, R. M. Smith, J. A. Saenz Renauld, H. K. Schnoes, A. L. Burlingame, and D. H. Smith, *J. Org. Chem. 34* (1969), 3858.

3. S. M. Kupchan, A. H. Gray, and M. D. Grove, *J. Med. Chem. 10* (1967), 337.

4. S. M. Kupchan and E. Abushanab, *J. Org. Chem. 30* (1965), 3931.

5. S. M. Kupchan, K. K. Chakravarti, and N. Yokoyama, *J. Pharm. Sci. 52* (1963), 985.

6. M. Tomita, H. Furukawa, S.-T. Lu, and S. M. Kupchan, *Tetrahedron Letters* (1965), 4309; *Chem. Pharm. Bull.* (Tokyo) *15* (1967), 959.
7. S. M. Kupchan, N. Yokoyama, and B. S. Thyagarajan, *J. Pharm. Sci. 50* (1961), 164.
8. S. M. Kupchan, T.-H. Yang, G. S. Vasilikiotis, M. H. Barnes, and M. L. King, *J. Amer. Chem. Soc. 89* (1967), 3075; *J. Org. Chem. 34* (1969), 3884.
9. S. M. Kupchan, A. P. Davies, S. J. Barboutis, H. K. Schnoes, and A. L. Burlingame, *J. Amer. Chem. Soc. 89* (1967), 5718; *J. Org. Chem. 34* (1969), 3888.
10. S. M. Kupchan, R. J. Hemingway, P. Coggon, A. T. McPhail, and G. A. Sim, *J. Amer. Chem. Soc. 90* (1968), 2982; S. M. Kupchan, R. J. Hemingway, and R. M. Smith, *J. Org. Chem. 34* (1969), 3898.
11. S. M. Kupchan, Y. Aynehchi, J. M. Cassady, A. T. McPhail, G. A. Sim, H. K. Schnoes, and A. L. Burlingame, *J. Amer. Chem. Soc. 88* (1966), 3674; S. M. Kupchan, Y. Aynehchi, J. M. Cassady, H. K. Schnoes, and A. L. Burlingame, *J. Org. Chem. 34* (1969), 3867.
12. S. M. Kupchan, J. C. Hemingway, J. M. Cassady, J. R. Knox, A. T. McPhail, and G. A. Sim, *J. Amer. Chem. Soc. 89* (1967), 465.
13. S. M. Kupchan, J. E. Kelsey, M. Maruyama, and J. M. Cassady, *Tetrahedron Letters* (1968), 3317; S. M. Kupchan, J. E. Kelsey, M. Maruyama, J. M. Cassady, J. C. Hemingway, and J. R. Knox, *J. Org. Chem. 34* (1969), 3876.
14. S. M. Kupchan, R. J. Hemingway, D. Werner, A. Karim, A. T. McPhail, and G. A. Sim, *J. Amer. Chem. Soc. 90* (1968), 3596; S. M. Kupchan, R. J. Hemingway, D. Werner, and A. Karim, *J. Org. Chem. 34* (1969), 3908.
15. L. Sequeira, R. J. Hemingway, and S. M. Kupchan, *Science 161* (1968), 789.
16. S. M. Kupchan, J. R. Knox, J. E. Kelsey, and J. A. Saenz Renauld, *Science 146* (1964), 1685.
17. S. M. Kupchan, R. J. Hemingway, and R. W. Doskotch, *J. Med. Chem. 7* (1964), 803.
18. S. M. Kupchan, M. Mokotoff, R. S. Sandhu, and L. E. Hokin, *J. Med. Chem. 10* (1967), 1025.
19. L. E. Hokin, M. Mokotoff, and S. M. Kupchan, *Proc. Nat. Acad. Sci. 55* (1966), 797.
20. S. M. Kupchan, R. J. Hemingway, and J. C. Hemingway, *Tetrahedron Letters* (1968), 149; *J. Org. Chem. 34* (1969), 3894.
21. A. E. Ruoho, L. E. Hokin, R. J. Hemingway, and S. M. Kupchan, *Science 159* (1968), 1354.

The Phytochemistry and Biological Activity of *Catharanthus lanceus* (Apocynaceae)

NORMAN R. FARNSWORTH*
Professor of Pharmacognosy and Head,
Department of Pharmacognosy and Pharmacology,
College of Pharmacy, University of Illinois
at the Medical Center, Chicago, Illinois

INTRODUCTION

The genus *Catharanthus* (Apocynaceae) is composed of six distinct species, i.e. *C. roseus* (L.) G. Don, *C. lanceus* (Boj. ex A. DC.) Pich., *Catharanthus trichophyllus* (Bak.) Pich., *C. longifolius* (Pich.) Pich., *C. pusillus* (Murr.) G. Don, and *C. scitulus* (Pich.) Pich. This genus has also been referred to incorrectly as *Vinca* or *Lochnera,* but the correctness of *Catharanthus* has recently and authoritatively been confirmed by Stearn (1).

Catharanthus trichophyllus, C. longifolius, C. lanceus, and *C. scitulus* are restricted in their distribution to Madagascar, *C. pusillus* is indigenous to India, and *C. roseus* is a pantropical species (2-4).

The medicinal importance of this genus of plants is attributed, at present, to the clinically useful anticancer alkaloids vincaleukoblastine (vinblastine, VLB) and leurocristine (vincristine, VCR), which are derived from *C. roseus.* However, four additional alkaloids, of the more than 66 that have been isolated from this species, have varying degrees of anticancer activity, i.e., leurosine (vinleurosine, VLR), leurosidine (vinrosidine), leurosivine, and rovidine (5).

The purpose of this paper is to review the current status of the phytochemistry of one of these species, *C. lanceus,* and outline the biolog-

*Formerly at the School of Pharmacy, University of Pittsburgh.

ical activities of those alkaloids isolated from this plant. Several recent reviews are available on all species of this genus, particularly dealing with *C. roseus* (3-13). No reviews have appeared in the scientific literature concerned only with *C. lanceus*.

Prior to the initiation of our work in this plant during 1959, the only phytochemical studies reported for *C. lanceus* were those by Janot and coworkers, who isolated and identified ajmalicine (III, δ-yohimbine), tetrahydroalstonine (IV), yohimbine (II, quebrachamine) and lanceine from the roots (14-16) (Fig. 1). Our studies have resulted in the isolation, to date, of 21 bases from *C. lanceus*, and we have confirmed the presence of all four alkaloids isolated by Janot et al.

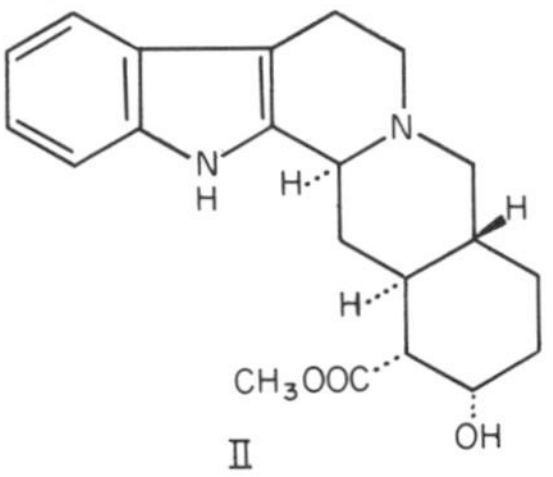

Fig. 1 β-Carboline Alkaloids of *Catharanthus lanceus* I, Pericyclivine; II, Yohimbine; III, Ajmalicine; IV, Tetrahydroalstonine.

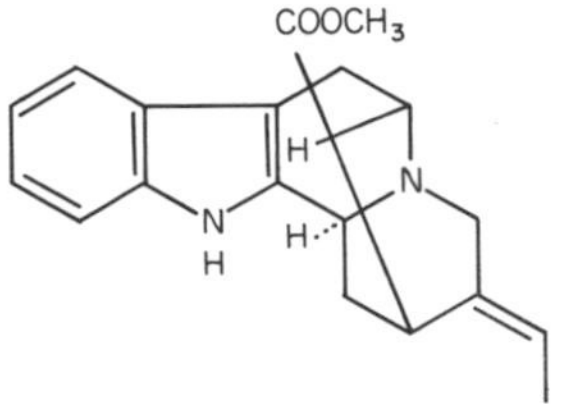

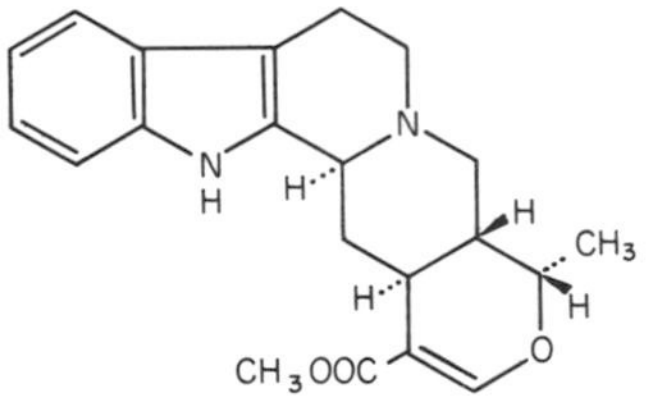

Description and Occurrence of Catharanthus lanceus. C. lanceus is a perennial herb with a sub-woody base and a smooth quadrangular stem. It attains a height of from 30 cm to 1 m, and the leaves are oblong, subsessile, and from 1.5 to 2.5 cm long and 0.5 cm to 1 cm wide. The flowers of *C. lanceus* are violet or rose-colored, having a linear-lobed calyx. It is indigenous to Madagascar and is most frequently found growing on the high plateaus of that country. Plants found growing among rocks and on dry slopes often exhibit ericoid leaves (2-4).

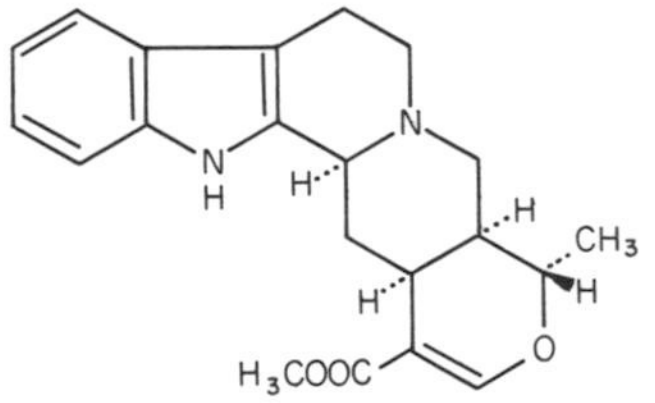

Folklore of C. lanceus. The leaves of *C. lanceus* have been used as an astringent (17), bitter (17), and emetic (17) in South Africa; the roots as a purgative (18) and vermifuge (19) in Madagascar; the aerial parts as a galactogogue (18) and vomitive (18) in Madagascar; and the whole plant as a remedy for dysentery (18) in Madagascar. Although the related *C. roseus* has been widely used as a folklore medicine in the treatment of diabetes (8), no reference to this use could be found for *C. lanceus.*

PHYTOCHEMICAL STUDIES

Non-alkaloids of C. lanceus. Choline has been detected in *C. lanceus* by means of paper chromatography (20), but has never been isolated from this plant. Sucrose has been isolated from the roots of *C. lanceus* (21), and tannins have been detected (17), but no other non-alkaloid entities are known to exist in this plant. A systematic phytochemical screening of the roots of *C. lanceus* in our laboratories has shown the presence of alkaloids, tannins, saponins, and unsaturated sterols, but tests for flavonoids and cardiac glycosides were negative (22).

Alkaloids of C. lanceus. The quantities of crude alkaloids have been estimated in various parts of *C. lanceus* grown in Madagascar. These vary from 0.20 to 1.4 percent in the roots (14, 16, 18, 24), to 0.10 to 0.15 percent in the stems (23), and 0.10 to 0.55 percent for mixtures of leaves and stems (16, 23, 24). It has been our experience to find 1.36 percent of crude bases in *C. lanceus* leaves (25) and 2.10 percent of crude bases in *C. lanceus* roots of Madagascan origin (26).

In our laboratories we have worked separately on the alkaloids of *C. lanceus* leaves and roots. Major emphasis has been on the leaf alkaloids, since bioassay results have shown antitumor activity in certain leaf alkaloid fractions. We have followed the systematic alkaloid fractionation scheme which was developed by Svoboda, and which has been so successful for his work on the *C. roseus* alkaloids (27). This scheme, with minor modifications used in our work, is shown in Fig. 2.

The yields of crude alkaloid in each fraction obtained by this method are presented in Table 1. Each of these fractions is exceedingly complex. However, most of the mixtures can be resolved using thin-layer chromatography (TLC) on silica gel G plates, with solvent systems composed of ethyl acetate-absolute ethanol (3:1), *n*-butanol-acetic acid-water (4:1:1), or methanol.

The visualization of *Catharanthus* alkaloids on TLC plates has been greatly aided by the use of ceric ammonium sulfate (CAS) as a spray reagent (28), which produces an array of colors with most indole alkaloids. The color reactions are somewhat associated with the u.v. chromophores

Fig. 2 Fractionation scheme used in the examination of alkaloids in *Catharanthus lanceus.*

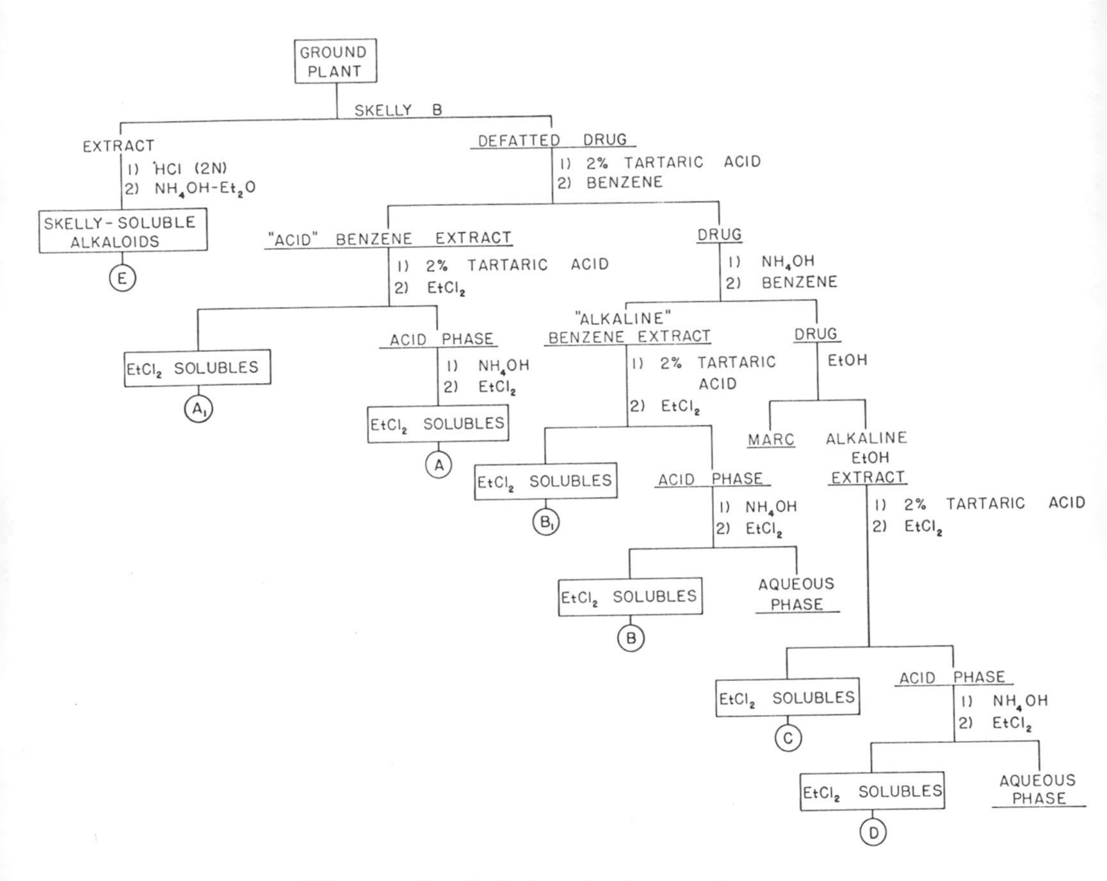

Table 1. Crude Alkaloid Yields from *Catharanthus lanceus*

Fraction[a]	Yield %[b]	
	Leaf	Root
A	0.27	0.58
A_1	0.19	0.29
B	0.43	0.51
B_1	0.22	0.04
C	0.08	0.25
D	0.12	0.33
E	0.05	0.10

[a]See Fig. 2.
[b]Expressed on the air-dried plant material.

in each alkaloid, but functional groups on the molecule can cause marked color changes. However, in most cases, the u.v. spectrum is directly related to the colors produced with CAS; e.g., 2-acylindoles (Fig. 3), oxindoles and quaternary bases will not react with the reagent. However, subsequent spraying of CAS-treated plates with modified Dragendorff's reagent (29) will serve to detect alkaloids of this class. All β-carbolines react to give a yellow or yellow-green color, but all α-methyleneindolines (Fig. 4) give an initial blue with a yellow or orange center which changes rapidly during

Fig. 3 2-Acylindole Alkaloids of *Catharanthus lanceus* V, Perivine (R=H); VI, Periformyline (R=CHO).

Fig. 4 α-Methyleneindoline Alkaloids of *Catharanthus lanceus* VII, Lochnerinine ($R=CH_3O$, $R_1=H_2$); VIII, Hörhammericine (R=H, R_1=OH); IX, Hörhammerinine ($R=CH_3O_1$, R_1=OH).

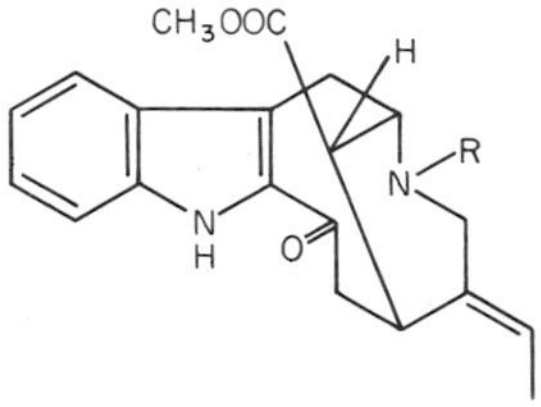

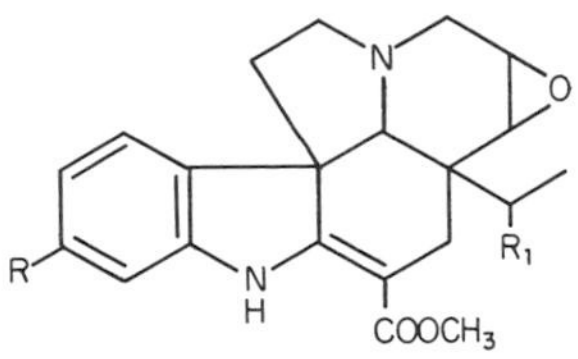

the first few minutes after spraying, usually to a lighter color. All dihydroindole alkaloids (Fig. 5), or bases having an *N*-methyl group, will give a red or orange color with the CAS reagent. Alkaloids with adjacent OCH_3 groups on the A ring of indole alkaloids will give light pink color reactions. We have observed that all α-methyleneindoline bases having an epoxide group attached to the D ring will give the characteristic blue color with an orange center, which fades after one hour, but which then gradually develops an emerald-green color. Lochnerinine (VII), lochnericine, hörhammericine (VIII) and hörhammerinine (IX), all contain this epoxy group, and they are the only *Catharanthus* alkaloids that give this characteristic reaction.

Fig. 5 Dihydroindole Alkaloids of *Catharanthus lanceus* X, Vindoline; XI, Vindolinine.

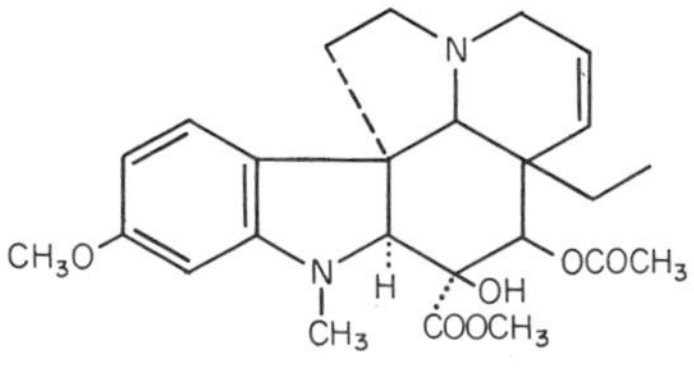

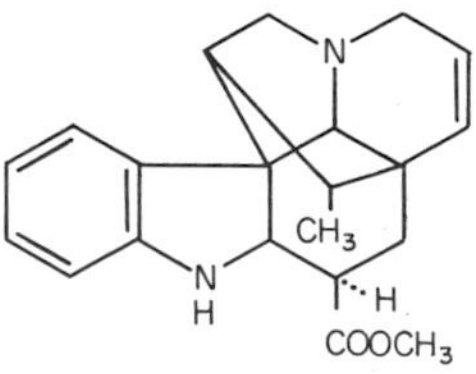

By interpretation of chromatograms of the seven crude alkaloid fractions from *C. lanceus* roots, and the seven from the leaves, we conservatively estimate that at least 100 different bases are present in this plant. However, we are reasonably sure that many of the minor bases are actually artifacts. For example, if we heat the alkaloid pericalline (XIII) on a water bath for a few minutes with either methanol or acetone, this base is converted partially into the alkaloid ammocalline (Fig. 6). We have isolated ammocalline from *C. lanceus* roots (30), and Svoboda has isolated ammocalline from *C. roseus roots* (31).

Fig. 6 Miscellaneous Alkaloids from *Catharanthus lanceus* XII, Catharanthine; XIII, Pericalline.

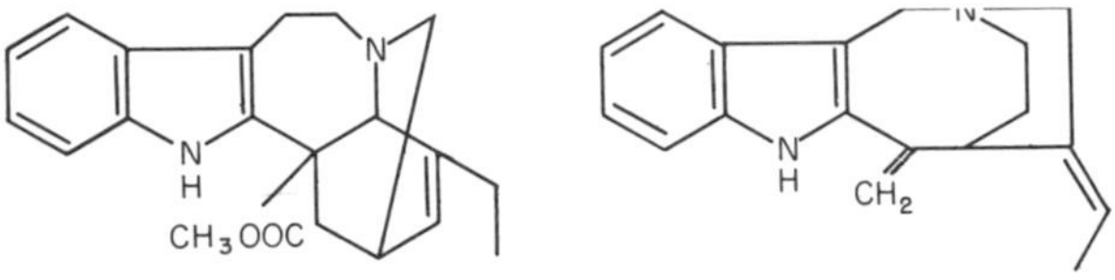

XII XIII

The dimeric alkaloid leurosine (XIV) is extremely unstable as the free base in any solvent, and its decomposition is accelerated by fluorescent light (Fig. 7). We have produced and isolated the major artifact resulting from this decomposition by allowing leurosine to be exposed to light in solution, and have also isolated the same artifact from the leaf (A) fraction, which is rich in leurosine (32).

In our efforts to isolate alkaloids from the 14 leaf and root crude alkaloid fractions, we initially attempted direct crystallization procedures.

Fig. 7 Structure of Leurosine (XIV).

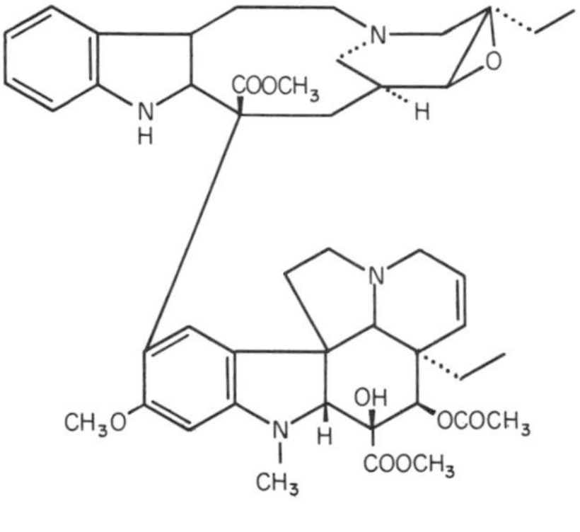

XIV

Typically, 5.0 gm of the fraction was dissolved in a minimum volume of hot benzene. If insoluble in this solvent, chloroform, ethyl acetate, methanol, ethanol or acetone were used in that order. The fraction, in solution, was then refrigerated for a minimum period of four days. If no crystallization occurred in that time, the solvent was removed, and the fraction was then dissolved in a minimum volume of the next solvent in the series, followed by refrigeration and so on. In this manner, we were able to isolate those alkaloids which were present in high concentration in any given fraction. We have isolated ajmalicine (III) from the root (A) fraction, the root (A_1) fraction, and the root (B_1) fraction in this manner, in yields of 11.85, 16.10 and 3.17 percent respectively, calculated in terms of the weight of the crude alkaloid fraction (33).

The leaf alkaloid (A) fraction has yielded 12.80 percent ajmalicine in this same manner, and the two major alkaloids of this plant, yohimbine (II) and perivine (V), are easily obtained from the leaf (B) fraction by initial treatment with benzene, which deposits crystalline yohimbine (24.70 percent) directly, due to its poor solubility in this solvent. Following the removal of yohimbine, the fraction is dried and dissolved in acetone. Perivine will crystallize directly from acetone at room temperature, or by chilling, within a few hours in a 15 percent yield. Thus, the three major alkaloids of *C. lanceus* are yohimbine, ajmalicine, and perivine, and all three of these bases can be obtained by direct crystallization procedures (33).

Next, we performed gradient *p*H separations on the crude alkaloid fractions in order to obviate tedious column chromatographic separations. This technique was developed by Svoboda (34), in his work with *C. roseus,* but he never reported applying it to crude fractions, only to column chromatography fraction cuts. In this manner we were able to again isolate ajmalicine (III) from the root (A) fraction in a yield of 15.90 percent, which was better than the direct crystallization technique. Similarly, this same alkaloid was obtained from the root (A_1) (3.20 percent), (B_1) (3.80 percent), (C) (0.53 percent) and (E) (4.02 percent) fractions. Yohimbine (1.0 percent was obtained from the root (B) fraction, although it could not be obtained by direct crystallization (33).

Gradient *p*H separation of the leaf (A) fraction yielded 0.20 percent of ajmalicine, and surprisingly enough, 2.30 percent of the active anticancer alkaloid leurosine (XIV), which was only obtained previously in our work in a 0.70 percent yield by means of column chromatography of the same fraction. Using gradient *p*H techniques, yohimbine and perivine were obtained from the leaf (B) fraction in comparable yields as the direct crystallization procedure of 27.0 and 20.0 percent respectively. Perivine and yohimbine were also isolated from the leaf (D) fraction in yields of 3.50 and 3.30 percent respectively, and neither of these bases could be obtained by the direct crystallization technique (33).

Thus, by direct crystallization and gradient *p*H techniques on the 14 crude alkaloid fractions, yohimbine, ajmalicine, perivine, and leurosine could be obtained in yields ranging from 2.30 to 27.0 percent. However, none of the remaining 17 alkaloids that we have isolated from *C. lanceus* were obtainable by these methods, although, interestingly enough, they were obtained in yields of from 0.04 percent (vinosidine) to 8.12 percent (tetrahydroalstonine (IV)), but most of these 17 bases were present as less than 2 percent of the crude alkaloid mixture.

Prior to initiating column chromatographic separations on *Catharanthus* alkaloid fractions, it therefore appears of some value to remove the major alkaloids either by direct crystallization procedures, or by the gradient *p*H technique, neither of which takes much time. Of course, with the latter technique, several fractions are then obtained. Each of these can be separately applied to a column, or they can be combined and separated on a single column.

As Svoboda found in his work, our major success has been realized through the column chromatographic separation of alkaloid mixtures on partially neutralized and partially deactivated (*ca.* activity II or III) alumina. We have used silicic acid-celite (5:1) at times, but this adsorbent does not have the capacity for resolving complex mixtures as well as alumina. The eluting solvents usually used are benzene, followed by benzene with increasing amounts of chloroform, chloroform alone, chloroform with increasing amounts of methanol, and finally stripping the column with methanol. We have never been able to isolate crystalline bases from column fractions past the point where chloroform was used as the eluant. Also, it should be pointed out that the fractions eluted with increasing ratios of methanol in chloroform contain a great deal of inorganic material. This material seems to vary depending on the fraction being separated, but sodium carbonate has been isolated and identified in pure form in some cases. Ammonium tartrate, which one would expect to be present in certain of the alkaloid fractions, has also been obtained quite frequently.

Following the grouping of chromatographic fractions, direct crystallization was made, using the normal alkaloidal solvents. If no crystalline

Table 2. A Comparison of Alkaloid Yields from *C. lanceus* Root Fractions (33)

Fraction[a]	Alkaloid	Yield[b]		
		Column chromatography	Direct crystallization	Gradient *p*H
A	Ajmalicine III	5.47	11.85	15.90
	Lanceine	1.47		
	Catharanthine XII	0.15		
	Cathalanceine	0.12		
	Perimivine	0.12		
	Vincoline	0.06		
	Ammocalline	0.05		
	Hörhammericine IX	0.05		
	Vinosidine	0.04		
	Pericalline XIII	0.03		
A_1	Ajmalicine	–	16.10	3.20
B	Yohimbine II	5.94	–	1.00
	Pericalline	5.66		
	Ajmalicine	1.11		
B_1	Ajmalicine	–	3.17	3.80
C	Ajmalicine	–	–	0.53
D	None isolated			
E	Ajmalicine	–	–	4.02

[a] See Fig. 2.
[b] Expressed as a percentage of the alkaloid fraction.

entities were obtained, the fraction was then separated by gradient *p*H, if sufficient material was available, or the fraction converted into sulfate salts, and crystallization again attempted. It is quite apparent from our work that many of the alkaloids of *C. lanceus* exist as oils and would never crystallize; hence salt formation is often advantageous. A good example of this is the alkaloid vindolinine (XI), which has never been obtained as the crystalline base. However, it crystallizes readily as the dihydrochloride.

A summary of all of our isolation work is presented in Tables 2 and 3, in which a comparison is made of the 21 alkaloids, and their yields by the various isolation techniques used.

The 21 alkaloids of *C. lanceus* can be classified into six groups on the

Table 3. A Comparison of Alkaloid Yields from *C. lanceus* Leaf Fractions (33)

Fraction[a]	Alkaloid	Yield[b] Column chromatography	Direct crystallization	Gradient pH
A	Leurosine XIV	0.70	–	2.30
	Yohimbine II	0.55		
	Perivine V	0.31		
	Ajmalicine III	0.23		
	Pericyclivine I	0.23		
	Hörhammerinine IX	0.18		
A_1	Tetrahydroalstonine IV	8.12		
	Lochnerinine VII	1.95		
	Catharine	0.73		
	Periformyline VI	0.22		
	Ajmalicine	0.06		
B	Yohimbine	24.70	24.70	27.00
	Perivine	15.00	15.00	20.00
	Vindolinine XI	0.04		
B_1	None isolated			
C	Yohimbine	0.40		
	Leurosine	0.17		
D	Perivine	2.10	–	3.50
	Yohimbine	–	–	3.30
E	Vindoline X	1.20		
	Lochnerinine	0.33		

[a] See Fig. 2.
[b] Expressed as a percentage of the alkaloid fraction.

Table 4. Classification and u.v. Absorption Maxima of *Catharanthus lanceus* Alkaloids

Compound	λ max nm.
I. Monomeric Alkaloids	
A. β-Carbolines	
Pericyclivine I	280, 290
Yohimbine II	223, 274, 290(sh)
Ajmalicine III	227, 292
Tetrahydroalstonine IV	226, 272, 291

Table 4 (continued)

B. 2-Acylindoles	
Perivine V	227, 237(sh), 314
Periformyline VI	239, 314
C. α-Methyleneindolines	
Lochnerinine VII	247, 312, 326
Perimivine	232, 302, 340
Hörhammericine VIII	299, 327
Hörhammerinine IX	245, 325
Cathalanceine	226, 296, 323
Lanceine	288, 300, 325
D. Dihydroindoles	
Vindoline X	212, 250, 304
Vincoline	244, 300
Vindolinine XI[a]	245, 300
E. Miscellaneous	
Catharanthine XII[a,b]	224, 282, 291
Pericalline XIII	207, 230(sh), 240(sh), 304
Ammocalline	218, 288
Vinosidine	226, 254, 259, 300
II. Dimeric Indole Alkaloids	
Leurosine XIV	214, 263, 287(sh), 296(sh)
Catharine	222, 265, 293, 310(sh)

[a] As hydrochloride.
[b] Free base has λ_{max} 226, 284, 292 nm.

basis of similarities in their basic nucleus, and their u.v. chromophore (Table 4).

The structures for 13 of the 21 *C. lanceus* alkaloids are known and the partial structure for the dimeric alkaloid catharine has been deduced in our laboratory (35). However, we do not at present know the structures for ammocalline, lanceine, perimivine, vincoline, vinosidine or cathalanceine. Several of these alkaloids are now being studied in our laboratories.

Although many of the *C. lanceus* alkaloids are found in related apocynaceous plants, periformyline (VI), cathalanceine, lanceine, hörhammericine (VIII) and hörhammerinine (IX) are restricted to this plant. The alkaloids pericyclivine (I), periformyline (VI), hörhammericine (VIII), hörhammerinine (IX), and cathalanceine were first discovered in our laboratories, perimivine being a co-discovery with Svoboda.

Table 5. Certain Physical Data for *Catharanthus lanceus* Alkaloids[a]

Alkaloid		Formula	m.p. 1°	pK'_a	$[\alpha]_D$	Ref.
Pericalline	XIII	$C_{18}H_{20}N_2$	196-202	8.05[b]		(26, 31)
Ammocalline	—	$C_{19}H_{22}N_2$	>335(d)	7.3[b]		(30, 31)
Pericyclivine	I	$C_{20}H_{22}N_2O_2$	232-233	6.75[b]	+5.2°[d]	(36)
Perivine	V	$C_{20}H_{24}N_2O_3$	181-183	7.6[b]	-121°[d]	(25)
Lanceine	—	$C_{20}H_{26}N_2O_3$	143-145	—	+64°[e]	(15, 37)
Vindolinine	XI	$C_{21}H_{24}N_2O_2$	212-214(d)(HCl)	3.3, 7.1[c]	-8[f]	(38, 39)
			210-212(d)(HCl)			
Catharanthine	XII	$C_{21}H_{24}N_2O_2$	179-181(d)(HCl)	6.8[c]	+65.9°[g]	(31, 39)
			126-128(base)		+29.8°[d]	
Ajmalicine	III	$C_{21}H_{24}N_2O_3$	252-254(d)	6.31[c]	-60°[d]	(26)
Tetrahydroalstonine	IV	$C_{21}H_{24}N_2O_3$	227-227.5(d)	5.83[c]	-107°[d]	(40)
Periformyline	VI	$C_{21}H_{22}N_2O_4$	206-209(d)	>4.0[c]	—	(40, 41)
Perimivine	—	$C_{21}H_{22}N_2O_4$	292-293(d)	insol.	-98.71°[d]	(26, 42)
Vincoline	—	$C_{21}H_{24}N_2O_4$	222-224(d)	6.1[b]	-37.5°[d]	(30, 42)
Yohimbine	II	$C_{21}H_{26}N_2O_3$	232-233(d)	7.60[b]	+47°[e]	(25)
Hörhammericine	VIII	$C_{21}H_{26}N_2O_4$	140-144	—	-403°[d]	(43)
Lochnerinine	VII	$C_{22}H_{26}N_2O_4$	164-166	—	-442°[d]	(40)
Vinosidine	—	$C_{22}H_{26}N_2O_5$	253-257(d)	6.80[b]	—	(31, 37)
			251-253(d)			
Hörhammerinine	IX	$C_{22}H_{28}N_2O_5$	209.5-211(d)	—	-381°[d]	(44, 45)
Vindoline	X	$C_{25}H_{32}N_2O_6$	154-155	5.5[c]	+42[d]	(25, 39)
			153-154	5.60[b]	—	
Cathalanceine	—	—	188-190(d)	4.50[b]	—	(26)
Leurosine	XIV	$C_{46}H_{56}N_4O_9$	200-202(d)	4.80, 7.10[b]	+59.80[d,h]	(25, 46, 47)
Catharine	—	$C_{46}H_{54}N_4O_{10}$	257-258(d)	—	—	(35)

[a]Arranged by increasing M.w.; [b]in 33% DMF; [c]in 67% DMF; [d]in chloroform; [e]in ethanol; [f]2.HCl, in water; [g]in methanol; [h]analysis on the hexahydrate.

The physical data for the 21 *C. lanceus* alkaloids are presented in Table 5, and the structures, where known, are grouped by class in I-XIV.

BIOLOGICAL ACTIVITY STUDIES

Hypotensive Activity of C. lanceus Alkaloids. All alkaloid fractions from *C. lanceus* have been evaluated in normotensive, anesthetized rats, for hypotensive activity (Table 6). The root (D) and (E) fractions, as well as the leaf (A_1), (C) and (E) fractions were found to elicit little or no hypotensive effects (48). The root (A_1) fraction gave a moderate hypotensive effect (48), and all remaining fractions, i.e. the root (A), (B), (B_1), (C), and the leaf (A), (B), (B_1) and (D) fractions, gave marked hypotensive activity of long duration, at doses ranging from 8-40 mg./Kg. (i.p.) (48).

Yohimbine (II), a potent α-adrenergic blocking agent, was shown to significantly reduce blood pressure in rats and in dogs, and this alkaloid

Table 6. Hypotensive Evaluation of *C. lanceus* Alkaloid Fractions (48)[a]

Fraction	Dose (mg/Kg)	Animals (no.)	Drop in b.p. (%) (range)	Duration of b.p. drop (min.) (range)
Root				
A	20-40	4	33-44	203-310+
A_1	20	6	28-44	28-67
B	10	2	47-56	350+
B_1	20-40	5	32-55	38-147+
C	40	2	27-37	70-120+
D	20	4	13-53	8-70
E	10-20	4	b	b
Leaf				
A	10	6	34-52	147-305+
A_1	5-20	6	b	b
B	8	6	42-73	260-378
B_1	8-16	6	22-36	56-222+
C	20-40	5	24-36	5-7
D	20	6	45-53	112-287+
E	10-20	4	0-50	2-3

[a]In normotensive rats, anesthetized with urethan (1.25 Gm/Kg, i.p.). Extracts were administered i.v.

[b]No hypotensive activity.

has been isolated from the root (B) fraction, and the leaf (A), (B), (C) and (D) fractions. However, no hypotensive alkaloids have as yet been obtained from the root (A), (B_1) or (C) fractions, or from the leaf (B_1) fraction, all of which were markedly hypotensive (48).

In an attempt to determine whether or not yohimbine was the sole hypotensive alkaloid in *C. lanceus,* we studied the leaf (B) fraction, which is rich in this alkaloid, and gives a pronounced hypotensive effect. Following column chromatographic separation of this fraction, all chromatographic cuts containing yohimbine, as evidenced by TLC, were removed. The remaining yohimbine-free cuts were then pooled and evaluated for hypotensive activity in rats. This fraction was found to elicit a high degree of hypotensive activity of long duration. Thus, it appears that there is still at least one highly active hypotensive alkaloid remaining in *C. lanceus* (48).

Leurosine (XIV) and perivine (V) gave only a transient hypotensive effect, whereas pericyclivine (I), pericalline (XIII), vindoline (X), tetrahydroalstonine (IX), ajmalicine (III), lochnerinine (VII) and periformyline (VI) were found to be devoid of hypotensive activity at doses ranging from 1.0 to 40 mg/Kg (i.p.). Pericalline was observed to elicit a pronounced analeptic effect in all test animals (48). None of the remaining alkaloids isolated from this plant have been available in sufficient quantity for hypotensive evaluation.

Antitumor Activity of C. *lanceus.* Our original interest in *C. lanceus* was aroused more than a decade ago when we obtained a small sample of the roots of this plant. At that time we were primarily interested in the hypotensive principles shown to be present in preliminary tests, but other types of biological activity were examined. Thus, we submitted a lyophilized aqueous extract of the roots of *C. lanceus* to the Clayton Biochemical Institute, University of Texas,* for antitumor screening. The studies were conducted using the RC mammary carcinoma in mice. To our surprise, this crude aqueous extract inhibited this tumor 70 percent at doses of 5.0 mg/Kg (i.p.), with no apparent evidence of toxicity on autopsy.

From this point we obtained a larger sample of *C. lanceus* roots, in addition to a quantity of leaf material. The leaf sample was of interest because of a knowledge that practically all of the antitumor activity of the related *C. roseus* was found in the leaf alkaloid (A) fraction. We prepared extracts of both the root and leaf samples according to the method

*Tumor evaluations were conducted by Mr. G. F. McKenna.

of Svoboda (27), and submitted all 14 alkaloid fractions to the Lilly Research Laboratories, Indianapolis, Indiana, for testing against the P-1534 leukemia in DBA/2 mice, a tumor system especially sensitive to the action of *C. roseus* antitumor alkaloids, and which was responsible for their discovery. To our disappointment, all 14 alkaloid fractions were inactive on replicate testing against this neoplasm.

To further complicate matters, identical alkaloid fractions were submitted to the Cancer Chemotherapy National Service Center (CCNSC), National Cancer Institute, for additional testing in DBA/2 mice infected with the P-1534 leukemia. It was then determined that of the 14 fractions, only the *C. lanceus* leaf alkaloid (A) fraction was active against this neoplasm. The results from evaluating all of the leaf and root fractions are presented in Table 7.

On the basis of these data, we placed major emphasis on the isolation of alkaloids from the leaf (A) fraction, using column chromatography, followed by bioassay of the chromatographic cuts to determine the fractions containing antitumor activity. A typical experiment of this type is shown in Table 8, where it can be seen that an unusual situation exists. Fractions 521-559 were devoid of activity against the P-1534 leukemia, but we subsequently isolated high yields of the active alkaloid leurosine (XIV) from these fractions. The activity elicited by fractions 560-759 was probably due to the presence of leurosine, which we were unable to isolate in crystalline form. However, one can see by inspection of the results from testing fractions 959-1399 that at least one more active alkaloid remains in the leaf (A) fraction. Experiments are now in progress to isolate this entity.

Another unusual finding in this work was noted during the investigation of the leaf alkaloid (C) fraction, which was inactive against the P-1534 leukemia. Column chromatography of this fraction resulted in the isolation of yohimbine (II), followed by significant amounts of the active alkaloid leurosine (XIV) (32).

We are unable to explain with any degree of certainty how an active antitumor alkaloid can be isolated from a crude fraction which has been shown to be devoid of activity. However, Svoboda has also encountered this phenomenon in his work with *C. roseus,* during studies leading to the discovery of leurocristine (34).

On the basis of these inconsistent bioassay results, we feel that there is justification for investigating all of the alkaloid fractions from *Catharanthus* species, rather than restricting studies to those fractions showing confirmed antitumor activity in animals.

Table 7. Activity of *C. lanceus*, the Leaf A Alkaloid Fraction, against the P-1534 Leukemia (48)[a]

Dose (mg/Kg)	Survivors ( ) of ( )		Animal Wt. Diff. (T/C)	Survival (days) (T/C)	% Activity[b] (T/C)
25.0	6	6	0.5	18.0 / 14.0	128
12.5	6	6	-0.5	24.0 / 14.0	171
6.25	6	6	0.2	21.0 / 14.0	150
3.13	6	6	0.7	20.0 / 14.0	142
3.13	6	6	-0.9	17.0 / 13.0	130
1.56	6	6	-0.7	15.0 / 13.0	115
0.78	6	6	0.1	14.0 / 13.0	107
0.39	6	6	-0.9	13.5 / 13.0	103

[a]All other extracts were inactive at the following dose rates tested: Leaf (A_1), 3.13 to 100.0 mg/Kg; Leaf (B), 3.13 to 50.0 mg/Kg; Leaf (B_1), 25.0-200.0 mg/Kg; Leaf (C), 6.25 to 50.0 mg/Kg; Leaf (D), 1.56 to 50.0 mg/Kg; Leaf (E), 25.0 to 200.0 mg/Kg; Root (A), 3.13 to 50.0 mg/Kg; Root (A_1), 3.13 to 100.0 mg/Kg; Root (B), 25.0 to 50.0 mg/Kg; Root (B_1), 3.13 to 100.0 mg/Kg; Root (C), 6.35 to 50.0 mg/Kg; Root (D), 3.13 to 50.0 mg/Kg; Root (E), 25.0 to 100.0 mg/Kg.

[b]Evaluation of the assay is such that an extract is considered active if it causes a prolongation of life in excess of (T/C) 125 percent.

Table 8. Column Chromatographic Separation of *C. lanceus* Leaf (A) Fraction and Activity against the P-1534 Leukemia

Fraction No.[a] (1000 ml. ea.)	Eluent	Fraction Wt. (Gm.)	P-1534 Activity (% prolongation/dose)[b]	Alkaloids Isolated	Yield (Gm.)
1-7	Benzene	0.10	not tested		
8-9		1.90	not tested		
10-12		0.28	not tested		
13-19		8.10	not tested		
20-60		40.90	not tested		
61-150		36.44	87 / 40.0	Ajmalicine	0.52
151-340	Benzene-chloroform (99:1)	22.02	100 / 40.0		
341-520	Benzene-chloroform (3:1)	32.00	100 / 40.0	Ajmalicine	1.70
521-550		12.60	120 / 20.0	Leurosine	3.20
				Pericyclivine	0.69
551-559		9.78	91 / 10.0	Leurosine	0.18
560-630		14.30	225 / 10.0	Hörhammerinine	0.88
631-735		16.00	216 / 40.0		
736-759		4.80	133 / 40.0		
760-850		13.20	87 / 40.0	Yohimbine	0.15
851-958		22.50	108 / 40.0	Perivine	0.64
				Yohimbine	2.03
959-1132	Benzene-chloroform (1:1)	27.50	150 / 10.0	Perivine	3.25
1133-1148	Chloroform	3.50	233 / 40.0		
1149-1200		8.00	254 / 40.0		
1201-1249	Chloroform-methanol (99:1)	14.50	229 / 40.0		
1250-1328		18.70	241 / 20.0		
1329-1386		5.00	208 / 40.0		
1387-1399	Chloroform-methanol (95:5)	4.89	137 / 40.0		
1400-1515	Chloroform-methanol (9:1)	9.75	112 / 40.0		
1516-1568		9.25	inactive		
1569-1630	Methanol	11.80	inactive		
1631-1800	Ethanol	51.70	inactive		

[a] A total of 500 Gm. of alkaloid (A) fraction was separated. [b] An activity of $\geqq$ 125 is considered to be active.

Of the 21 alkaloids that we have isolated from *C. lanceus,* only 12 were obtained in sufficient quantity for P-1534 leukemia evaluation. The dimeric base leurosine (NSC-90636) is the only *C. lanceus* alkaloid found to date that has antitumor properties (48). This alkaloid was originally isolated from *C. roseus* by Svoboda (46), but to date clinical trials for leurosine have been disappointing (49, 50).

Alkaloids which were inactive against the P-1534 leukemia are yohimbine, tetrahydroalstonine, ajmalicine, pericalline, perivine, periformyline, vindoline, lochnerinine, pericyclivine, and catharanthine, which are all monomeric bases. The dimeric alkaloid catharine was similarly devoid of antitumor properties (48).

Since all of the alkaloids that we have isolated from *C. lanceus* in quantities insufficient for P-1534 leukemia evaluation are monomeric (hörhammericine, hörhammerinine, vindolinine, lanceine, cathalanceine, perimivine, vincoline, vinosidine), it is not likely that they would have any activity against this neoplasm. Such activity is well established as being associated only with certain of the dimeric *Catharanthus* alkaloids.

Cytotoxicity of C. lanceus. We have used Eagle's 9 KB carcinoma of the nasopharynx in cell culture as in indicator of cytotoxicity for certain alkaloid fractions and isolated compounds from *C. lanceus.* The criteria for cytotoxicity are such that plant extracts must exhibit an $ED_{50} \leqq 15.0\ \mu g/ml$, and pure compounds an $ED_{50} \leqq 1.0\ \mu g/ml$, in order to be considered as active cytotoxic agents (51).

Cytotoxicity is not always an effective or reliable means for predicting *in vivo* antitumor activity, or for determining clinical efficacy. However, the antitumor alkaloids of *Catharanthus* are all cytotoxic. Thus, we have found the 9 KB assay advantageous as a bioassay to guide us in our isolation studies. It has been particularly useful since it is a rapid assay, and because only small amounts of extract or pure compound are necessary for testing.

Table 9 gives the bioassay result from testing the 14 *C. lanceus* root and leaf crude alkaloid fractions. It can be clearly seen that all of the fractions contain cytotoxic material except the root (B_1), (C), (D) and (E) fractions, and that all of the leaf fractions show significant cytotoxic activity.

All 21 of the *C. lanceus* alkaloids have been evaluated against the 9 KB carcinoma and only leurosine sulfate ($ED_{50}\ 1.0 \times 10^{-5}$) and lochnerinine ($ED_{50}\ 1.0 \times 10^{-2}$) are active cytotoxic agents.

Diuretic Activity of C. lanceus Alkaloids. When evaluated for diuretic

Table 9. Cytotoxicity Testing of *C. lanceus* Alkaloid Fractions (48)

Fraction	ED_{50}(μg/ml)
Root	
(A)	2.7
(A_1)	2.1
(B)	2.7
(B_1)	21
(C)	22
(D)	18
(E)	24
Leaf	
(A)	not tested
(A_1)	<1.0
(B)	1.9×10^{-3}
(B_1)	3.8
(C)	2.4
(D)	0.25
(E)	1.3

activity in saline-loaded rats, and compared with chlorothiazide (+2) and dihydrochlorothiazide (+3), the following results were obtained with several *C. lanceus* alkaloids: vindolinine dihydrochloride (+3), catharanthine hydrochloride (+2), vindoline (0), perivine sulfate (0) and tetrahydroalstonine (0). Ajmalicine (-1) gave an antidiuretic effect. All alkaloids were administered orally at doses of 5.0 and 50.0 mg/Kg, except for vindoline, which was given at 1.0 and 10.0 mg/Kg (42, 52).

Hypoglycemic Activity of C. *lanceus Alkaloids.* Several of the *C. lanceus* alkaloids have been examined for hypoglycemic activity in normal rats at a dose level of 100 mg/Kg (p.o.). Ajmalicine and perivine were inactive, and catharanthine hydrochloride and tetrahydroalstonine gave questionable results which appear to be manifest in a delayed onset. Vindoline gave only a slight hypoglycemic effect, and leurosine sulfate, as well as vindolinine dihydrochloride were rated as the most active of those alkaloids evaluated. The hypoglycemic action of the latter two alkaloids was intermediate between that of acetohexamide and tolbutamide (53).

Antiviral Activity of C. *lanceus Alkaloids.* A routine screening of all available *Catharanthus* alkaloids for antiviral activity in cell culture has revealed that pericalline, perivine, periformyline and vindolinine were inhibitory for the polio type III virus. Perivine, in addition, was inhibitory for the vaccinia virus. Alkaloids that were not inhibitory for either the polio type III or the vaccinia viruses, were ajmalicine, leurosine, vinosidine, ammocalline, catharine, lochnerinine, perimivine, tetrahydroalstonine and vincoline (54). The significance of these findings has not yet been explored.

SUMMARY

Catharanthus lanceus has yielded 21 crystalline bases, 19 being monomeric indoles, and two dimeric indoles. The structures for 14 of these alkaloids are known, five being deduced in our own laboratory.

Alkaloids from *C. lanceus* have been shown to elicit a variety of biological activities, including analeptic, anticancer, antidiuretic, antiviral, cytotoxic, diuretic, hypoglycemic, and hypotensive. These activities are summarized in Table 10.

Table 10. Summary of Biological Activities of *C. lanceus* Alkaloids

Reported activity	Alkaloids
Analeptic	Pericalline
Anticancer	Leurosine
Antidiuretic	Ajmalicine
Antiviral	Pericalline, Perivine, Periformyline, Vindolinine
Cytotoxic	Leurosine, Lochnerinine
Diuretic	Catharanthine, Vindolinine
Hypoglycemic	Vindolinine, Leurosine, Catharanthine, Tetrahydroalstonine
Hypotensive	Yohimbine (α-adrenergic blocking agent)

Our studies are continuing on the alkaloids of the leaves and the roots of this plant, primarily in an attempt to isolate additional anticancer agents, which are known to be present. This research effort, spanning a period of some ten years, has necessitated a truly interdisciplinary approach, which has involved the cooperation of pharmacognosists, organic chemists, biochemists, medicinal chemists, physical chemists, botanists, pharmacologists, biologists, and others.

Acknowledgments. The work reported herein was supported, in part, by research grants HE-06162, CA-08228, CA-12230, and FR-05455 from the National Institutes of Health, U.S. Department of Health, Education and Welfare, Bethesda, Md.; from the Jane Olmstead Thaw Fund, University of Pittsburgh; from the American Cancer Society (institutional grant funds); and from Eli Lilly and Company, Indianapolis, Ind.

Since the work reported in this review has involved the cooperation of many individuals, the author gratefully acknowledges the following: Drs. D. J. Abraham, R. N. Blomster, J. P. Buckley, F. J. Draus, H. H. S. Fong, M. Gorman, I. Hassan, A. N. Masoud, G. A. McKenna, W. A. Meer, A. G. Sharkey, Jr. and G. H. Svoboda; graduate students Messrs. W. W. Brown, W. D. Loub, R. E. Martello, E. M. Maloney and N. A. Pilewski; and technical assistants L. V. Cammarato, D. Damratoski and R. E. Rhodes.

We are especially grateful to the Cancer Chemotherapy National Service Center, National Cancer Institute, and to Dr. J. L. Hartwell and Miss B. J. Abbott, for their cooperation in having our extracts and isolates evaluated for anticancer activity.

References

1. W. T. Stearn, "*Catharanthus roseus,* the correct name for the madagascan periwinkle," *Lloydia 29* (1966), 196-200.

2. M. Pichon, "Classification des Apocynacées. IX. Rauwolfiées, Alstoniées, Allamandées, Tabernemontanoidées," *Mem. Mus. Hist. Nat.* XXXII, fasc. *6* (1948), 153-252.

3. R. Paris and H. Moyse, "Les Pervenches indigènes et exotiques. I. Étude botanique," *J. Agric. Trop. Botan. Appl. 4* (1957), 481-489.

4. N. G. Bisset, "The occurrence of alkaloids in the Apocynaceae," *Ann. Bogorienses 3* (1958), 105-236.

5. G. H. Svoboda, "The current status of research on the alkaloids of *Vinca rosea* Linn. (*Catharanthus roseus* G. Don)." *Excerpta Med.* International Congress Series 106 (1966), 9-27.

6. R. Paris and H. Moyse, "Les Pervenches indigènes et exotiques. II. Étude chimique," *J. Agric. Trop. Botan. Appl. 4* (1957), 645-653.

7. R. Paris and H. Moyse, "Les Pervenches indigènes et exotiques. III. Étude pharmacologique, *ibid. 5* (1958), 35-43.

8. N. R. Farnsworth, "The pharmacognosy of the periwinkles: *Vinca* and *Catharanthus,*" *Lloydia 24* (1961), 105-138.

9. J. Trojánek, "Novější poznatky o alkaloidech rodu *Vinca* a *Catharanthus,*" *Csl. Farm. 11* (1962), 508-518.

10. O. Strouf and K. Kavkova, "Alkaloidy rodu *Vinca* a *Catharanthus,*" *Chem. Listy 56* (1962), 987-1028.

11. G. H. Svoboda, I. S. Johnson, M. Gorman, and N. Neuss, "Current status of

research on the alkaloids of *Vinca rosea* Linn. (*Catharanthus roseus* G. Don)," *J. Pharm. Sci. 51* (1962), 707-720.

12. I. S. Johnson, J. G. Armstrong, M. Gorman, and J. P. Burnett, Jr., "The *Vinca* alkaloids: a new class of oncolytic agents," *Cancer Res. 23* (1963), 1390-1427.

13. N. Neuss, I. S. Johnson, J. G. Armstrong, and C. J. Jansen, "The *Vinca* Alkaloids," in *Advances in Chemotherapy* (New York, Academic Press, 1964), pp. 133-174.

14. M.-M. Janot, J. LeMen, and Y. Hammouda, "Présence de la yohimbine (québrachamine) dans les racines du *Lochnera lancea* Boj. (ex A. DC.) K. Schumm., pervenche de Madagascar," *Ann. Pharm. Franc. 14* (1956), 341-344.

15. M.-M. Janot, J. LeMen, and Y. Gabbai, "Alcaloïdes du *Lochnera lancea* Boj. (ex A. DC.) K. Schumm. III. Présence de la tetrahydroalstonine et d'un alcaloïde nouveau: la lancéine," *ibid. 15* (1957), 474-478.

16. M.-M. Janot and J. LeMen, "Présence de la δ-yohimbine dans *Lochnera lancea* Boj. (ex A. DC.) K. Schumm. ou *Vinca lancea* Boj. (ex A. DC.)," *Compt. Rend. 239* (1954), 1311-1313.

17. T. S. Githens, *Drug Plants of Africa* (Philadelphia, Penn., University of Pennsylvania Press, 1949).

18. R. Pernet, "Les Plantes médicinales Malgaches. *Mem. Inst. Sci. Madagascar* [Ser. B.] *8* (1957), 7-15.

19. L. Aldaba and L. Oliveros-Belardo, "Preliminary chemical study of the leaves of the Philippine variety of *Lochnera rosea* (L.) Reichb.," *Rev. Filipina Med. y Farm. 29* (1938), 259-293.

20. R. Paris and H. Moyse-Mignon, "Caractérisation de la choline par chromatographie sur papier chez quelques plantes médicinales," *Ann. Pharm. Franc. 14* (1956), 464-469.

21. N. A. Pilewski, "Studies on *Catharanthus lanceus* Root Alkaloids," M.S. Thesis, University of Pittsburgh, Pittsburgh, Penn., 1963.

22. R. E. Martello, "A Phytochemical Investigation of *Catharanthus lanceus* Roots," M.S. Thesis, University of Pittsburgh, Pittsburgh, Penn., 1964.

23. R. Pernet, G. Meyer, J. Bosser, and G. Ratsiandavana. "Les *Catharanthus* de Madagascar," *Compt. Rend. 243* (1956), 1352-1353.

24. R. Paris and H. Moyse, "Examen en électrophorèse sur papier des alcaloïdes totaux de diverses espèces de pervenches," *ibid. 245* (1957), 1265-1268.

25. W. D. Loub, N. R. Farnsworth, R. N. Blomster, and W. W. Brown, "Studies on *Catharanthus lanceus.* IV. Separation of leaf alkaloid fractions and the isolation of leurosine, perivine, yohimbine and vindoline," *Lloydia 27* (1964), 470-479.

26. R. N. Blomster, R. E. Martello, N. R. Farnsworth, and F. J. Draus, "Studies on *Catharanthus lanceus.* V. Preparation of root alkaloid fractions and the isolation of ajmalicine, yohimbine, pericalline, perimivine and cathalanceine," *ibid. 27* (1964), 480-485.

27. G. H. Svoboda, N. Neuss, and M. Gorman, "Alkaloids of *Vinca rosea* Linn. (*Catharanthus roseus* G. Don). V. Preparation and characterization of alkaloids," *J. Am. Pharm. Assn., Sci. Ed. 48* (1959), 659-666.

28. N. R. Farnsworth, R. N. Blomster, D. Damratoski, W. A. Meer, and L. V. Cammarato, "Studies on *Catharanthus* alkaloids. VI. Evaluation by means of thin-layer chromatography and ceric ammonium sulfate spray reagent," *Lloydia 27* (1964), 302-314.

29. R. Munier and M. Macheboeuf, "Microchromatographie de partage sur papier des alcaloïdes et de diverses bases azotees biologiques. III. Exemples de separations

de divers alcaloïdes par la technique en phase solvente acide (familles de l'atropine, de la cocaine, de la nicotine, de la sparteine, de la strychnine et de la corynanthine)," *Bull. Soc. Chim. Biol. 33* (1951), 846-856.

30. N. R. Farnsworth, H. H. S. Fong, and R. N. Blomster, "*Catharanthus* alkaloids. XII. Isolation of catharanthine, ammocalline and vincoline from *Catharanthus lanceus* roots," *Lloydia 29* (1966), 343-347.

31. G. H. Svoboda, A. T. Oliver, and D. R. Bedwell, "Alkaloids of *Vinca rosea* (*Catharanthus roseus*). XIX. Extraction and characterization of root alkaloids," *ibid. 26* (1963), 141-153.

32. N. R. Farnsworth. Unpublished data.

33. N. R. Farnsworth, R. N. Blomster, A. N. Masoud, and I. Hassan, "*Catharanthus* alkaloids. XIV. Comparison of alkaloid isolation techniques for *Catharanthus lanceus*," *Lloydia 30* (1967), 106-110.

34. G. H. Svoboda, "Alkaloids of *Vinca rosea* (*Catharanthus roseus*). IX. Extraction and characterization of leurosidine and leurocristine," *ibid. 24* (1961), 173-178.

35. D. J. Abraham, N. R. Farnsworth, R. N. Blomster, and R. E. Rhodes, "Structure elucidation and chemistry of *Catharanthus* alkaloids. II. Isolation and partial structure of catharine, a dimeric indole alkaloid from *C. lanceus* and *C. roseus*," *J. Pharm. Sci. 56* (1967), 401-403.

36. N. R. Farnsworth, W. D. Loub, R. N. Blomster, and M. Gorman, "Pericyclivine, a new *Catharanthus* alkaloid," *ibid. 53* (1964), 1558.

37. R. N. Blomster, N. R. Farnsworth, and D. J. Abraham, "*Catharanthus* alkaloids. X. Isolation of lanceine and vinosidine from *Catharanthus lanceus* roots," *ibid. 56* (1967), 284-286.

38. E. M. Maloney, H. H. S. Fong, N. R. Farnsworth, R. N. Blomster, and D. J. Abraham, "*Catharanthus alkaloids.* XV. Isolation of vindolinine from *C. lanceus* leaves," *ibid. 57* (1968), 1035-1036.

39. M. Gorman, N. Neuss, G. H. Svoboda, A. J. Barnes, Jr., and N. J. Cone, "A note on the alkaloids of *Vinca rosea* Linn. (*Catharanthus roseus* G. Don). II. Catharanthine, lochnericine, vindolinine and vindoline," *J. Am. Pharm. Assn., Sci. Ed. 48* (1959), 256-257.

40. E. M. Maloney, N. R. Farnsworth, R. N. Blomster, D. J. Abraham, and A. G. Sharkey, Jr., "*Catharanthus lanceus.* VII. Isolation of tetrahydroalstonine, lochnerinine and periformyline," *J. Pharm. Sci. 54* (1965), 1166-1168.

41. D. J. Abraham, N. R. Farnsworth, R. N. Blomster, and A. G. Sharkey, Jr., "Structure elucidation of periformyline, a novel alkaloid from *C. lanceus*," *Tetrahedron Letters* (1965), 317-319.

42. G. H. Svoboda, M. Gorman, and R. H. Tust, "Alkaloids of *Vinca rosea* (*Catharanthus roseus*). XXV. Lochrovine, perimivine, vincoline, lochrovidine, lochrovicine and vincolidine," *Lloydia 27* (1964), 203-213.

43. R. N. Blomster, N. R. Farnsworth, and D. J. Abraham, "*Catharanthus* alkaloids. XIX. Isolation of hörhammericine from *C. lanceus* (Apocynaceae) roots," *Naturwissenschaften 55* (1968), 298-299.

44. N. R. Farnsworth, W. D. Loub, R. N. Blomster, and D. J. Abraham, "*Catharanthus* alkaloids. XVIII. Isolation of hörhammerinine, a new alkaloid from *C. lanceus* (Apocynaceae)," *Z. Naturforsch. 23b* (1968), 1061-1063.

45. D. J. Abraham, N. R. Farnsworth, W. D. Loub, and R. N. Blomster, "Structure elucidation and chemistry of *Catharanthus* alkaloids. IV. Structures of hörhammericine and hörhammerinine," *J. Org. Chem. 34* (1969), 1575-1576.

46. G. H. Svoboda, "A note on several new alkaloids from *Vinca rosea* Linn. I.

Leurosine, virosine, perivine," *J. Am. Pharm. Assn., Sci. Ed. 47* (1958), 834.
47. D. J. Abraham and N. R. Farnsworth, "Structure elucidation and chemistry of *Catharanthus* alkaloids. III. Structure of leurosine, an active anticancer alkaloid," *J. Pharm. Sci. 58* (1969), 694-698.
48. N. R. Farnsworth, R. N. Blomster, and J. P. Buckley, "*Catharanthus* alkaloids. XIII. Antineoplastic and hypotensive activity of alkaloid fractions and certain alkaloids from *Catharanthus lanceus,*" *ibid. 56* (1967), 23-27.
49. S. D. Gailani, J. G. Armstrong, P. P. Carbone, C. Tan, and J. F. Holland, "Clinical trial of vinleurosine sulfate (NSC-90636): a new drug derived from *Vinca rosea* Linn.," *Cancer Chemother. Rept. 50* (1966), 95-103.
50. M. E. Hodes, R. J. Rohn, W. H. Bond, and J. Yardley, "Clinical trials with leurosine methiodide, an alkaloid from *Vinca rosea* Linn.," *ibid. 28* (1963), 53-55.
51. Anonymous. *Cancer Chemother. Rept. 25* (1962), 1-66.
52. M. Gorman, R. H. Tust, G. H. Svoboda, and J. LeMen. "Alkaloids of *Vinca rosea* (*Catharanthus roseus*). XXVI. Structure-activity studies of some alkaloids and their derivatives," *Lloydia 27* (1964), 214-219.
53. G. H. Svoboda, M. Gorman, and M. A. Root, "Alkaloids of *Vinca rosea* (*Catharanthus roseus*). XXVIII. A preliminary report on hypoglycemic activity," *ibid. 27* (1964), 361-363.
54. N. R. Farnsworth, G. H. Svoboda, and R. N. Blomster, "*Catharanthus* alkaloids. XXII. Antiviral activity of selected *Catharanthus* alkaloids," *J. Pharm. Sci. 57* (1968), 2174-2175.

The Ordeal Bean of Old Calabar: The Pageant of *Physostigma venenosum* in Medicine

BO HOLMSTEDT

Department of Toxicology
Swedish Medical Research Council
Karolinska Institutet
Stockholm, Sweden

INTRODUCTION

Many drugs have played a role not only in the cure and alleviation of disease but also as tools in elucidating physiological and pharmacological mechanisms. Among the latter may be mentioned atropine, curare, muscarine, nicotine, and, not least, physostigmine. Physostigmine, also called eserine, is an alkaloid contained in the Calabar bean, *Physostigma Venenosum Balf.* It is an open question which of the above-mentioned alkaloids has contributed most to pharmacology. One thing is sure—we could not have advanced in our understanding of basic mechanisms without any one of them. The story of the Calabar bean and physostigmine and its role in medicine is perhaps less well known than curare. In this paper the author has tried to recount its colorful history from African poison ordeals to the arrival of modern carbamate insecticides.

DISCOVERY OF THE WEST COAST OF AFRICA, INCLUDING CALABAR

In 1469 when the appetites of the Portuguese traders had been whetted by Henry the Navigator, a Lisbon merchant, Fernão Gomes, persuaded the King of Portugal to grant him a five-year monopoly of the trade of the Guinea coast beyond the Cape Verde Islands on condition that he explore a hundred leagues (nearly 400 miles) of new coastline each year. Gomes's contract was subsequently extended until 1475, by which time his ships had reached Fernando Po and crossed the equator. In so doing they had discovered a country in which gold and gold dust were evident in such abundance that they gave it the name of the Gold Coast.

The delta region of the Niger and other rivers in the East proved too unhealthy to permit permanent European occupation. It was found that the slave trade of the delta could be conducted by visiting ships with less risk from the island of São Thomé and Fernando Po, where Portuguese had begun to settle about 1493.

By the end of the sixteenth century the Portuguese were no longer the only Europeans who wished to establish permanent contact with West Africa. First the Dutch, then the English and French and some lesser European nations were anxious to break in on the Portuguese monopoly and establish trading bases of their own on the shores of West Africa. A new era was beginning in which these western European nations were to regard West Africa primarily as a source of slave labor for the plantations they were setting up in the West Indies and other tropical parts of the New World, and soon they were to compete bitterly among themselves for trade in slaves.

The Dutch first set about to capture the Portuguese trading posts in West Africa, and by 1642 all the Portuguese forts on the Gold Coast, which in the seventeenth century were deemed essential for the conduct of the slave trade, were in Dutch hands. The European traders secured their slaves by buying them at the coast from African merchants who obtained the bulk of them from the interior. Since, in practice, it was impossible to ensure that slaves and ships both arrived at the coast at a steady rate, depots of some kind were needed where stocks of slaves and of the trade goods that were offered in exchange for them could be kept. Sometimes the ships remained for a long time at the trading posts.

Between about 1640 and about 1750, a great number of European forts and trading-posts were established on the shores of West Africa. The coast east of the Gold Coast received early attention, since the prevailing winds and currents required ships leaving the Gold Coast to proceed eastward, close to the shore, to about as far as Fernando Po before they could set course for Europe or America. By the end of the seventeenth century the coast between the mouth of the Volta and Badagri was already known as the Slave Coast, and during the following century slave-traders of many nations became active in this part of Africa, commonly called the Oil Rivers (1). One of the most prominent of the trading-posts was Old Calabar, which probably got its name from

Fig. 1 The West Coast of Africa, including Calabar.

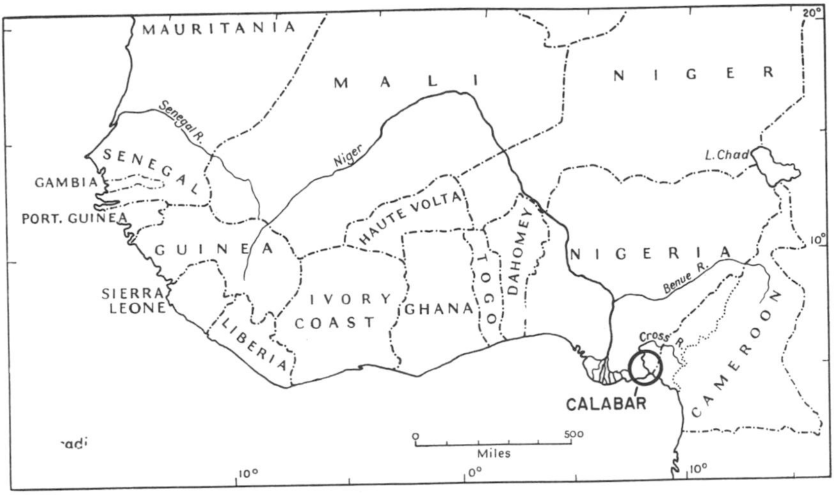

its earliest frequenters, the Dutch. In any case, the name first appears on Dutch maps of the seventeenth century (Fig. 1) (see essay by Simmons in (2).

The Efik People of Old Calabar. The Oil Rivers, the stretch of the West African coast which included the Niger Delta and the estuary of the Cross and Calabar rivers, took its name from the trade in palm oil which developed in place of the notorious slave trade that was gradually suppressed by the British in the first half of the nineteenth century. The main rivers and the innumerable creeks here provided a network of waterways into the interior which both the slave- and oil-traders commanded with their fleets of large canoes.

The Efik people of Old Calabar—originally a small branch of the Ibibio-speaking peoples—lived on the Cross River estuary in what is now the Calabar Province of Nigeria (Fig. 2). They were famous in the eighteenth and nineteenth centuries among the trading peoples of the Oil Rivers of West Africa. From their coastal settlements, such as Duke Town and Creektown, these African traders dealt with their European counterparts who lived in their ships that were moored close by. In return for slaves and oil they obtained supplies of European goods, and not least the firearms which enabled them to dominate the hinterland (2). Originally fishermen, the Efiks found their opportunity in the slave trade. Their communities grew and multiplied during the seventeenth and eighteenth centuries with the success of this traffic.

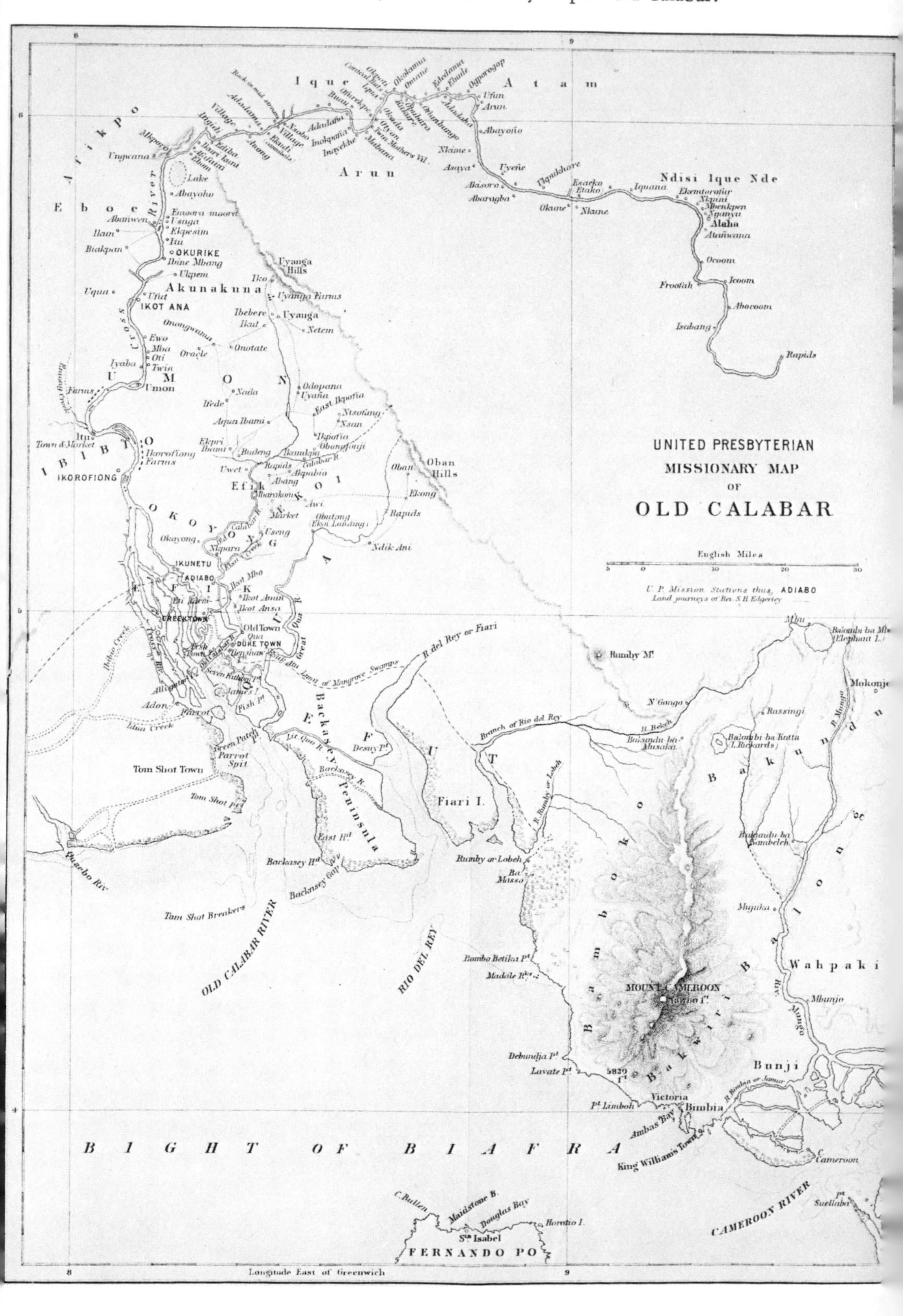

Fig. 2 United Presbyterian missionary map of Old Calabar.

At Old Calabar, and in the other trading centers of the Oil Rivers, there developed, through intercourse with the English traders and seamen, a jargon mainly English in its vocabulary which was to become in the eighteenth century a fairly standardized pidgin English. The value of written records, especially of business transactions, was also recognized by the Efik traders, and some of them were effectively instructed in writing by Englishmen from the ships. Some of this written documentation has survived, and the Diary of Antera Duke, which dates from 1785, is particularly famous (2).

For many generations, the only occupation considered noble by Calabar men was that of trading with the mbakara "rulers" or Europeans. Acting as middle men, they made great profits which they shared with the Qua and other tribes, who supplied them with farm produce and meat. A cash economy came to Calabar before it came to any other part of the Coast, and it came without restraint and in its worst guise. Each Calabar family kept its own slave community, which performed every function for the household—even, at times, large-scale trading. After the abolition of the slave trade, domestic slaves remained for some time; and it was with this subject-population, who worked the outlying plantations, that the great houses of Calabar continued to prosper in the palm oil trade. The great number of imported laborers working on Calabar land is still a feature of the area.

Every major Efik settlement had a senior *obong*, or chief, who always ranked high in the Egbo (Leopard) Society and was usually the head of the highest of its grades in the town. Many of them carried the European title of duke or king. The chief enforced laws in his role as head of the society, mediated or adjudicated in disputes, led armed forces in time of war, and arranged peace pacts with neighbors.

The Egbo, or Leopard, Society was the most important male association and consisted of several grades, each possessing a distinctive costume. There were further subdivisions within the higher grades. Under the aegis of the chief and the important elders of the town, the Leopard Society promulgated and enforced laws, judged important cases, recoverd debts, protected the property of members, and constituted the actual executive government of the Efik. It enforced its laws by capital punishment or fines for individuals, and by trade boycotts against European traders or other Efik towns (3).

What Egbo provided, and what was so conspicuously lacking in the Efik political system, was an executive staff able to carry out the orders and decisions of the senior members of the society. It had agents who, whether as recognized members of the society ("officers of the society") or disguised as Egbo spirits ("Egbo runners"), carried out the orders of the society or of its senior members. With the weight of the society behind them, these agents could be resisted by an individual or a local group only at his or their peril.

The religious and recreational functions of the Old Calabar Egbo did not differ noticeably from those of present-day societies in the area. The only special feature was the establishment of an Egbo day in the eight-day Efik week. On this day members of the society met together while the Egbo runners patrolled the streets (Fig. 3) and the uninitiated had to remain indoors and avoid coming into contact with Egbo (Jones in (2)).

To most Efik people, Egbo represented a mysterious overlordship more significant than any king or government. Great power lay in the hands of its leaders, and there can be no doubt that it was the local equivalent of the power of the Church in medieval Europe. Sometimes it worked with and sometimes against the secular power. But, as is often the case with African societies of a similar nature, powerful individuals often sat in the councils of both the religious and the secular bodies. Much the same thing happened in Europe during the later centuries.

In Calabar, this dual sanction was developed to a most thorough degree, and during the nineteenth century the fear of Egbo was exploited by officials of the traditional secular government who were invariably members of the secret society. Private ends were ensured by individuals working through government, and the rule of Egbo became openly oppressive. Heavy punishments were given for any minor offenses which happened to cause inconvenience.

It is not surprising, therefore, to find Old Calabar society suffering from many superstitions, including an intense fear of witchcraft. But accusations of witchcraft were not lightly made; they were followed by counter-accusations, and accusers as well as accused might find themselves compelled to submit to the deadly esere (Calabar bean) poison ordeal.*

*In the Efik language, *esere* is pronounced with stress on the first syllable (*ésere*). The word, as far as I have been able to find out, stands for both the bean and the poison ordeal.

Fig. 3 Ceremonial costume of the Egbo Society.

The custom of strangling some of the wives and slaughtering the slaves of a great man at his funeral was common throughout the Oil Rivers; the bigger the man the greater the number of both. In view of the great increase of wealth at the beginning of this period, the tendency for conspicuous waste may have been partly responsible for the ever-increasing number of victims, but it does not explain the manner of their execution and the lust for killing that activated the relatives of the kings of Old Calabar in the early part of the century. Slaves and wives were not the only persons to die on these occasions. Free-born men and women accused of witchcraft underwent the esere ordeal in remarkably large numbers. Ekpenyong Eyamba (Mr. Young), recorded in a journal the names of fifty persons who had died by esere ordeal on Duke Ephraim's death in 1847; at Archibong I's, in 1852, the victims were never counted. The increased number of deaths by the esere ordeal on the latter occasion may have had some relation to the absence of human sacrifices, but it also illustrates the constantly increasing use of the ordeal as a political weapon. Ordinary folk might believe in the genuineness of these witchcraft accusations; the more sophisticated recognized them as a deliberate method of removing dangerous rivals, weakening powerful houses, and settling old scores.

In addition to the human sacrifices there were other local customs appalling to early visitors. One of them was Egbo Iquo, which permitted the young men of the town to take an Egbo out, seize, strip, and indecently assault any woman wearing a dress or cloth in the street. Another custom compelled widows to remain in their houses in filth and wretchedness after the death of their husbands. Sometimes they were kept for years in this state of misery. Furthermore, as in many other parts of the world, twins that were born were killed and the mother expelled from the village.

The most mysterious of these gruesome customs was, however, *esere,* the poison ordeal, first observed by William Freeman Daniell (1818-1865), a British Army medical officer. Daniell was born in Liverpool in 1818. In 1841, he became a member of the Royal College of Surgeons of England and entered the medical service of the British Army. He served on the pestilential coast of West Africa and returned to England in 1863. Later, he accompanied the "Expeditionary Force" to China and was present at the taking of Pekin. A short time afterward he traveled to West India; on his next homecoming his health was broken, and he died in 1865 (4).

While in foreign service, Daniell collected and described many native plants. His publications are concerned with the plants, trees, shrubs, and their fruits, which were indigenous to the countries he visited. One of his treatises deals with the Matemfé, or miraculous fruit of the Sudan, which was afterward named *Phrynium danielli,* Benn., as a compliment to him.

Daniell first observed the use of the Calabar bean in native judicial procedure about 1840. A communication read by him before the Ethnological Society of Edinburgh, in 1846 (5), contains the first known record of its application:

> The Rio Calbary, or Old Callebar, formerly designated "Oude Calburgh" by its earliest frequenters, the Dutch, is one of the largest and most important of the rivers in the intertropical regions of Western Africa. It is situated nearly in the central portion of the Bight of Biafra, between the river Bonny and the Rio Del Rey; its embouchure being in Lat.4° 32′ N. and Long. 8° 25′ E. At the commencement of this century, it constituted one of the ordinary marts for the slave trade; but in proportion as this odious traffic declined, a more lucrative, if not extensive, commerce with this country has imperceptibly taken its place; our cotton and other home manufactures being received in exchange for exports of native produce. The entrance of this river is 10 miles in breadth, but contracts in size as it proceeds toward the interior, dividing, at 35 or 40 miles from its mouth, into two divergent branches; the first, or the one of the greatest magnitude, known as Cross River, flows from the northward for several hundred miles.
>
> The government of the Old Callebar towns is a monarchical despotism, rather mild in its general character, although sometimes severe and absolute in its details. The king and chief inhabitants ordinarily constitute a court of justice, in which all country disputes are adjusted, and to which every prisoner suspected of capital offences is brought, to undergo examination and judgment. If found guilty, they are usually forced to swallow a deadly potion, made from the poisonous seeds of an aquatic leguminous plant, which rapidly destroys life. This poison is obtained by pounding the seeds and macerating them in water, which acquires a white milky colour. The condemned person, after swallowing a certain portion of the liquid, is ordered to walk about until its effects become palpable. If, however, after the lapse of a definite period, the accused should be so fortunate as to throw the poison from off the stomach, he is considered as innocent, and allowed to depart unmolested. In native *parlance* this ordeal is designated as "chopping nut." Decapitation is also practised, but not so much amongst criminals as the former process, being more employed for the immolation of the victims at the funeral obsequies of some great personage.
>
> Many cruel and superstitious ceremonies occur upon the death of any influential personage, whether male or female. They mourn for some weeks, which is indicated by their binding a black silk handkerchief across the forehead, and neither washing their body nor changing their clothes; being therefore literally in sackcloth and ashes during the allotted period. Two or three days elapse after the inhumation of the body, when several guns and muskets are fired off, and a proportionate quantity

of slaves decapitated to accompany the deceased into the next world. Wives, friends, and confidential servants alike share the same fate, if the departed individual be a man of consequence. Upon the death of Duke Ephraim, one of the former kings of Old Callebar, some hundreds of men, women, and children, were immolated to his manes—decapitation, burial alive, and the administration of the poison-nut, being the methods resorted to for terminating their existence.*

Daniell's account may be compared to that of Donald Simmons, an anthropologist, who conducted ethnological research in Calabar Province in 1952-53. Simmons writes the following (2, 3):

The Efik of Calabar Province consult diviners to ascertain the cause of an unfortunate event, the verity of an allegation, the guilt or innocence of an individual, or the perpetrator of a heinous dead. Unlike some West African tribes which reportedly rely on one or two divinatory methods, the Efik possess several forms of divination.

Certain distinctions between divination, ordeal, and omen may be made. Divination is an attempt to discover something that cannot be ascertained through empirical experience or ordinary linguistic means. An ordeal is a form of divination in which the accused risks bodily injury.†

The Efik believe that the *esere* or Calabar bean possesses the power to reveal and destroy witchcraft. A suspected person is given eight of the beans ground and added to water as a drink. If he is guilty, his mouth shakes and mucus comes from his nose. His innocence is proved if he lifts his right hand and then regurgitates. If the poison continues to affect the suspect after he has established his innocence, he is given a concoction of excrement mixed in water which has been used to wash the external genitalia of a female. If a person dies from the ordeal his body is usually thrown into the forest after the eyes have been removed. A type of clay pot called *eso ntibe* which

*An early account of the use of the Calabar bean may be found in Holman's *Voyage to Old Calabar* (edited with notes by Donald C. Simmons, American Association for African Research, Publ. no. 2, Calabar, 1958). Holman visited Old Calabar in 1828 and wrote (p. 23):

The brig *James,* from Liverpool, arrived this afternoon. About eight in the evening, a Calabar man was brought on board from the Kent's oilhouse; he wanted to be secreted until we sailed, as he wished to make his escape; he said, his master wanted to cut his head off, or to make him chop nut, i.e. to oblige him to eat a poisonous nut which produces speedy death, because he had free-mason (meaning witchcraft), and that his master had been sick ever since he had last flogged him.

Thomas J. Hutchinson also mentions the effects of the Calabar bean: "Him do dis," said one of the Kalabar gentlemen, describing to me its effects; and in the words as well as the action suited to them there was a graphic power impossible for me to transfer to paper; "him do dis, soap come out of him mout, and all him body walk," —a most perfect description of the frothing from the mouth, and the convulsive energy of the whole frame (*Impressions of Western Africa. With Remarks on the Diseases of the Climate and a Report on the Pecularities of Trade up the Rivers in the Bight of Biafra,* by Thomas J. Hutchinson, Esq., Her Britannic Majesty's Consul for the Bight of Biafra and the Island of Fernando Po [London, Longman, Brown, Green, Longmans, & Roberts, 1858]), Frank Cass & Co. Ltd., 1970.

†For an excellent general account of ordeals and ordeal poisons, see (6).

has several holes and is normally used for drying shrimps is put over its head. Corpses of witches were sometimes buried with the face to the ground, or the corpse was burnt and the ashes buried. In this way the Efik hoped to prevent the return of the *ekpo ifot* or 'witch ghost' to hinder and 'wreak havoc' among the living. If trouble continues to occur in a compound the people may suspect a recently deceased relative of being an undetected witch. The relatives exhume the corpse to inspect the process of decay. If it appears fresh, with little sign of decay, the deceased is believed to have been a witch and the corpse is burnt.*

INFLUENCE OF THE UNITED PRESBYTERIAN MISSION AND THE BRITISH CONSUL ON LIFE IN CALABAR

August 1, 1838, was the day of emancipation fixed for the slaves of the West Indian islands. Many expected a time of great crisis, but the event passed peaceably. Among others, the Scottish Missionary Society had begun mission work in Jamaica in 1800. When the slaves were freed the daring thought came to the missionaries that among their converts might be found men willing and able to take the Gospel back to their own people in Africa.

The Reverend Hope Masterton Waddell (Fig. 4a) had the opportunity of reading Sir T. F. Buxton's book, *The Slave Trade and Its Remedy,* which dealt fully with the whole subject. When the Presbytery met in July 1841 they were convinced that the time had come to go forward (7).

Earlier, the missionaries had discussed Calabar as a site. It happened that just at this time a Queen's ship was there to make a treaty with the chiefs for abandoning the slave trade and receiving missionaries. Letters from two of the Efik notables on that occasion largely on account of their desire for schooling seemed an added call to the Jamaican missionaries. To Commander Raymond of the naval vessel King Eyamba wrote:

> Now we settle treaty for not sell slaves, I must tell you something I want your Queen to do for we. Now we can't sell slaves again, we must have too much man for country, and want something for make work and trade, and if we could get seed for cotton and coffee we could make trade. Plenty sugar cane live here, and if some man come teach we way for do it we get plenty sugar too, and then some man must come for teach book proper and make all men saby God like white men, and then go on for same fashion.

The missionaries got in touch at once with Captain Beecroft, Governor of Fernando Po, the Spanish island lying off the mouth of the Calabar

*Interestingly enough, the Efiks, according to Simmons (3), have a second "ordeal poison" called *mbiam.* As far as I know, it has never been investigated chemically. Nothing is known about its source or possible toxicological action.

Fig. 4

(a) Hope M. Waddell (1804-1895).

(b) John Hutton Balfour (1808-1884), from a lithograph by L. Ghémar in the Wellcome Institute. Courtesy of the Wellcome Trustees.

Fig. 5 Calabar River as seen from Duketown. Photograph by the author.

River. His reply was that he had interviewed the chiefs of Calabar and that they eagerly awaited the missionaries' coming (7).

Waddell set sail for Scotland in January 1845, and, after a tedious voyage, in which he was shipwrecked on Grand Cayman, he reached his destination and began to press the claims of Africa. In the first year, £4000 were raised for its support. Mr. Jamieson, a Liverpool merchant, gave the use of a brigantine, the *Warree,* and £100 a year for working expenses. Expectation ran high. The *Warree* set sail on January 6, 1846. The missionary party consisted of the Reverend Hope Waddell, Mr. Samuel Edgerley, printer and catechist from Jamaica, Mrs. Edgerley, experienced as a teacher, Andrew Chisholm, carpenter, a brown man, Edward Miller, teacher, and George, Mr. Waddell's colored boy. On April 10, 1846, after more than three months of weary voyaging, they dropped anchor off Duke Town. This marked an important turning point for Old Calabar.

The evening of their arrival, the missionaries went ashore to pay their respects to King Eyamba of Duke Town. His "palace" was a two-story iron house which had been constructed for him in Liverpool. Its furnishings were beautiful, varied, and to a large extent inappropriate. In his yard were seen mahogany chests of drawers, puncheons of rum, hogsheads of tobacco, iron pots, bales of Manchester cloth, crates of earthenware, and, although there were neither roads nor horses, two four-wheeled carriages. Eyamba himself was a large, coarse-faced and coarse-natured man, a keen trader with a commercial interest in the Gospel (8).

In a few days, arrangements were made in regard to the site of the mission station.

"I look long time for you," said Eyamba, "glad you come now to live here; look about and chose what place you like for make house. . . I glad you come. That palaver done."

Next Sabbath a Bible from friends in Scotland was presented to the king. He received the missionaries in state in an elegant apartment. Eyamba, in satin hat and feathers and waist cloth, and beads and brass, strutted in front of the large mirrors which hung about around the room. Two peacocks did likewise. By-and-by Eyamba sat down on a chair of solid brass under a canopy. Four sofas were provided for the missionaries. The Bible was presented, the object of the mission explained, and Eyamba, proud of his good fortune and vaguely expectant of better fortune, thanked the missionaries and the God who sent them (9).

It was not long before the missionaries started schools, produced a written version of the Efik language, and eventually suppressed various Efik customs particularly repugnant to Europeans. In this they had considerable help from the British Consul for the Bight of Biafra, a post first established in 1849. The Consul was resident on the nearby island of Fernando Po and acted during part of this period as its governor. He could be appealed to in serious disputes between European and native interests when he duly appeared in a British naval vessel to adjudicate. His judgments were accepted without question as having behind them, so far as the inhabitants of Old Calabar were concerned, the sanction of British naval power. This power had been effectively displayed in the movement for the suppression of the slave trade (2).

From 1861 to 1863, Richard Burton was the British Consul. He spent very little time on the island but visited a great number of places on the mainland of Africa. During all these trips he accumulated notes enough to fill four two-volume studies totaling 2,500 pages. He also found time to compile a collection of native proverbs, his 450-page *Wit and Wisdom in West Africa.* The writing of almost 3,000 pages in three years provides one of the most extraordinary records of industry and observation in the annals of the British Foreign Office (10). Burton disliked missionaries, and was one of the best observers and recorders of native customs in all history. In his *Wanderings in West Africa* (11) he has given a glowing and vivid description of Bonny, but he apparently missed nearby Old Calabar. What a tragedy for posterity that this incomparable narrator has not given us an account of the customs of the Efik people of a hundred years ago.*

The authority of the British Consul gradually came to play a steadily increasing part in Old Calabar politics until eventually replaced by the British Protectorate of 1884-85. In 1878 Consul Hopkins visited Calabar, and on the instruction of Lord Salisbury, then Foreign Secretary, drew up a treaty with the king and chiefs of Duke Town to put an end to the murderous customs of the country. The agreement consisted of fifteen articles, of which the following two concern the Calabar bean:

Article three: Any person administering the *esere* bean, whether the person taking it dies or not, shall be considered guilty of murder, and shall suffer death.

Article four: Every person taking the *esere* bean wilfully, either for the purpose of committing suicide, or for the purpose of attempting to prove their innocence of any crime of which they may have been accused, shall be considered guilty of attempted murder, and shall be fined as heavily as the circumstances will permit, and shall be banished from the country.

In spite of this agreement, occasional instances still occur when individuals desire to show their innocence of any imputation of witchcraft by *esere.*

Nigerian law forbids the use of the Calabar bean, and possession of the bean entails fine and imprisonment. However, some Efik keep one bean in their pocket book or with a cache of coins in order to prevent

*The early British consuls at the Bight of Biafra were an odd mixture of traders, bush lawyers, and medical men. Burton, in spite of his frequent absenteeism, played a role in reforming the only existing institution for hearing grievances between Africans and Europeans. Most noteworthy was the clause of the new regulation which abolished trust at Old Calabar and the Cameroons and made the consul a final arbiter of disputes (See C. W. Newbury's Introduction to *A Mission to Gelele, King of Dahome* by Sir Richard Burton [London, Routledge & Kegan Paul], 1966).

witches from 'drinking' the money; the rainbow (Efik *akpatre*) signifies a witch is 'drinking' money from the house located at the point where the rainbow intersects the ground (3).

When the present writer visited Calabar in 1963 he had a good deal of difficulty finding anyone willing to show him the plant from which the bean is obtained. Also he was strongly warned not to take pictures of the big drum in front of the palaver house of the still existing Egbo Society.

STUDIES OF THE CALABAR BEAN IN EDINBURGH DURING THE NINETEENTH CENTURY

Long before Consul Hopkins and the still very active Church of Scotland mission had put an end to the *esere* ordeal, the story of the Calabar bean had entered its second and scientific phase. This, in the beginning, took place in Edinburgh.

The Edinburgh Medical School was started about the year 1670 by a group of doctors who decided to lay out an herb garden and to start a college of physicians. Most drugs were made from plants and the only way to get reliable drugs was to grow the plants. A Professor of Botany was appointed in 1676 to teach the subject to medical students, and this arrangement continued for 64 years. In 1738 Dr. Charles Alston was appointed Professor of both Botany and Materia Medica and started giving courses of lectures on these two subjects. Materia Medica was a large and important subject in those days, since the doctor was responsible for the preparation of his own medicines (12).

The medical men of Edinburgh continued to be good botanists. Among them was John Hutton Balfour (1808-1884) (13). Balfour, seen in Fig. 4, was of Scottish ancestry, and a number of his forebears were famous scientists. He was born in Edinburgh and studied at the universities of St. Andrews and Edinburgh. He had prepared first for ordination in the Church of Scotland but gave this up in order to become a botanist and a distinguished biologist of the new school. Balfour studied in continental medical schools and started medical practice in Edinburgh in 1834. He was prominent in establishing the Botanical Society of Edinburgh in 1836 and the Edinburgh Botanical Club two years later. In 1840 he began to give extra-academical lectures on botany in Edinburgh, and in 1841 he became Professor of Botany at Glasgow and gave up medical practice. In 1845 he transferred to Edinburgh as Professor of Botany. Not much else is known about him. He was known to his students as "Old Woody Fibre."

In 1845 Balfour was serving as keeper of the Royal Botanic Gardens and the Queen's botanist for Scotland. He was the first in Edinburgh to hold classes for practical instruction in the use of the microscope. He is mostly known as a teacher and writer of textbooks. His original work is not extensive—the best-known is his "Description of the plant which produces the Ordeal Bean of Calabar" (14).

It is only natural that a man with Balfour's background should have close contact with the missionaries of Old Calabar; among them were some of his own pupils. The Scottish missionaries reported their findings, including their observations of the Calabar bean and its origin, in *The Missionary Record* which started its publication in 1846. Hope Waddell himself brought beans to Edinburgh, where they grew but never flowered in the garden of which Balfour was keeper. From his observations there and from accounts and specimens sent from Calabar he was able to give the first botanical description of the plant.* The following is abstracted from the publications of Balfour (14) and Fraser (15):

> The following is a description of the plant the *Physostigma venenosum* (Balfour) (Fig. 6), found in the neighbourhood and to the west of Calabar Proper, in the territory of a tribe called Eboe (Ibio, Aboi, Abo, or Ibo), who inhabit a region extending westward from the source of the Niger. The plant is described as a *runner*, climbing on the bushes and trees in its neighbourhood; and this character was well shown in the plants which have been cultivated in the Edinburgh Botanic Garden. Its habitat is on the sides and edges of streams, thriving best on swampy river banks. The ripe beans are frequently dropped into the stream, and carried down to Calabar in considerable quantity, so that the natives obtain their supply principally from the banks of the river, irrespective of what is used judicially by the 'idol-priests.' From this source also was derived one of the parcels of the bean for which I am indebted to the Rev. John Baillie. The plant is perennial, probably producing fruit only after some years. The fruit ripens at all seasons of the year, in common with that of many other tropical plants; but the most abundant crop is produced in the rainy season from June to September inclusive. Natur. order-Leguminosae. Suborder-Papilionaceae. Tribe-Euphaseolae.

*Professor Daryll Forde in a communication to the author mentions the following:

"In 'Christian Missions in Nigeria 1841-1891' by J. F. A. Ajayi (London, Longmans, 1965) the author states on p. 279 that five of the eleven volumes of Hope Waddell's private diaries are in the National Library of Scotland, Edinburgh. Hope Waddell is not mentioned in 'The Church Missionary Record'—the periodical of the Church Missionary Society (Anglican). There is another periodical 'The Missionary Register,' running from 1813 to 1856, being a cooperative effort between various non-conformist missionary societies. The British Museum Library is the only location for a complete run of The Missionary Register and there are partial runs in the Bodleian Library, Oxford, and in the Cambridge University Library."

Fig. 6 *Physostigma venenosum* Balf. *1.* Young pod of *Physostigma venenosum,* with three ovules. *2.* Full-grown pods of *Physostigma venenosum.* *3.* Seed of ordeal bean seen laterally. *4.* The same, showing the sulcate and extended hilum on the convex edge. All figures are natural size. Drawings from Balfour's original publications.

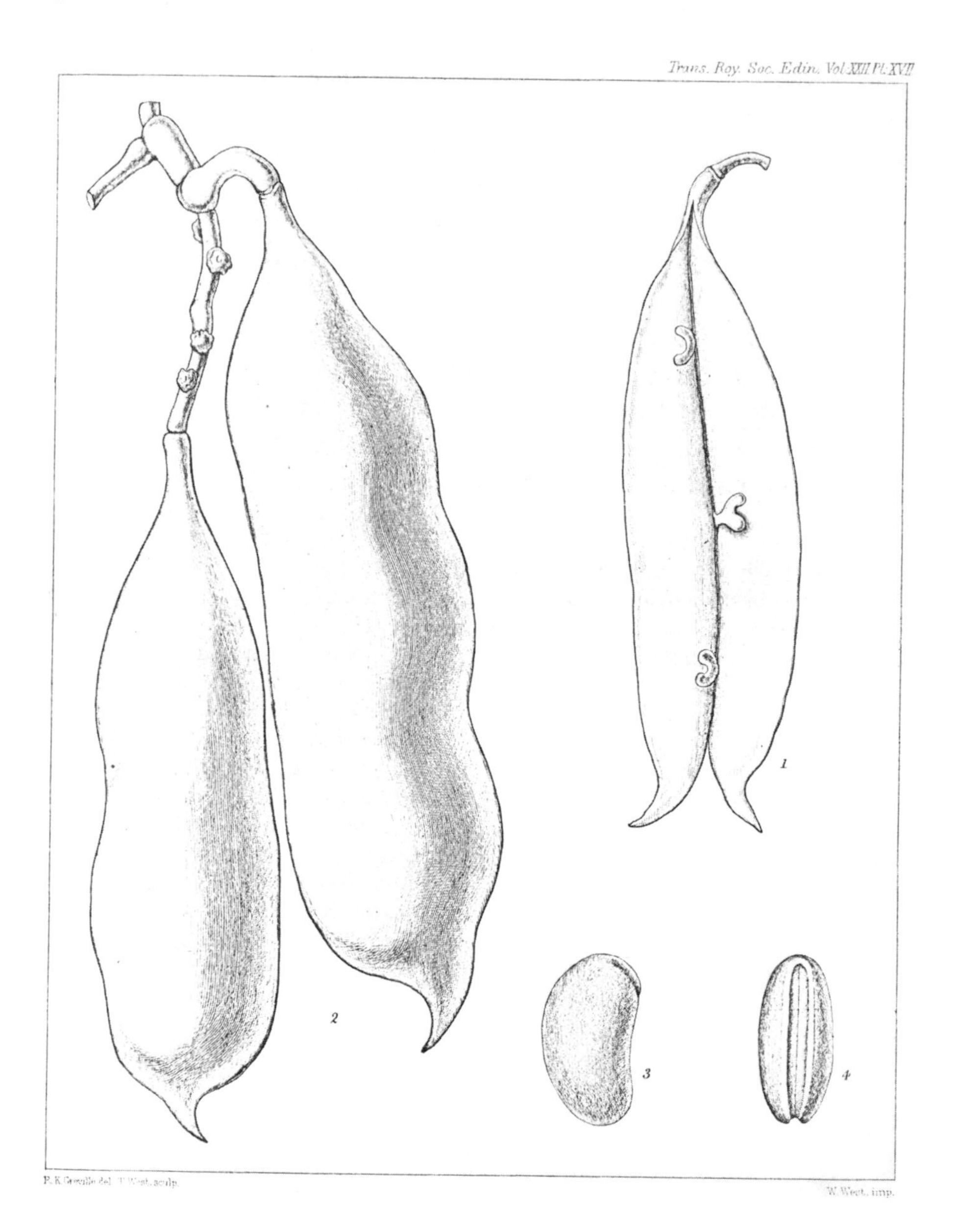

Genus and Species-Physostigma venenosum. It has generic characters closely resembling those of *Mucua* and *Phaseolus,* but is separated from the former by the characters of the flower and pod, and from the latter by its seed. It has accordingly been placed by Professor Balfour in a separate genus, *Physostigma,* and is itself the only known species, *venenosum.*

Generic Characters: Root, spreading with numerous fibrils, often having small succulent tubers attached. Inflorescence, axillary; on pendulous multifloral racemes; rachis of each raceme zig-zag and knotty. Calyx, campanulate, four cleft at apex, the upper division being notched and its segments ciliated. Corolla, papilionaceous; veined with a pale pink, having a purplish tinge, and curved in a crescentic manner. Stamens, ten, diadelphous. Pistil, more than one. Stigma, blunt, covered by a remarkable ventricular sac or hood, which extends along the upper part of the convexity of the style, having a resemblance to an "admiral's hat set in a jaunty manner." Legume, dark brown and straight, when mature, about seven inches in length, elliptico-oblong, with an apicular curved point, and with outer and inner integument easily separable. Seeds two or three, separated from each other by a woolly substance.

Characters of the Seed or Bean. The part of the plant of interest on account of any known properties is the *seed* or *bean.*

Synonyms. Eséré nut; the bean of the Etu esere; chop nut; the ordeal bean of Calabar.

Form. Irregular reniform, having the appearance of a somewhat flattened fusiform body bent on one of its edges. It has two margins, a shorter or concave, and a longer or convex, and two flattened surfaces. Extending along the convex margin is a sulcus, having a minute aperture near one of its extremities.

Color. As obtained from Calabar, the beans have a grey color, and are incrusted with earthy matter. This is readily removed by washing, and a somewhat shining integument is exposed of various shades of brown, ranging from a light coffee to an almost perfect black.

Dimensions. The average length is $1\frac{1}{16}$ of an inch, and varies from 1 to $1\frac{1}{2}$ inches. The average breadth is $\frac{12}{16}$ of an inch, and varies from $\frac{10}{16}$ to $\frac{14}{16}$. These measurements are the extremes in each direction, and the sides slope from the greatest breadth to the comparatively narrow extremities. The average thickness or breadth from one flattened side to the other is $\frac{8}{16}$, the maximum $\frac{11}{16}$, and the minimum $\frac{6}{16}$ of an inch.

The toxicological studies of *Physostigma venenosum* start with Sir Robert Christison (1797-1882) (16). He was born and educated in Edinburgh, where he graduated in the medical faculty in 1819. Somewhat later he went to Paris for further studies, especially in chemistry, under several prominent teachers. He also came in contact with the famous toxicologist M. J. B. Orfila. At the early age of twenty-four, Christison was appointed to the newly established chair of Medical Jurisprudence in Edinburgh, the first of its kind in Great Britain. He held this position for ten years, whereupon he was transferred to the Professorship of Materia Medica and Therapeutics, which he occupied until 1877. Simultaneously he was Professor of Clinical Medicine. His scientific activity had reference to the different aspects of his academic position. Thus, he worked especially in the field of toxicology. Among other things, he worked on arsenic, lead, opium, and hemlock. He ascertained that the action of hemlock and its alkaloid, coniine, were identical, coming to the conclusion after analyzing them that the action on the spinal cord was "the counterpart of the action of *nux vomica* and its alkaloid strychnia."

In later life Christison became interested in cocaine and carried out experiments on himself at the age of 78. He must have been an energetic old man, since these experiments involved walking 15 miles at four miles an hour and climbing 2900 feet up a mountain. He found that cocaine removed extreme fatigue and made him temporarily less hungry and thirsty, but eventually his appetite returned.

Christison's best-known contribution to toxicology is, however, an experiment on himself with the calabar bean which he apparently obtained from Waddell, and of which he gave the following account (17):

> It appears that many persons think it an easy task to investigate experimentally the physiology of poisoning. But they are assuredly mistaken. A long apprenticeship must be passed before any one can observe with accuracy the phenomena of the action of poison.
>
> These cautions are prefatory to the remark, that it is a matter of great nicety to apprehend the deceptively simple manifestations of the action of the ordeal-bean on the lower animals. Scarcely do signs of uneasiness appear after a fatal dose has been given, when the animal becomes in quick succession languid, prostrate, flaccid, immovable; respiration, now faint, speedily ceases; and death is complete. It may thus appear to die insensible and comatose. But that is not the case. So long as the power of expression remains, amidst the swiftly advancing languor, signs of sensation may be elicited. Or we might infer from the phenomena that it dies of paralysis of the voluntary and respiratory muscles. But this too is in all probability not the fact. For, on dissection immediately after respiration ceases, the heart is found in a state of paralysis; and it is evident that a quickly increasing paralysis of the heart not only explains the mode of death, but might likewise account for the antecedent muscular weakness and flaccidity.
>
> These effects were well exemplified in the first experiment I tried, when twenty-one grains of fine powder, made into an emulsion with two drachms of water, were secured in a cavity in the subcutaneous, cellular tissue of the flank of a rabbit. For three minutes there was no appreciable change. But the animal then evidently became weaker, especially in the hind legs. Its feebleness quickly increased, and was attended with slight irregular twitches of the muscles of the trunk and extremities, and occasional twitching of the head backwards. But sensation remained; for the animal struggled a little when held up by the ears, and resisted attempts to shove it from behind. In four minutes, when put upon the side, it lay in that position; which the rabbit always vehemently resists so long as it is able. The trunk and extremities immediately afterwards became quite flaccid. Respiration ceased in five minutes certainly; probably indeed sooner; but the precise time could not be fixed,

owing to continuance of slight muscular twitches. The chest being immediately opened, the heart was seen pulsating slowly, feebly, and inefficiently for ten minutes; and when its cavities were then perforated, the left side gave out a much brighter blood than the right, showing that the circulation, owing to paralysis of the heart, had not been maintained after respiration had ceased. The muscles of voluntary motion contracted at this time vigorously under the stimulus of galvanism, and continued to do so twenty-five minutes after death. . .

Having ascertained the mode of death from the action of the ordeal-bean, I did not consider it advisable to study further the details of its action by means of experiments on animals, because I had been fully informed as to this in a more precise manner by an experiment made with the bean in my own person. I shall conclude this notice with an account of what I experienced; and I trust the details will not appear needlessly minute, as they seem to me to establish an action of a very singular kind in the case of this poison, and one of which we might discover other instances among known poisons, had we equally precise opportunities of determining the true phenomena.

Having some doubts whether I had obtained the true ordeal-poison, as it tasted so like an eatable leguminous seed, I took one eighth of a seed or six grains, about an hour after a very scanty supper. During an hour that I passed in bed reading, I could observe no effect whatever, and next morning I could still observe none. I am now satisfied, however, that a certain pleasant feeling of slight numbness in the limbs, like that which precedes the sleep caused by opium or morphia, and which I remarked when awake for a minute twice or thrice during the night, must have been owing to the poison.

On getting up in the morning I carefully chewed and swallowed twice as much, viz. the fourth of a seed, which originally weighed forty-eight grains. A slight giddiness, which occurred in fifteen minutes, was ascribed to the force of the imagination; and I proceeded to take a warm shower bath; which process, with the subsequent scrubbing, might take up five or six minutes more. The giddiness was then very decided, and was attended with the peculiar indescribable torpidity over the whole frame which attends the action of opium and Indian hemp in medicinal doses. Being now quite satisfied that I had got hold of a very energetic poison, I took immediate means for getting quit of it, by swallowing the shaving water I had just been using, by which the stomach was effectually emptied. Nevertheless I presently became so giddy, weak, and faint, that I was glad to lie down supine in bed. The faintness continuing great, but without any uneasy feeling, I rung for my son, told him distinctly my state, the cause, and my remedy—that I had no feeling of alarm, but that for his satisfaction he had better send for a medical friend. Dr. Simpson, who was the nearest, reached me in a few minutes, within forty minutes after I ate the seed, and found me very prostrate and pale, the heart and pulse extremely feeble and tumultuously irregular; my condition altogether very like that induced by profuse flooding after delivery; but my mental faculties quite entire, and my only sensation that of extreme faintness, not, however, unpleasant. Dr. Simpson judged it right to proceed at once for Dr. Douglas Maclagan as a toxicological authority, and returned with him in a very few minutes.

In his absence, feeling sick, I tried to raise myself on my elbow to vomit, but failed. I made a second more vigorous effort, but scarcely moved. At once it struck me—"This is not debility, but volition is inoperative." In a third effort I was more nearly successful; and in the fourth, a resolute exercise of the will, I did succeed.

But I could not vomit. The abdominal muscles acted too feebly; nor were they much aided by a voluntary effort to make them act. I then gave up the attempt, and fell back, comforting myself with the reflection that vomiting was unnecessary, as the stomach had been thoroughly cleared. At the same time the sickness ceased, and it never returned. There were now slight twitches across the pectoral muscles. I also felt a sluggishness of articulation, and, to avoid any show of this, made a strong effort of the will to speak slowly and firmly, through fear of alarming my son, who was alone with me.

Dr. Maclagan, on his arrival, thought my state very like the effects of an overdose of aconite. Like Dr. Simpson, he found the pulse and action of the heart very feeble, frequent, and most irregular, the countenance very pale, the prostration great, the mental faculties unimpaired, unless perhaps it might be that I felt no alarm where my friends saw some reason for it. I had, in fact, no uneasy feeling of any kind, no pain, no numbness, no prickling, not even any sense of suffering from the great faintness of the heart's action; and as for alarm, though conscious I had got more than I had counted on, I could also calculate, that, if six grains had no effect, twelve could not be deadly, when the stomach had been so well cleared out.

Presently my limbs became chill, with a vague feeling of discomfort. But warmth to the feet relieved this, and a sinapism over the whole abdomen was peculiarly grateful when it began to act. Soon afterwards the pulse improved in volume, but not in regularity. I was now able to turn in bed; and happening to get upon the left side, my attention was, for the first time, directed to the extremely tumultuous action of the heart, which compelled me to turn again on the back, to escape the strange sensation. Two hours after the poison was swallowed, I became drowsy, and slept for two hours more; but the mind was so active all the while, that I was not conscious of having been asleep. On awakening, the tumultuous action of the heart continued. In an hour more, however, I took a cup of strong coffee; after which I speedily felt an undefinable change within me, and on examining the condition of the heart, I found it had become perfectly and permanently regular.

For the rest of the forenoon, I felt too weak to care to leave my bed; and on getting up, after a tolerable dinner, I was so giddy as to be glad to betake myself to the sofa for the evening. Next morning, after a sound sleep, I was quite well.

Christison's work on the Calabar bean was followed up by Thomas Richard Fraser (1841-1920) who at the age of twenty-one discovered that an extract of the bean acted on many different organs, such as the pupils, the central nervous system, the heart, the glands, the voluntary muscles, and the intestines. He became Christison's assistant and in 1877 succeeded him as Professor of Materia Medica and Therapeutics. The following year he also became physician to the Royal Infirmary, the combination of materia medica with practical medical work being still maintained (12, 16).

The following is quoted from Fraser's early work on the Calabar bean (15):

The two varieties of symptoms following the administration of the bean may be harmonized in the same way. It exerts a special influence on the spinal cord; when

Fig. 7

(a) Robert Christison (1797-1882), from an engraving in the Wellcome Institute. Courtesy of the Wellcome Trustees.

(b) Sir Thomas Richard Fraser (1841-1919), from an album of the International Medical Congress (1881) at the Wellcome Historical Medical Museum.

(c) Douglas Argyll Robertson (1837-1909).

(d) Frederick Jolly (1884-1904).

(a) (b)

(c) (d)

this is limited in extent and energy, the only marked effect is paralysis, and death is caused by the extension of this paralysis to the muscles of respiration, causing death by *asphyxia.* When, on the other hand, this spinal action is more extensive and energetic, the heart is affected, its contractions cease, and death occurs by *syncope.*

The kernel or embryo of the bean of *Physostigma venenosum* has the following actions:

1. It acts on the spinal cord by destroying its power of conducting impressions.
2. This destruction may result in two well-marked and distinct effects: (a) In muscular paralysis, extending gradually to the respiratory apparatus and producing death by *asphyxia;* (b) In a rapid paralysis of the heart, probably due to the extension of this action to the sympathetic system, thus causing death by syncope.
3. A difference in dose accompanies this difference in effect.
4. This action does not extend to the brain proper *pari passu* with the action on the spinal cord. The functions of the brain, may however, be influenced secondarily.
5. It also produces paralysis of muscular fibre, striped and unstriped.
6. It acts as an excitant of the secretory system, increasing more especially the action of the alimentary mucous glands.
7. Topical effects follow the local application of various preparations; they are destruction of the contractility of muscular fibre, when applied to the muscles, and contraction of the pupil, when applied to the eyeball. . .

When the extract is applied to the eyeball it immediately causes a copious secretion of tears, and in about five minutes a distinct contraction of the pupil confined to the side of the application. In about thirty minutes after the application the pupil becomes a mere speck, but still retains a certain degree of mobility. It continues in this state for twelve or fourteen hours, but greater or less degree of contraction of the pupil may persist for five or six days. A slight headache and dimness of vision with myopia in the affected side, are almost always produced, but these only continue for one or two hours at the commencement . . . When the extract is applied to the edges and outer surface of the eyelids, there is produced, in addition to the contracted pupil, a degree of immobility of the eyelids. . . No effect was produced on the pupil by friction on the temples or over the eyebrows with any preparation of the kernel.

The antagonism between physostigmine and atropine which Fraser found to exist with respect to their action on the pupil and the heart rate also made it possible that a physiological antagonism might be found for other effects. This led Fraser to perform a careful study in which he showed that the lethal effect of physostigmine could be prevented by atropine (18).

This was outlined in a lecture which dealt with drug antagonism (19) and which has been carefully commented upon by Gaddum.

Fraser pointed out that some antidotes destroy the poison in the stomach but that others act after absorption. He emphasizes the fallacies which complicate the interpretation of clinical observations on the effects of antidotes in those cases where it is not possible to know whether the patient would have died if no antidote had been given. He then describes his own work on the antagonism between physostigmine and atropine when both are injected subcutaneously in rabbits. This antagonism was itself a new concept at that time, but Fraser was interested in its quantitative aspect. After a given dose of physostigmine, there is a range of doses of atropine that will save life. If too little is given, the rabbit dies of physostigmine poisoning, and if too much, it dies of the effects of atropine. As the dose of physostigmine is increased this range gets smaller and if the dose of physostigmine is more than 3½ the lethal dose, life cannot be saved by any quantity of atropine.

Fraser plotted the results on a graph, with atropine on a horizontal scale and physostigmine on a vertical scale, and drew a line which separated combinations of doses which led to death from those which led to recovery. No one else made graphs like this for about half a century, and then Dr. S. Loewe reintroduced them and called them isobols. They still provoke discussion.

Fraser's experiments showed that small doses of atropine saved life by opposing the lethal action of physostigmine, but he also obtained evidence that small doses of physostigmine caused death by increasing the lethal action of atropine. As pointed out by Gaddum (12), unthinking persons may perhaps find this result surprising, but it was just the kind of thing that Fraser expected, and his experiments were designed to detect it. He points out that as both atropine and physostigmine possess a number of separate actions, it was not unreasonable to anticipate that several of them are not mutually antagonistic; in some cases the two drugs might be expected to work together.

At the end of this lecture Fraser compared the use of antidotes with the treatment of disease; rabbits can be killed by excess of antidote, and patients may be killed by excessive treatment, but, in both cases, careful work can determine the best dose (12).

Fraser's work on the pupil of the eye with the Calabar extract was followed by that of Douglas Argyll Robertson, 1837-1909 (see McKay, 20a,b). Robertson was born in Edinburgh and died at Gondal, India. His full name was Douglas Moray Cooper Lamb Argyll Robertson. His father, John Argyll Robertson, was a well-known lecturer in surgery in the Extra-Academical School of Edinburgh. The distinct preference of Robertson's father for ophthalmic surgery greatly influenced his son, who from the beginning of his career devoted himself to this science. From 1867 to 1870 he was Assistant Ophthalmic Surgeon to the Edinburgh Royal Infirmary, in 1870 he was appointed full Ophthalmic Surgeon and in 1882 sole Surgeon until 1897, after which he was Consulting Surgeon. At the time of Robertson's entrance into his chosen career, the founders of modern ophthalmic surgery, Bowman and von Graefe, were at the height of their fame. Robertson was for a time their pupil in Berlin.

Argyll Robertson published several papers dealing with eye diseases. One of them dealth with the pupillary condition generally known as the "Argyll Robertson phenomenon" and contributed to making him immediately famous. Sir Anderson Citchett said, about this sign of neurosyphilis, "It was better to be an Argyll Robertson pupil than to *have* one."

On a hint from his friend, Sir Thomas Fraser, Robertson turned his attention to the Calabar bean as an important agent in the treatment of eye conditions. His paper, "On the Calabar bean as a new agent in ophthalmic medicine," appeared in 1863 in both the *Edinburgh Medical Journal* and the *London Ophthalmic Journal* (21).

The need of a miotic was recognized by the early ophthalmologists, and as late as 1861, Stellwag von Carion, in his famous textbook, lamented the fact that "we possess no means of effecting contraction of the pupil" (4).

The following is an abstract from Robertson's paper of 1863 (21):

> For more than a year past I have recognized the numerous advantages that would flow from the discovery of a substance which, when applied to the conjunctiva, should produce effects exactly opposite to those well known to result from belladonna or atropine; which should stimulate the muscle of accommodation and the sphincter pupillae as the above-named remedies paralyze them. With the view of discovering such an agent, I endeavoured to ascertain from experiments of my own, and from the writings of previous observers, whether any of the common vegetable principles possessed this property. These investigations were, however, productive of no satisfactory results, until my friend Dr. Fraser informed me that he had seen contraction of the pupil result from the local application of an extract of the ordeal bean of Calabar. I resolved to investigate the action of this substance, and, above all, to ascertain whether it exerted any influence on the accommodation of the eye. With some difficulty I got a few Calabar beans, and, with the view of obtaining their active principles in a convenient form, prepared from them three extracts of varying strengths.

Argyll Robertson then proceeded to describe a series of detailed experiments on his own eyes and finally concluded:

> I have narrated these experiments somewhat in detail, so as to elucidate, as far as possible, the method of action of this new agent and its energy. These experiments prove that the local application of the Calabar bean to the eye induces,—*first,* A condition of shortsightedness. That this is present, and the cause of the indistinctness of distant vision cannot be doubted, as it is relieved by the use of concave glasses. The

fact that objects appear larger and nearer than natural may be atributed to the induced myopia. And, *second,* It occasions contraction of the pupil, and sympathetically dilatation of the pupil of the other eye. We further observe that atropine possesses the power of counteracting its effects, and, *vice versa,* that it is capable of overcoming the effects produced by atropine. The first symptom noticed is dimness of distant vision, and shortly after the pupil becomes contracted; the symptoms also subside in the same order, first the derangement of accommodation, and then the affection of the pupil.

Let me now say a few words as to the method of action of the Calabar bean. In respect to its effects on the pupil they might be produced either by causing contraction of the circular fibres of the iris, or by paralyzing its radiating fibres. I am inclined to believe that the contraction of the pupil is due to increased action of the sphincter pupillae, and this chiefly on the ground that the other effects produced by the Calabar bean can only be explained by an induced contraction of the ciliary muscle—the muscle of accommodation; and as the sphincter pupillae and ciliary muscle are both supplied by the ciliary nerves, I think the most feasible explanation of the action of the Calabar bean on the eye is to regard it as a stimulant to the ciliary nerves. In favour of this view we have the feeling of straining in the eye shortly after the physiological effects are produced. The alteration, too, in the accommodation of the eye exhibits much of the character of a spasmodic action; thus we find in experiment *third,* after the second application of the Calabar bean, that the extent of distinct vision is limited to 3 inches, viz., from 6 to 9 inches from the eye, but an hour after distinct vision extends to any distance beyond 5 inches. It has also been observed that the accommodation of the eye is not usually affected in cases where contraction of the pupil is due to lesion of the sympathetic. . .

As regards the cases in which this substance may be applied in practice, it is applicable in all instances where atropine is used to render the examination of the eye more perfect or more simple. This includes two classes of cases, those in which dilatation of the pupil is either necessary or desirable to aid ophthalmoscopic examination, and those in which paralysis of the ciliary muscle is necessary, in order to ascertain the state of the accommodation of the eye.

As seen from the above, Argyll Robertson did not realize the value of Calabar-bean extract as a miotic in glaucoma but did suggest that it be applied after atropine preparatory to examination of the eyes. He also proposed its use for the relief of photophobia and paralysis of the ciliary muscle. Moreover, he believed it to be beneficial in cases of ulceration of the margin of the cornea leading to perforation, and even when prolapse of the iris had occurred.

Ludwig Laqueur (1877) was the first opthalmologist to use the chief alkaloid of the Calabar bean, physostigmine or eserine, therapeutically in glaucoma (22). He noticed that it lowered intraocular pressure temporarily or permanently. From that time on, eserine has been used more and more extensively and occupies a prominent place in the treatment of glaucoma. For a time, certain organophosphorus compounds were great competitors, but since it has been recently found that the latter substances may produce cataracts, many ophthalmologists are reverting to eserine (23).

NINETEENTH-CENTURY STUDIES ON THE CALABAR EXTRACT AFTER PUBLICATION OF FRASER'S THESIS IN 1863

Much work was done on the Calabar bean and its alkaloid content during the remainder of the nineteenth century, and an overwhelming number of papers were published. From our present vantage point it is interesting to observe that many valuable observations were recorded and predictions made that later were proved to be true. For example, Rudolf Lenz (24) in a careful study of blood pressure, pulse recording, using A. Fick's instrumentation, reached the following conclusion.

> The Calabar bean acts upon the last part of the nervous system in the heart, the sympathetic ganglia, in such a way that the inhibition is potentiated. Only this can explain the action of injections of Calabar extract. Cutting of both vagi had no effect.

Lenz's observation was confirmed and extended by Arnstein and Sustschinsky (25), who undertook an investigation to see whether Calabar extract acted on the nerves of the heart. They showed that it did not paralyze the sympathetic nerves but that it actually increased the action of the vagus upon stimulation. They also showed that this potentiation of the action of the vagus could be antagonized by atropine and then produced again by a second injection of Calabar extract.

There was also a discussion of the mechanism of death following physostigmine poisoning, and there were proponents of both circulatory and respiratory failure (26-28). Bauer (28) proved that open airways in experimental animals made them tolerate much higher doses and suggested that spasm on the bronchial muscles may contribute to the cause of death.

All these individual observations did not merge into a general picture of the toxicology of cholinesterase inhibitors, including the Calabar extract, until some 100 years later. Death due to this group of compounds is complex and involves central effects (convulsions and paralysis of respiration), respiratory effects (increased secretion and bronchospasm), and neuromuscular block with early involvement of respiratory muscles. The effects on the circulatory system involves bradycardia, decreased cardiac output, and peripheral vascular phenomena.

Study of the part played by the cardiovascular system has confirmed what many nineteenth-century authors reported (15a, 24, 26-29). When death, due to a large dose, comes quickly, the circulation is still relatively unimpaired when respiration fails. When death is delayed, it is impossible to make such a distinction, since mounting depression and final failure then involve both systems equally. Death appears to be primarily asphyxial in some instances and primarily cardiovascular in others; and in some cases failure of both seems to coincide (30).*

A most important pharmacological finding was made appropriately at the turn of the century by Pal (31), who was engaged in the studies of the action of physostigmine on the gut and, like many other pharmacologists at the time, used curarized animals. Upon injection of physostigmine, Pal observed return of the respiration paralyzed by curare and thus established physostigmine's antidotal effect on the latter drug (Fig. 8).

Fig. 8 Observation by J. Pal, 1900. Respiration of diaphragm in curarized animal is resumed at 10 hrs 52 min (4 min after injection of 0.0025 physostigmine salicylate).

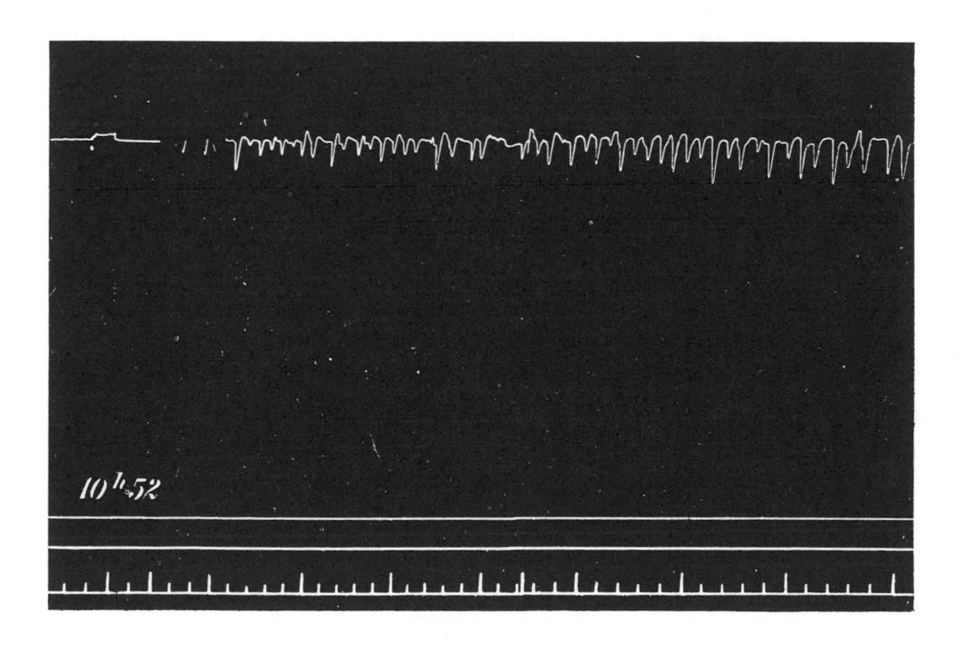

In addition to these pharmacological experiments, the papers by Fraser (15) and Argyll Robertson (21) precipitated a large number of clinical articles about the use of Calabar extract in various diseases. Among these

*Fraser's original paper, quoted above, stated this very clearly. Early attempts to reconcile the divergent findings are to be found in Roeber (32) and Harnack and Witkowski (33).

more or less uncontrolled studies may be mentioned the alleged curative effects of the Calabar bean in the treatment of chorea, tetanus, colic, cholera Asiatica, and tic, as well as its use as a lactagogue, in epilepsy and progressive paralysis of the insane, and in other nervous affections (34-37).* Cases of human intoxications were also reported (38, 39). Thus, for example, 46 cases, of which one proved fatal, occurred in Liverpool in 1864 (40):

> On the afternoon of Thursday, August 11 (1864), forty-two children, poisoned by eating the 'Calabar ordeal bean,' were admitted into the Southern Hospital under the care of Dr. Cameron, and, on the following day, four more cases. Of these forty-six cases, one only proved fatal. It appears that the beans were obtained from amongst a heap of rubbish—the sweepings of a ship which had recently brought a considerable quantity of them as part of her cargo from the West Coast of Africa. These sweepings had been inadvertently deposited on some waste ground near the Hospital, and in the centre of a very populous neighbourhood.

Dr. Cameron seems to have been entirely unfamiliar with the work in Edinburgh because the cases were treated with "emetics and a plentiful supply of warm water and brandy."

Kleinwächter (41) at the ophthalmic clinic in Prague, however, was familiar with Argyll Robertson's experiments with eserine and atropine on the eye and had the courage to use Calabar extract systemically as an antidote in atropine poisoning. His is the earliest published reference to this fact: he was called to a prison to see three prisoners who had drunk a solution of atropine under the impression that it was some kind of liquor. He gave Calabar extract to the one who seemed to be the most severely poisoned and kept the others as controls. In this way he obtained convincing evidence that Calabar extract and atropine were antagonistic.† In another paper from the same year, this time from St. Petersburg, it was observed that the cases of peroral intoxications with the Calabar bean did not necessarily have miosis (42), an observation which has subsequently been confirmed.

*A recent investigation carried out with physostigmine in Huntington's chorea demonstrates, that the alkaloid can improve some of the symptoms in this disease, thus confirming what was observed already in 1875 (Sten Magnus Aquilonius: Svenska Läkarsällskapets Förhandlingar, November 1970).

†Interestingly enough, physostigmine has been suggested as an antidote against modern chemical warfare agents with an atropine-like action (*International Defense Review*/Interavia/II/1969, pp. 170-174).

ALKALOIDS OF THE CALABAR BEAN

In 1863 Thomas Fraser separated from the kernel (the spermoderm of the bean), and later from the excrement of a lepidopterous insect which feeds on the kernel (15b), an amorphous active principle with the properties of a vegetable alkaloid (15a). He proposed the name eserine for this alkaloid, derived from eséré as the ordeal poison is called in Calabar. Some time later he obtained a purer, crystalline form of the alkaloid.

The main alkaloid present in the seeds of the Calabar bean was first isolated in a completely pure form in 1864 by Jobst and Hesse (43), who called it physostigmine (I). One year later it was obtained in crystalline form by Vée, who called it eserine (44). Both names are still used to designate the base. Other alkaloids have been reported as occurring in the seeds, such as calabarine (33), later claimed to have been a mixture of decomposition products (45), eseridine (46, 47), eseramine (45), isophysostigmine (48), physovenine (49) and geneserine (50a,b).

Physostigmine (eserine), $C_{15}H_{21}O_2N_3$ is a tertiary amine readily oxidized by oxygen in the presence of aqueous potassium hydroxide to the red compound rubreserine which is pharmacologically inactive (47, 51, 52). It is readily hydrolyzed by alkali to yield methylamine, carbon

Fig. 9

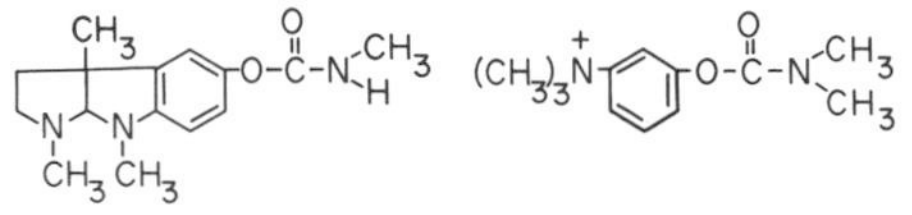

I Physostigmine II Neostigmine

dioxide, and a base, eseroline ($C_{13}N_{18}ON_2$) (45, 53). Methylamine and carbon dioxide are present in the molecule as a carbamate group -$OCONHCH_3$ since, when heated with sodium ethylate in the absence of air, physostigmine yields methyl urethane ($CH_3NHCO_2C_2H_5$) and eseroline, while its oxidation with potassium permanganate yields methyl isocyanate (54).

The elucidation of the structural formula of physostigmine proved difficult. A complete description of the many blind alleys and the painstaking work leading to final synthesis has been given by Marion (55). It may be mentioned here that several groups of investigators devoted years and even decades to this topic; for example, Polonovski, 1893-1923 (50a,b, 53, 54, 56a,b,c), and King and Robinson, 1932-1935 (57). The chemical structure of the ring system (I) was established by Stedman and Barger in

1925 (58). Appropriately, this work was carried out in Edinburgh where so many studies on the Calabar bean had been made. The final proof of the structure was obtained through the complete synthesis achieved by Julian and Pikl in 1935 (59). They developed a simple route of synthesis which had the advantage over those already described in that it gave rise to the same isomer as the natural base.

The property of physostigmine to produce miosis and other symptoms is attributable to the urethane group, since it is absent in its hydrolysis product, eseroline. This observation has prompted the preparation of a number of substituted urethanes, the best known being neostigmine (II) (see Fig. 9) which was synthesized by Aeschliman and Reinert in 1931 (60).

NEUROHUMORAL TRANSMISSION, PHYSOSTIGMINE, AND CHOLINESTERASE

The main alkaloid of the Calabar bean was destined to play an important role in the elucidation of neurohumoral transmission—that is, how impulses are transmitted from one nerve to another or from a nerve to an end organ by means of a hormone—in this case acetylcholine. The powerful biological activity of synthetic acetylcholine was discovered as early as 1906 (61). Readers interested in the detailed sequence of events in the discovery of neurohumoral transmission are referred to Gaddum (12, 62), Holmstedt and Liljestrand (16), Lembeck and Giere (63), Dale (64), Feldberg (65) and Koelle (66).

An important advance in the understanding of the mechanism of action of physostigmine was made in 1906 by Anderson (67), who found that if the nerves of the pupil were cut, eserine still caused constriction of the pupil, but that if the nerves were cut and allowed to degenerate, this effect disappeared. The explanation of these observations is presumably that the physostigmine could only act when the nerves were there to supply a transmitter and that the freshly cut nerves still did this. No such transmitter was known at the time, but in the next year Dixon (1871-1931) produced some evidence for this effect. He expressed views on the action of drugs which were an extension of Elliott's suggestion regarding the liberation of adrenaline and was of the opinion that muscarine might be responsible for the action of vagus stimulation on the heart (68).

> If physiological activity is brought about by the chemical combination of an animal alkaloid with some substance in the activated tissue, as it is, probably, in the

> two examples given (adrenaline and secretin), why should not such a procedure be responsible for all forms of activity? That is to say, when a muscle contracts, when a gland secretes, or a nerve ending is excited, the cause in each case may be due to the liberation of some chemical substance, not necessarily set free in the circulation as in the case of secretin, but more likely liberated at the spot upon which it is required to act.
>
> In order to test the validity of this reasoning, I investigated the action of the vagus nerve upon the heart. Animals were killed by pithing; they were bled, and the vagus nerves were then placed on the electrodes and excited for half an hour. The heart was next extirpated, placed in boiling water for ten seconds and extracted with alcohol. The alcoholic extract was evaporated to dryness and taken up once again with 100% alcohol. This was again evaporated off on the water bath, and a few drops of normal saline solution added. The solution so obtained was found to have the power of inhibiting the frog's heart, and, like muscarine, the effect was completely antagonized by atropine. Moreover, the substance disappeared from the solution if it were allowed to stand in the laboratory 24 hours.
>
> Hearts treated in an identical manner, but in which the vagus nerve had not been excited, also gave a supply of this inhibitory substance, but in a smaller degree than in the excited heart. I interpret these experiments to mean that some inhibitory substance is stored up in that portion of the heart to which we refer as a 'nerve ending,' that when the vagus is excited this inhibitory substance is set free, and by combining with a body in the cardiac muscle brings about the inhibition. If cardiac inhibition is brought about in this way, drugs must act by liberating the inhibitory hormone. Atropine either prevents the liberation of the hormone, or saturates the substance in the end organ upon which it acts. The former seems the most probable explanation.

The proof of a transmitter working in this way was not to come until 15 or 20 years later.

Loewi and Mansfeld in 1911 (69) confirmed what other authors had previously observed, namely, that physostigmine increased the sensitivity of peripheral organs to electric stimulation. They investigated the action of the salivary glands and the bladder. From the experiments it was concluded that physostigmine made the end organ sensitive to subthreshold stimulation. The authors thereupon proceeded to establish whether autonomic organs have a peripheral nervous resting tonus. Experiments on the eye had already been positive. A resting tonus could be shown for the salivary glands but not for the bladder. The action of physostigmine was discussed toward the end of the paper, but at this time Loewi would not commit himself to say whether or not physostigmine sensitized the end organ to nervous or chemical stimulation. In 1912 Loewi again worked with eserine. He repeated the experiments of Arnstein and Sustschinsky from 1867 (25) and recorded the effect of stimulation of the vagus and of an injection of 2 mg of pilocarpine on a rabbit's blood pressure, and then injected 2 mg of eserine. The effect of the vagus was enormously increased but that of pilocarpine was not (70).

In 1914 Dale (1875-1968) published an epoch making paper entitled: "The action of certain esters and ethers of choline and their relation to muscarine" (71). In it, he outlined the now well-known muscarine-like and nicotine-like actions of the autonomic nervous system. Among other things Dale concluded:

> The question of a possible physiological significance, in the resemblance between the action of choline esters and the effects of certain divisions of the involuntary nervous system, is one of great interest, but one for the discussion of which little evidence is available. Acetylcholine is, of all the substances examined, the one whose action is most suggestive in this direction. The fact that its action surpasses even that of adrenaline, both in intensity and evanescence, when considered in conjunction with the fact that each of these two bases reproduces those effects of involuntary nerves which are absent from the action of the other, so that the two actions are in many directions at once complementary and antagonistic, gives plenty of scope for speculation. On the other hand, there is no known depot of choline derivatives, corresponding to the adrenine depot in the adrenal medulla, nor, indeed, any evidence that a substance resembling acetylcholine exists in the body at all. Reid Hunt found evidence of the existence of a substance in the suprarenal gland which was not choline itself, but easily yielded that base in the process of extraction. If acetylcholine, however, or any substance of comparable activity, existed in the suprarenal gland in quantities for chemical detection, its action would inevitably overpower that of the adrenine in a gland extract. The possibility may, indeed, be admitted, of acetyl-choline, or some similarly active and unstable ester, arising in the body and being so rapidly hydrolysed by the tissues that its detection is impossible by known methods. Such a suggestion would acquire interest if methods for its experimental verification could be devised.

Dale was obviously impressed by the fact that the actions of acetylcholine in the body were very brief, and commented: "It seems not improbable that an esterase contributes to the removal of the active ester from the circulation and the restoration of the original condition of sensitiveness." This was the first mention of an esterase in connection with acetylcholine.

Hunt (72, 73) had shown that both vascular effects of acetylcholine—the "muscarine-like" action normally produced by small doses, and the "nicotine-like" action of larger doses given after atropine—were intensified by eserine. He had earlier (74) found that the parasympathetic effects of homologues of acetylcholine were similarly increased by eserine.

Other evidence has definitely related the enhancement by eserine of the effects of acetylcholine on the esteric structure of the latter. Fühner (75) found that soaking the plain muscle of a leech in a very weak solution of eserine (1 ppm.) did not alter the stimulant action of pilocarpine or of choline, but increased that of acetylcholine a million-fold. He suggested that the effect of eserine was to inhibit hydrolysis of acetylcholine by the tissues. Fühner's technique was later developed by Feldberg (76) into a bioassay for acetylcholine:

> At that time I was keen to have a sensitive test preparation for acetylcholine. Looking through the literature I came across the paper by Fühner in which he used the acetylcholine-physostigmine synergism on the leech muscle as the basis for assaying physostigmine in its smallest amounts. It struck me at once that if Fühner was right, this synergism might also provide a sensitive and specific method for detecting acetylcholine. So why not reverse Fühner's procedure and test the leech muscle with physostigmine and then test for acetylcholine? I had a young coworker, Dr. Minz, working in my lab and suggested to him that he test the specificity of the eserinized leech muscle to acetylcholine, and to try out as many substances as possible on this preparation, particularly those which might be present in our perfusate, or in the blood under experimental conditions. He published the method in 1932 in the *Archives of Experimental Pathology,* acknowledging my suggestion (76).

The name of Otto Loewi (1873-1961) is forever linked with the series of papers on humoral transmission published between 1921 and 1926. He has himself given a vivid account of the beginning of these studies (77):

> The night before Easter Sunday of that year (1920), I awoke, turned on the light and jotted down a few notes on a tiny slip of thin paper. Then I fell asleep again. It occurred to me at six o'clock in the morning that during the night I had written down something most important but I was unable to decipher the scrawl. The next night, at three o'clock, the idea returned. It was the design of an experiment to determine whether or not the hypothesis of chemical transmission that I had uttered seventeen years ago was correct. I got up immediately, went to the laboratory, and performed a simple experiment on a frog heart according to the nocturnal design. I have to describe this experiment briefly since its results became the foundation of the theory of chemical transmission of the nervous impulse.
>
> The hearts of two frogs were isolated, the first with its nerves, the second without. Both hearts were attached to Straub cannulas filled with a little Ringer solution. The vagus nerve of the first heart was stimulated for a few minutes. Then the Ringer solution that had been in the first heart during the stimulation of the vagus was transferred to the second heart. It slowed and its beats diminished just as if its vagus had been stimulated. Similarly, when the accelerator nerve was stimulated and the Ringer from this period transferred, the second heart speeded up and its beats increased. These results unequivocally proved that the nerves do not influence the heart directly but liberate from their terminals specific chemical substances which, in their turn, cause the well-known modifications of the function of the heart characteristic of the stimulation of its nerves.

The first of the series of papers was published in 1921 (78). It consisted of four pages with no references and is truly one of the most remarkable communications in the history of science. It proved for the first time that chemical transmission from nerves to end organ existed.

In the series of papers that followed, Loewi narrowed step by step the number of possible compounds that could be the chemical equivalent of

Vagusstoff. As early as the second paper, he clearly pointed toward a choline ester (79); and in the tenth appears a footnote reading as follows: "In view of the extremely high activity of the vagal substance it might be thought that it is acetylcholine."

The famous eleventh paper in the series (80) introduces the effects of enzyme inhibitors, primarily eserine, as further proof of the similarity between the transmitter and acetylcholine. In Loewi's own words, it is the first time that the pharmacological action of an alkaloid has been defined in terms of inhibition of an enzyme:

It is known that physostigmine and ergotamine sensitize the heart to vagal stimulation. This sensitization manifests itself in various ways; e.g., in the presence of the sensitizing substances, the effect considerably outlasts stimulation. Since the effect of vagal stimulation is determined by the vagal substance which is formed or released during stimulation, we investigated whether physostigmine and ergotamine sensitized the heart to the action of the vagal substance. This was indeed the case. Physostigmine and ergotamine increased the intensity of action of the vagal substance, and prolonged especially its duration. Physostigmine and ergotamine, however, did not sensitize the heart to the action of all substances with a vagus-like action, e.g. they had no effect on the action of muscarine or choline, whereas they sensitized the heart to both acetylcholine and vagal substance to the same extent. In these experiments vagal substance and acetylcholine, therefore, behaved once more in an analogous manner.

We proposed in our tenth paper that the administration of vagal substance or acetylcholine produce only a very short-lasting effect on the heart since both were speedily metabolized. We therefore investigated whether the long duration of action of vagal substance and acetylcholine, which is seen when physostigmine or ergotamine are given beforehand, was due to an inhibition of their metabolism.

These experiments showed that the metabolism of vagal substance and acetylcholine by heart extracts was indeed inhibited by physostigmine or ergotamine. This is further evidence for an analogous behavior of the two substances. The concentration of physostigmine and ergotamine which produce an inhibition *in vivo* are the same as those which sensitized the intact heart. There can hardly be any doubt, therefore, that the sensitization was the consequence of an inhibition of the metabolism of vagal substance and acetylcholine. It is at once obvious that this explained the increase in intensity and duration of action which was observed after sensitization. Our results also readily explain why the sensitized heart responded to stimuli which were ineffective in the normal heart. A weak stimulus forms only a minute amount of vagal substance which in normal heart is readily metabolized, and therefore produces no effect. Sensitization arrests the metabolism of the vagal substance and consequently an effect becomes noticeable.

The experiments described above only reveal the mechanism responsible for sensitization to vagal stimulation. It is known that physostigmine also produces sensitization to other types of stimulation, e.g. it sensitizes the striped muscle to electrical stimulation and the muscle in the skin of the leech to stimulation by barium. It is naturally problematical whether the sensitization observed in these cases is due to a mechanism similar to that which accounts for sensitization of the heart. Whatever the answer might be it does not affect our results which, as far as we know, show for the first time that alkaloids can produce a *visible* action by affecting an enzymic reaction. As a result of this the familiar observation that alkaloids are effective in very small doses is, at least in our case, more readily understood.

Dale (81) has commented on Loewi's discovery in the following way:

There was one feature, however, of Loewi's first published account of this crucial experiments, which I found more difficult to understand. In referring to the two different transmitters, which he found to appear in the saline contents of the heart, in response to the stimulation of its inhibitor vagus nerve and of its augmentor-accelerator sympathetic nerve-supply, he referred to them merely in terms of their respective origins, as "Vagus-stoff" and "Accelerans-stoff." He had thus appeared to ignore, or to be unaware of the fact, that there were two good chemical candidates, the respective sympathomimetic and parasympathomimetic, [the] activities of which had long entitled them, in theory, to be considered for the transmitter roles, if once the real existence of such a chemical transmitter function could be demonstrated. At our next encounter I ventured to rally him on this reticence. "Why," I asked him, "did you not say frankly that your vagus-substance might well prove to be something like acetylcholine?" He replied that, of course, he believed it to be acetylcholine, but that, having recently, in an entirely different connexion, committed himself too rashly to a conclusion which he had subsequently to withdraw, he was abnormally sensitive to the danger of publishing another speculative suggestion, lest it might prove also to be premature. Most fortunately, however, this attitude, which seemed to me to be one of excessive caution, did not prevent him . . . from pursuing the comparison between the properties and behavior of acetylcholine and those of his "vagus-substance" . . . with a succession of collaborators . . . And . . . succeeded in establishing the identity of the parasympathetic transmitter with acetylcholine.

In summary, Loewi and Navratil (80) showed that extracts of the heart and various other tissues inactivated "Vagus-stoff" and, also, acetylcholine *in vitro,* and that the inactivation was inhibited by eserine. The conclusion that this was due to an enzyme was disputed by Galehr and Plattner (82) and others, who found that the results obtained with enzyme from red blood corpuscles differed from the results obtained with enzyme from serum; but Engelhart and Loewi (83) were able to explain most of the anomalies and came to the conclusion that the effects really were due to an enzyme. In the same year, Matthes (84) produced other evidence in support of this conclusion, and proved that the combination between the eserine and the enzyme was reversible by dialysis. As mentioned earlier, the chemical structure of eserine had been established at this time by Stedman and Barger (58). Dr. and Mrs. Stedman followed this work with a series of fundamentally important studies of the relationship between chemical constitution and pharmacological action. At the time, this led to the conclusion that the action depended on competition between the alkaloids and acetylcholine (85). In 1932,

Stedman, Stedman, and Easson (86) showed that an enzyme present in horse's serum acted specifically on cholineesters and proposed that it should be called cholineesterase.

Later work has shown that there is more than one cholineesterase (for a review see [66]), and recently the enzyme has been purified and crystallized (87a). The progress made has also led to different interpretations of the mechanism of inhibition of the enzyme.

Based on classical approaches, it was thought for many years that inhibition by eserine and other carbamates was simply a question of reversible complex formation between the carbamate and the cholineesterase.

The reasons for believing that this was the operative mechanism were as follows: If one measures the amount of inhibition obtained, it can be shown to be proportional to the amount of inhibitor used, and this suggests a reversible component is involved. Secondly, if one dialyzes the enzyme or dilutes it greatly following inhibition, one can virtually recover the enzyme activity back again. On these grounds, it seemed clear that simple complex formation was involved. Stedman (87b,c) noted that the pharmacological activity of methylcarbamates was dependent on the carbamate ester grouping, and appeared to be related to the rate of hydrolysis. He proposed that one of the products liberated on hydrolysis in the nerve or at the nerve ending might be responsible for the activity of these compounds.

Hobbiger (88) suggested that the inhibition of choline esterase by physostigmine might involve carbamylation, but that the extent to which this occurred was uncertain. Wilson et al., in 1960 and 1961 (89, 90), produced convincing evidence that this was so, suggesting that carbamylation of the enzyme occurs most probably at a serine residue to give a carbamylated enzyme. This is analogous with the case of the organophosphates, which for many years have been thought to inhibit enzymes precisely in this way (91). The evidence that Wilson offered to support his model was that the nature of the inhibited enzyme is dependent on the carbamylating group. Thus, all *N*-methyl carbamates give the same kind of inhibited enzyme, as judged by ability to recover from inhibition.

Wilson suggested that cholineesterase inhibitors could be of two types: a) acid transferring inhibitors (organophosphates and carbamates); or b) prosthetic inhibitors (compounds that bind reversibly as exemplified by tensilon and other quaternary ammonium bases).

However, long before modern biochemistry revealed the exact mode of inhibition by eserine, its use led to many important advances in our knowledge of cholinergic nerves, that is, nerves that act by liberating a cholineester (64). The mere fact that the actions of certain nerves were found to be increased by eserine was used in support of the theory that these nerves were cholinergic (92, 93).

Acetylcholine was shown to be the transmitter not only of postganglionic parasympathetic but also of some postganglionic sympathetic fibers, for instance, of those innervating the sweat glands in cats. Thus the transmitter function of acetylcholine was not limited to only one of the two anatomical divisions, parasympathetic or sympathetic, of the autonomic nervous system. However, the discovery of the nicotinic transmitter function of acetylcholine entailed a much wider implication —it came as a real shock. The finding that acetylcholine was the neuromuscular transmitter at the motor endplate of striated muscle and the synaptic transmitter in a sympathetic ganglion went against all the, often ingenious ideas which had been propounded up to that time by those working in this field, because it had been taken for granted that such transmission processes were physical—purely electrical events. How many today realize or recall the complete change in concept that took place as a result of the discovery of the nicotinic transmitter function of acetylcholine (65)?

The likelihood that acetylcholine was the transmitter increased considerably when Dale and Dudley (94) reported its isolation from mammalian tissue (the spleen), but a direct test of this theory was only made possible by the use of eserine. This was first achieved in experiments on a perfused sympathetic ganglion (95). In the absence of eserine, nothing was detected, but in its presence, stimulation of the nerves to the ganglion led to the appearance of acetylcholine in the perfusate.*

The identification of the ester was made by means of parallel quantitative assays of the same solutions by different methods. Such bioassay can distinguish between acetylcholine and other similar substances such as propionylcholine (98), and the evidence was in favor of the view that the observed effects were due to acetylcholine. There is reason to believe that propionylcholine and acetylcholine are both present in extracts of

*Kibjakow (96) first used the perfused isolated ganglion. He did not use eserine in the perfusate, and it is questionable whether he ever got any acetylcholine in the effluent upon stimulation. Feldberg seems to have been largely responsible for the introduction of eserine under these experimental conditions (64, 65, 97).

ox spleen (99). As pointed out by Gaddum (62), it is still possible that other esters besides acetylcholine are liberated in autonomic ganglia, and modern chromatographic methods may eventually decide this point, but there is still no reason to doubt that acetylcholine is the main ester liberated in the superior cervical ganglion of a cat. The specific tests on which this conclusion depends were also used in the work proving that acetylcholine is liberated at the motor nerve endings in voluntary muscle (100).

Early and important experiments with physostigmine on the central nervous system and the blood-brain barrier were carried out by Schweitzer et al. (101). Recently, the presence of acetylcholine in the central nervous system has been proved beyond any doubt by modern chemical methods (102).

USE OF PHYSOSTIGMINE AND RELATED COMPOUNDS IN MYASTHENIA GRAVIS

In contrast to the important role physostigmine played in the discovery of cholinergic transmission and in the treatment of glaucoma, it has come to be of major use only in one systemic disease, namely *Myasthenia gravis.* The merit of having initiated the study of this disease should be attributed to Guillermo Erb (103). Muscular fatigability in the patients described by him was particularly localized in the muscles of the neck, in chewing and swallowing, and in the ocular muscles. The extremities were unaffected, which justified Erb's hypothesis that these disturbances resulted from upper bulbar lesions.

In 1891 Goldflam presented new observations (103). He insisted principally on the periods of remission in myasthenia, and also on the fatigability of striated muscle which ultimately might produce the whole symptomatology. In recognition of these studies, *Myasthenia gravis* has also been called Erb-Goldflam's disease.

In a series of studies between 1891 and 1895, Friedrich Jolly (1844-1904) pursued investigations on muscular reaction to electrical stimulus verifying that the fatigability was manifested as much in voluntary movements as with electric stimulation (104, 105). He called this special characteristic of the skeletal muscles the myasthenia reaction and named the disease "pseudo-paralytic myasthenia gravis," which is the name actually most used.

Friedrich Jolly was born on November 24, 1844, in Heidelberg. His father, Philipp Jolly, a Professor of Physics at the University, was not only a distinguished teacher but also a personality of importance (106). In the 1850's Friedrich Jolly moved with his parents to Munich where his father belonged to a circle of famous scientists, authors, and artists. From early youth on, therefore, the son was exposed to a diversity of influences and inspiration. His first wish was to become an engineer but he finally chose medicine.

He began his studies in Munich, and was said from the very beginning to be an exemplary student. He later moved to Würzburg where he joined a "Club of Youths" which consisted of alert, scientific, and sociable members and was for some time its chairman. In Würzburg he carried out research in the laboratory of Adolf Fick, the physiologist. In 1873 he went on a trip to England and Scotland for the purpose of studies and on his return accepted the chair at Strassburg University.

In 1874 he married Anna Böhm, the sister of his friend Rudolf Böhm (pharmacologist in Leipzig). After 17 years of arduous activity in Strassburg, Jolly was called to Berlin where the quantity of his work steadily increased from year to year.

It is easy to understand why Jolly, with his inclination for physics from early youth, in his first important paper under the leadership of Fick devoted himself to experiments dealing with intracranial pressure and blood circulation in the brain. Jolly's favorite method of procedure was to use experimental physiology to obtain knowledge about conduction and reaction of nerves and muscles in normal and pathological conditions. He repeatedly returned to this field, as evidenced by his many investigations of the electrical reaction of the neuromuscular apparatus. Jolly made many other contributions to neurology and psychiatry but this is what mainly concerns us here. He showed that an abnormal twitch to electric stimulus occurred during the abnormal state of muscle tonus in Thomsen's disease and in *Myasthenia gravis pseudoparalytica.* The myasthenia reaction was proven to be a condition of abnormal fatigue of muscle activity in response to tetanic stimulation. These results were presented by Jolly at a lecture on December 5, 1894, to the Berlin Medical Society (104): He introduced his subject as follows:

> I should like to draw your attention to the pathological picture that has been described, it is true, by various authors in recent years, but which nevertheless is not generally known and which, in my opinion, claims not only a certain theoretical but also a considerable practical interest. The case which I wish to demonstrate to you is the second of this strange sickness which has come to my own observation. In the year 1890/91 I observed the first case here in the Charité and demonstrated it at the time to the Society of the Charité physicians.

Jolly then proceeded to describe the symptoms of the patients and his technique of electrical stimulation. He commented as follows on the action of a tetanus applied to the muscle for a longer time, whether through the nerve or directly:

> We find the same phenomenon as when the muscle is fatigued voluntarily or when the patient is subjected to exercise. Upon the tetanic stimulation the response becomes less complete and even during the stimulation a decrease is more and more noticeable. A stage is soon reached at which every time that the current is applied only a contraction of short duration follows . . . It is practical to give a special name to this particular kind of reaction to stimulation that I have described and I propose for it the name of *myasthenic reaction.*
>
> I should like to stress once more that I found and expressly emphasized the myasthenic reaction also in the first case which came under my observation in the winter of 1890/91 and that it was this first observation that led me to seek out the second case. The peculiarity consists of the fact that the major part of the voluntary muscles of the patient exists in a state of abnormal fatigue brought about by normal innervation and through tetanic stimulation, resulting in a rapid discontinuance of the contractility. . .
>
> In connection with this disease as documented in the case I have presented and showing an abnormal muscle condition, an alteration in the chemistry of the muscle is likely to cause such a contraction phenomenon. In this instance it is of great interest to recall that we know of certain alkaloids which bring about quite similar conditions of the muscle during the duration of the poisoning. It has been repeatedly said that veratrine physostigmine, digitoxin, and other alkaloids bring about a state of the muscle resembling that of the Thomsen's disease. With regard to the *myasthenic condition* existing in today's case (which I should like to call the opposite of the myotonic condition in the Thomsen's disease), I have been informed by Geh. Rath Böhm in Leipzig that here also a roughly analogous alkaloid effect is known. Protoveratrine, a substance related to veratrine, under certain conditions creates a similar fatigue of the muscles*. . . The analogies, however, allow us to conclude that in these conditions also a disturbance of the chemical processes of the muscle contraction (muscle chemistry) exists. . .
>
> With regard to the observations I have made showing an electric fatigue of the muscles, it is only natural to think about drugs which could bring about pharmacologically the opposite of the myasthenic reaction, i.e. the myotonic condition. The alkaloids (veratrin, physostigmine, etc.) which may be considered are, however, not free from side-effects on the nervous system and their use should therefore be restricted to hospitals only.

It is quite apparent here that Jolly was considerably influenced by his brother-in-law, Rudolf Böhm (1844-1926). It may be mentioned that

*Fick's and Böhm's paper, published in the year 1872, on "The effect of veratrine on the muscle fibres" (107), presents a myogram of veratrine-treated muscles showing a very great similarity to those from myotonics. Reference also can be made to work by Walker-Overend: "On the influence of curare and veratrine on striated muscles" (108), and to the work of Thomas Watts Eden on "The effects of protoveratrine" (109).

Böhm graduated from Würzburg at the time Jolly was working there. He started as a psychiatrist, but soon switched over to pharmacology (16).

Jolly's study is most interesting, because in addition to the description of the myasthenic reaction, he established clearly the completely opposed characteristics of *Myasthenia gravis* to the so-called Thomsen's disease. He also mentioned the possibility of the therapeutic use of physostigmine, based on the fact that this alkaloid by its pharmacologic action can produce the opposite of the myasthenic reaction—that is, the myotonic condition.

More than forty years elapsed before this concept was applied by Mary Walker and proved of vast importance in the study and treatment of the malady. Mary Walker received her degree at the University of Edinburgh in 1913. (It seems that this university and physostigmine are forever linked together.) She joined the Royal College of Physicians in 1932, and for her work on myasthenia she was awarded the gold medal of medicine from the University of Edinburgh in 1935. She published a few other papers, mainly during the 1930's. During World War II she served in the British Royal Army Medical Corps in Salonika. Very little is known about her since then, except that she became an honorary member of the Medical Advisory Board of the Myasthenia Gravis Foundation in 1955 and that she received the Gene Hunter prize of the Royal College of Physicians in 1963. She now lives in retirement in Scotland.

There was still no idea how or where the myasthenic phenomenon was produced when, in 1934, Mary Walker first used physostigmine, and later prostigmine, for treatment of the disease, thus establishing a new fundamental stage in the knowledge of the myasthenic syndrome. Dr. Walker recognizing that myasthenia resembled curare poisoning, made successful use of physostigmine, whose antagonistic action to curare had been known for a long time (31). The result was rightly called "the miracle at St. Alfeges," the hospital in Greenwich, England, where Dr. Walker was then working (110).

Walker's original paper (111) reads as follows:

> The abnormal fatigability in *Myasthenia gravis* has been thought to be due to curare-like poisoning of the motor nerve-endings or of the "myoneural junctions" in the affected muscles. It occurred to me recently that it would be worth-while to try the effect of physostigmine, a partial antagonist to curare, on a case of *Myasthenia gravis* at present in St. Alfege's Hospital, in the hope that it would counteract the effect of the unknown substance which might be exerting a curare-like effect on the myoneural junctions. I found that hypodermic injections of physostigmine salicylate did have a striking though temporary effect.

Mrs. M., aged 56, had had a previous attack of *Myasthenia gravis,* lasting about six months, 14 years ago. Gastric ulcer four years ago. Non-specific infective arthritis seven months ago, now improved. Towards the end of last February she found that she was unable to hold her shopping bag, and that her head used to fall forwards when she knelt to do the hearth. She had to remain in bed after 18 March and had difficulty in sitting up. Her jaw then began to droop, she had to hold it up with her hand, and the left eyelid began to drop. Speech became indistinct when she was excited, swallowing was difficult, and fluid sometimes regurgitated through her nose. She was admitted to the hospital on 28 March and a few days later weakness came on in the middle and ring fingers of both hands. The weakness is much increased by excitement, and is lessened by rest. It becomes worse as the day goes on. There is no wasting, and the tendon reflexes are all present. The masseters respond slightly or not at all to faradism; a myasthenic reaction has been obtained in the left deltoid. Radiograms show obsolete pulmonary tuberculosis. The thymus is not enlarged.

On 11 April treatment with hypodermic injections of physostigmine salicylate, gr 1/60th once a day, was begun. In from half an hour after the injection the left eyelid "goes up," arm movements are much stronger, the jaw drops rather less, swallowing is improved, and the patient feels "less heavy." The effect wears off gradually in from 2-4 hours. With injections of gr 1/50th the improvement is greater, and it lasts for 4-5 hours. Still greater improvement, lasting for 6-7 hours, followed an injection of gr 1/45th, but the patient felt rather faint and trembly, her "inside seemed all on the work," and she felt as if "something were going to happen." These feelings did not completely disappear till an hour after the injection. In all, 26 injections of physostigmine salicylate have been given. The effect is not quite uniform; on two occasions injections of gr 1/45th and gr 1/60th failed to produce any obvious effect. She feels better and more cheerful since the injections were begun.

Given by the mouth, physostigmine salicylate gr 1/60th produced no obvious effect, but an hour after gr 1/30th, slight improvement occurred. No improvement followed control injections of water, pilocarpine gr 1/20th, strychnine gr 1/30th, adrenaline gr 5, ephedrine gr 1/2, or acetylcholine 0.05 and 0.01 g.

I think that this effect of physostigmine on *Myasthenia gravis* is important, though it is only temporary, for it improves swallowing and might tide a patient over a respiratory crisis. It supports the opinion that the fatigability is due to a poisoning of the motor end-organs, or "myoneural junctions," rather than to an affection of the muscle itself. It may be significant that physostigmine inhibits the action of the esterase which destroys acetylcholine. I have not had the opportunity of treating another case to confirm the findings. The administration of other drugs whose action resembles that of physostigmine is under consideration. It is also possible that physostigmine might be of some service in botulism and in cobra poisoning, in both of which a curare-like poisoning of the "myoneural junctions" of the respiratory muscles has been stated to be the main cause of death.

In a second paper (112) Mary Walker also used prostigmin:

D. C., female, aged 40. History: In spring 1930 she noticed undue fatigability of the arms and drooping of the right upper eyelid. In July 1930 she was admitted to the Middlesex Hospital with diplopia, and was discharged, appreciably better, within a month. January 1931, readmitted with impairment of all movement of the left eye and weakness of the arms and legs, which became worse towards the end of the day.

Control injections of ephedrine, lobeline, femergin, and water produced no effect on the muscular weakness. Physostigmine salicylate, 1 mgm. approximately, removed the ptosis but made the patient feel sick and faint and disinclined to move. Atropine, given with the physostigmine, counteracts these ill-effects, without altering the action on the motor nerve-endings.

Physostigmine salicylate was given in a previous case of the disease which was admitted to St. Alfege's Hospital in April 1934, because it was thought that as the muscles in *Myasthenia gravis* behave like muscles poisoned by curare, physostigmine, an antagonist to curare, might also counteract the unknown substance which might be exerting a curare-like effect on the motor nerve-endings in *Myasthenia gravis*. The patient tolerated it well, and its effect on the weakness and fatigability was the same as that of prostigmin, 1.0 mgm.

The advantages of prostigmin (Roche)—a synthetic drug analogous to physostigmine and with similar actions—over physostigmine are that it has a less depressing effect on the heart, less often causes nausea and vomiting, and is probably safer in large doses; 4 mgm. have been given without ill-effect, though in other cases the same dose has caused severe diarrhoea and cardiac and respiratory distress. The disadvantage is its expense, the price of an ampoule containing 0.5 mgm. of the drug being ninepence.

Dr. P. Hamill writes:

Whatever may be the mechanism of the weakness and fatigability of the muscles in myasthenia gravis, physostigmine, and its ally, prostigmin, overcome it. The resulting voluntary movements are normal in type. If it is the case that nervous impulses set free acetylcholine—or some analogous substance—at the nerve-endings, and that in myasthenia the supply is deficient, physostigmine, by delaying its destruction, would compensate for the lack.

On this hypothesis, defective innervation, whether resulting from some disability of the anterior-horn cells or impaired conduction of nerve-fibres, as in neuritis, should also be corrected by injection of physostigmine and prostigmin. This appears to be the case, the drugs having been used in cases of peripheral neuritis and in spinal-cord lesions with beneficial results. Muscular power is increased. As additions to massage in such cases, the drugs may help in maintaining nutrition of the paralyzed muscles and thus accelerate their recovery. The economic aspects are important. A wide field of usefulness in neuromuscular disorders is thus opened and is now being studied; the results will be communicated at a later date.

Actually, prostigmine, as pointed out by Viets (1890-1969)* (110), had been used in myasthenia once before by Lazar Remen in Germany. His short note (113) on the subject in 1932 remained, however, entirely forgotten: "*Therapy:* Experiments with prostigmine injections. One hour after the injection, the patient could stretch out his hands, open

*B. Spector, "Henry Rouse Viets, 1890-1969, neurologist, medical historian, librarian," *Bull. Hist. Med., 44* (1970), 173-175, and *New Eng. J. Med.*, Aug. 14, 1969, 387-388.

his eyes wider and eat food. This improvement remained for only one half to one hour. The patient could not stand two injections because of dizziness and attacks of vomiting. The patient remained in this condition some months up to March 1932."

The report of Mary Walker's second case awakened interest, and confirmatory observations filled the contemporary literature for many months thereafter. As stated by Viets (110), success could no longer be denied, and "the Miracle at St. Alfege's" became an integral part of medicine. In 1935, Viets and Schwab (114) introduced neostigmine as a diagnostic test and in their publication stated the following:

> The earlier work on the physostigmine-curare antagonism gave a suggestion that physostigmine or one of its derivatives might be of value in the treatment of *Myasthenia gravis,* a disease characterized by weakness of muscles similar to the paralysis of curare poisoning. Such has proved, in part, to be the case. The rapid response to the derivative, prostigmine, however, has furnished us with a new diagnostic test for *Myasthenia gravis,* while its therapeutic action is still uncertain, except for temporary relief of symptoms. A diagnostic schema is helpful in recording the action of the drug. While the diagnostic test is of some use to medicine, it is to be hoped that prostigmine or another preparation may in the near future furnish us with a form of therapy capable of prolonged effect and thus be more suitable in the treatment of *Myasthenia gravis.*

The authors concluded:

> Prostigmine, a derivative of physostigmine, causes a marked, but temporary, remission of the muscular weakness seen in *Myasthenia gravis.* The effect is so marked that the drug is suggested as a diagnostic test for the disease. Other diseases, either of nerves or muscles, respond only slightly to prostigmine. Prostigmine is injected subcutaneously, with an appropriate dose of atropine to counteract any intestinal stimulation. Three cc. of prostigmine (Roche) and atropine, 1/100 grain, has proved to be satisfactory as a diagnostic test.

A Myasthenia Gravis Clinic was started in Boston in 1935. Soon a research unit was attached to it that was devoted, as time passed, to clinical investigation, pathology, and thymic surgery. Since then, several conferences on an international level have been devoted to *Myasthenia Gravis* (115).

The basic cause of *Myasthenia gravis* is still obscure. As regards the pathophysiological changes, there are two main contradictory opinions. One of them is that the defect is localized presynaptically—that is, it lies either in the production of acetylcholine or in its liberation from the nerve terminals. According to the second opinion, the liberated acetylcholine is unable to stimulate the defective receptors in the muscle endplate, or that these receptors are blocked. Abnormal cholinesterase activity has also been considered. Up to the present time, experimental attempts at producing a neuromuscular block similar to that seen in *Myasthenia gravis* have been unsuccessful.

Prostigmine is still the drug of choice in the treatment, but many recent pharmacological studies have been concerned with the drugs mentioned by Jolly, namely, physostigmine and the Veratrum alkaloids (116).

CARBAMATE INSECTICIDES

In 1863 Thomas Fraser obtained from Old Calabar a shipment of beans which had been collected because they showed indications of attack by an insect. The beans had been partly eaten and contained traces of excrement, cocoons, and caterpillars. In fact, Fraser was able to follow the complete metamorphosis of the insect and to establish its identity as the moth *Deiopeia pulchella* (order-*Lepidoptera*).

Fraser was much astonished that any animal could ingest the Calabar bean without suffering and wrote (15b):

> The Ordeal-bean of Calabar is a poison of extreme activity: hitherto no living being has been known to be able to resist its action; and, from my knowledge of its properties, I confess to having been sceptical of the existence of any animal form which could be fairly subjected to its influence and still retain its hold on life. It appeared of importance to determine, as exactly as possible, the connection between this caterpillar and the kernel of the bean; as, supposing the kernel to be received into its alimentary system, the existence of a special assimilative selection might be shown, or it could be determined if the caterpillar were proof against this deadly poison.

Fraser used the excreta of the insect to perform certain experiments. He found that a solution of it contracted the pupil and that it was highly toxic and killed higher animals. Also, he found that the caterpillar itself was resistant to the injection of the active principle of the bean which he had isolated. Fraser concluded:

> (1) The caterpillar of *Deiopeia pulchella* feeds on the virulent poison contained in the kernel of the seed of *Physostigma venenosum;* and that (2) This caterpillar is unaffected by the poisonous principle of the kernel-eserine.
>
> The bearing of the second conclusion on our ideas of vital action should not be overlooked. A somewhat analogous case is furnished by *Anthonomus druparum* which feeds on the kernel of *Prunus cerasus;* and the poisonous properties of this kernel are well known to depend on the hydrocyanic acid [glycoside] it contains. Here, then, our difficulties are increased: *Deiopeia pulchella* is unaffected by one poison, but is rapidly killed by hydrocyanic acid; and this latter occurs in the food of another insect, *Anthonomus druparum.* If life be "the sum total of the function which resists death," we have in these examples two organisms, each furnished with an exceptional potency of one or more of these death-repelling functions, or having

bestowed on each, for a necessary purpose, a special, almost unrecognized, and certainly uninvestigated alexipharmic (antipoison). Unfortunately, we have no knowledge of those intimate and primary structural changes which accompany every vital action, and our acquaintance with the perversions of such changes is quite as unsatisfactory.

Physostigmine is indeed a poor insecticide, as has been shown by modern testing. Present-day activities in this field are directed toward finding compounds that show *selective toxicity:* that is, that are highly toxic to insects and either non-toxic or only moderately toxic to man and wild life (117). Before and after World War II, cholineesterase inhibitors of the organophosphorus type were developed as insecticides (see Holmstedt in ref. [66]), and the older types of cholineesterase inhibitors were then completely neglected.

The use of carbamates as insecticides started with their development as insect repellents. Dr. H. Gysin, a senior research chemist at Geigy Co., synthesized a number of cycloaliphatic carbamates as potential insect repellents and passed them on to the Biological Department for testing. The biologists Wiesman and Lotmar observed, however, that one of these substances, 5, 5-dimethyl dihydroresorcinol dimethylcarbamate, which was later given the common name dimetan (III) (Fig. 10), showed only a weak repellent effect, but showed a promising insecticidal activity. As a consequence, a considerable number of carbamates analogous to

Fig. 10

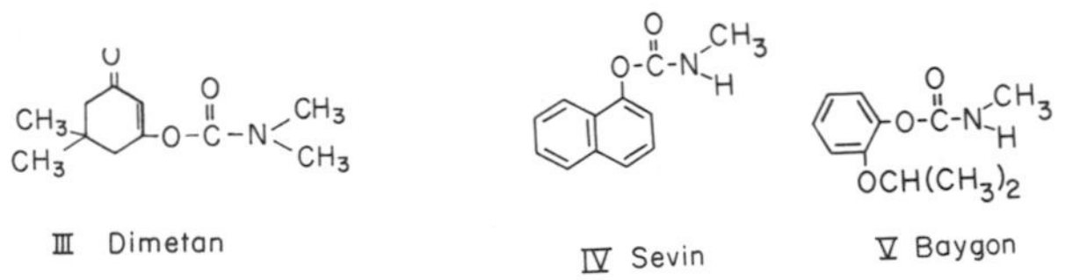

III Dimetan IV Sevin V Baygon

dimetan were synthesized and tested. Heterocyclic carbamates were found to be superior to the cycloaliphatic carbamates, one compound in this series even displaying systemic insecticidal activity (118a,b, 119).

The first marketed products in the carbamate field were the *N, N*-dimethylcarbamates of a variety of heterocyclic enols. The versatility of the carbamates was expanded by the introduction of the *N*-methylcarbamates of which the 1-naphtyl ester, Carbaryl or Sevin, (IV) (Fig. 10) is one of the most widely used broad spectrum insecticides. Its remarkable success is attributable to its low acute and chronic toxicity to mammals, and its environmental degradability. Therefore, it has replaced DDT for a number of uses, particularly where contamination of animal forage or the environment is a possibility. Carbaryl is presently recommended for more than 149 uses on 46 crops (120).

Another compound, Baygon (V), (Fig. 10) has become a standard item for household pest control and is exceptionally promising as a residual spray in malaria eradication programs. Still another carbamate, Zectran, is very effective on a wide range of pests, on ornamentals, and the home garden.

Various insect species react in different ways to substances having an anticholineesterase action. One explanation would be that different cholineesterases exist in the nervous system of different species. Derivatives of carbamic acid, similar to neostigmine, may inhibit the cholineesterase in some insect species, but not in others—which implies that the action of these substances is specific and that they are therefore suitable for use as selective insecticides. It is interesting to note that neither physostigmine (as mentioned previously) nor neostigmine has insecticidal properties, though both are well known as highly effective cholineesterase inhibitors and are closely related to the insecticides of the urethane group described above. On the other hand, all hydroaromatic and heterocyclic urethanes which have so far been found to have insecticidal properties have also proved to be highly inhibitory toward various esterase systems, from which it would seem that esterase-inhibition is an essential factor of insecticidal action. But it is still not clear why some cholinesterase inhibitors are not insecticides (118a,b, 119).

Carbamates, like the better-known organophosphorus insecticides, demonstrate all degrees of systemic action. The factors determining effective systemic activity are (1) sufficient water solubility to enter the roots of plants, together with appropriate lipophilicity to penetrate through waxes of plant cuticles; (2) sufficient stability in the plant environment to survive translocation and to permit accumulation in the leaves in toxic amounts; (3) high intrinsic insecticidal activity.

Detailed knowledge of pesticide metabolism in animals is essential for the registration of all new pesticides. However, for systemic insecticides which are absorbed and translocated within the plant environment, it is equally important to understand the chemical modifications which may take place within the "chemical factory" of the plant (121).

The use of carbamate insecticides is likely to be of significant potential value in protecting cotton plants, for example. Results are also impressive with the potato, certain seed crops, and ornamentals, and should lead to eventual wider use of these compounds on crops such as these. The persistence and desirable solubility of the carbamates should stimulate their use for protecting an even wider variety of crops from pests. For example, it has been shown that spreading granulated Temik on the soil surface prior to irrigation carries the toxicant down to the root zone of tree crops, where it undergoes absorption and translocation to the foliage (120).

In the program for testing and evaluating new insecticides by the World Health Organization (WHO), seven vigorous stages must be gone through to evaluate new insecticides both for their effectiveness and for any hazards involved in their application. Baygon has been found to be a promising residual insecticide against adult malaria vectors and has passed all stages of the WHO program (122).

Several other compounds, all monomethylcarbamates, have been tested by WHO (122) and have been found to be "direct inhibitors" of cholineesterase, since the unmetabolized compounds are capable of inhibiting the enzyme. The spontaneous reactivation of carbamylated enzyme is much more rapid than in the case of the phosphorylated enzyme, and, unless special precautions are taken, determination of blood cholineesterase in people or animals exposed to carbamates are likely to give results that are too high. The symptoms of intoxication with carbamates therefore disappear comparatively rapidly, and atropine treatment is usually not necessary by the time the patient reaches the place where the antidote is at hand.

The low toxicity to man, as well as other properties, make other carbamates, besides Baygon, promising for use as residual insecticides for control of adult mosquitos. Several have been tried by the WHO Insecticide Testing Unit on a village scale in Nigeria. We also have the curious information that modern pesticides have been used by African witch doctors as test poisons (123). The circle of history of the Calabar bean can thereby be considered as closed.

CONCLUSION

Alkaloids with strong cholineesterase inhibiting properties are a rare occurrence in the Plant Kingdom. In 1951 one was found in the Caucasian snowdrop, *Galanthus woronowii* Lozinsk (family Amaryllidaceae), a plant which grows in Georgia, USSR.

Galanthus woronowii contains galanthamine (VI) (Fig. 11), a potent cholineesterase inhibitor (124). When made into a carbamate derivative

Fig. 11

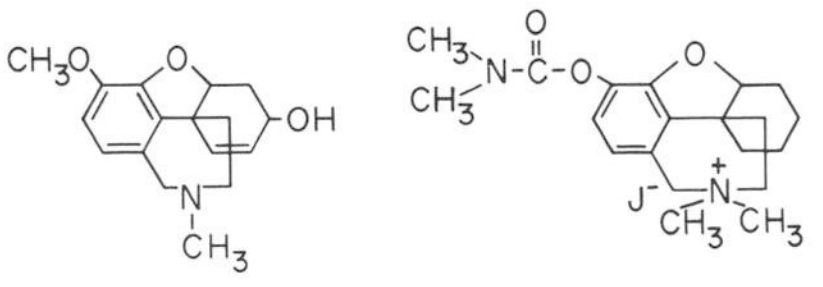

Galanthamine
VI

Deoxydemethyllycoramine
N,N-dimethylcarbamate
methiodide
VII

and quaternized (VII) (Fig. 11), it becomes even more potent (125a-d). Galanthamine has been used in *Myasthenia gravis* and other neurological diseases. Two new prospective sources of it, both snowdrops, *Leucojum aestivum* L. which is often found on the Caucasian part of the Black Sea coast, and *Ungernia victoris* Vved, which grows in the southern parts of Central Asia, have been suggested (126).

Galanthamine was discovered as a result of an extensive Russian screening program. It thus took more than 100 years from the discovery of physostigmine to find another group of alkaloids with strong choline esterase inhibiting activity.

Again, one is struck by the screening abilities of so-called primitive people. If attention had not been focused early on the ghastly poison ordeal of the Efik people of Old Calabar, it seems very likely that our present knowledge of many basic mechanisms in physiology and biochemistry would have been considerably retarded.

Physostigmine was the first alkaloid proven to act through inhibition of an enzyme. Knowledge of its action made possible the development of a successful bioassay for the neurohumoral transmitter, acetylcholine. Indeed, the alkaloid has contributed most to the understanding of neurohumoral chemical transmission and mapping out of cholinergic nerves, in relation to the heart, ganglionic transmission, neuromuscular transmission, vasodilator fibres, and sweat fibres. It also contributed much to the understanding of the function of the blood brain barrier.

Physostigmine has also been the subject of classical experiments with atropine and curare on pharmacological antagonism. It played an important role in the understanding of the kinetics of the enzyme, cholinesterase, and has contributed to the elucidation of the configuration of the active center of this enzyme.

Finally, physostigmine, although replaced by synthetic compounds in many cases, has had several uses in clinical medicine: In ophthalmology, as a miotic and in glaucoma; in *Myasthenia gravis;* in intestinal paralysis; in decurarization in anaesthesiology; and as an antidote in poisoning with atropine and atropinelike compounds. It can also be considered as the prototype of the carbamates used as insecticides.

ACKNOWLEDGMENT

This work has been supported by Stiftelsen Gustav och Tyra Svenssons Minne, Karolinska Institutet, Stockholm 60, Sweden, and grant MH 12007-03, National Institute of Mental Health, U.S. Public Health Service. The author wishes to express sincere thanks to his friends Paul and Helga Nya of Duketown Calabar.

References

1. J. D. Fage, *An Introduction to the History of West Africa,* 3rd ed. (Cambridge, Eng.: Cambridge University Press, 1962).
2. D. Forde (ed.), *Efik traders of Old Calabar* (London, Publ. for the International African Institute by the Oxford University Press, 1956).
3. D. C. Simmons, "Efik divination, ordeals and omens," *S. W. J. Anthropol. 12* (1956), 223-228.
4. F. H. Rodin, "Eserine: Its history in the practice of ophthalmology." *Am. J. Ophthalm. 30* (1947), 19-28.
5. W. F. Daniell, "On the natives of Old Callebar, W. Coast of Africa." *Proceedings of the Ethnological Society 40* (1846), 313-327.
6. L. Lewin, *Beiträge zur Giftkunde. Herausgegeben von Dr. Louis Lewin,* vol. II (Berlin, Gottesurteile durch Gifte und andere Verfahren, Verlag von Georg Stilke, 1929).
7. D. M. McFarlan, *Calabar* (London, Th. Nelson and Sons, Ltd., 1957).
8. H. M. Waddell, *Twenty-nine Years in the West Indies and Central Africa. A Review of Missionary Work and Adventure, 1829-1858* (London, 1863); 2nd ed. with intro. by G. I. Jones (Frank Cass & Co. Ltd., 1970).
9. W. Dickie, *Story of the Mission in Old Calabar* (Edinburgh, Offices of United Presbyterian Church, 1894).
10. F. M. Brodie, *The Devil Drives—A Life of Sir Richard Burton* (New York, W. W. Norton, 1967).
11. R. Burton, *Wanderings in West Africa from Liverpool to Fernando Po,* 2 vols. (London, F. R. G. S. Tinsley Bros. 1863).
12. J. H. Gaddum, "The pharmacologists of Edinburgh," *Ann. Rev. Pharmacol. 2* (1962), 1-9.
13. J. D. Comrie, *History of Scottish Medicine,* 2nd ed. (London, Wellcome Historical Medical Museum. Baillière, Tindall & Co., 1932), vol. II.
14. J. H. Balfour, "Description of the plant which produces the ordeal bean of Calabar," *Trans. Roy. Soc. Edinb. 22* (1861), 305-312.

15a. T. R. Fraser, "On the characters, actions and therapeutic uses of the bean of Calabar," *Edinb. Med. J. 9* (1863), 36-56, 123-132, 235-248. See also 15b.
15b. T. R. Fraser, "On the moth of the esere, or ordeal-bean of Old Calabar," *Ann. Nat. Hist.* (3rd ser.) *13* (1864), 389-393.
16. B. Holmstedt and G. Liljestrand, *Readings in Pharmacology* (London, Pergamon Press, 1963).
17. R. Christison, "On the properties of the ordeal bean of Old Calabar," *Monthly J. Med.* (London) *20* (1855), 193-204.
18. T. R. Fraser, "On atropia as a physiological antidote to the poisonous effects of physostigma," *Practitioner* (London) *4* (1870), 65-72.
19. T. R. Fraser, "The antagonism between the actions of active substances," *Brit. Med. J.* (1872), 457-487.
20a. G. Mackay, "Argyll Robertson-Nachruf," *Klin. Monatsblätter f. Augenheilkunde 47* (1909), 308-312; see also 20b.
20b. G. Mackay, "Obituary of Douglas Argyll Robertson," *Brit. Med. J. 1* (1909), 191-193.
21. A. D. Robertson, "The Calabar bean as a new agent in ophthalmic medicine," *Edinb. Med. J. 8* (1863), 193-204.
22. L. Laqueur, "Ueber Atropin und Physostigmin in ihre Wirkung auf den intraoculären Druck. Ein Beitrag zur Therapie des Glaucoms," *Arch. Ophthal. 23* (1877), 149-176.
23. U. Axelsson, "Glaucoma miotic therapy and cataract. Studies on Echothiophate (Phospholine iodide) and Paraoxon (Mintacol) with regard to cataractogenic effect," *Acta Ophthal. (Kbh.) Suppl. 102* (1969).
24. R. Lenz, "Versuche über die Einwirkung der Calabarbohne auf den Blutkreislauf" (Zurich, Inaugural-Dissertation, 1864).
25. C. Arnstein and P. Sustschinsky, "Über die Wirkung des Calabar auf die hemmenden und beschleunigenden Herznerven," *Zbl. med. Wiss.* (1867), 625-628.
26. G. Harley, "A brief account of the literary history, botanical characters, and therapeutical properties of the ordeal bean of Old Calabar," *Brit. Med. J.* (1863), 262-265.
27. J. Tachau, "V. Versuche über die Wirkung des Calabarbohnenextractes," *Arch. d. Heilkunde (von Wagner in Leipzig) Sechster Jahrg.* (1865), 69-78.
28. F. L. Bauer, "Einige Resultate von Versuchen über die Wirkung des Calabargiftes," *Zbl. med. Wiss. 37* (1866).
29. W. Westerman, "Untersuchungen über die Wirkungen der Calabarbohne." *Druck von C. Mattiesen. Diss. Dorpat* (1867).
30. B. Holmstedt, "Pharmacology of organophosphorus cholinesterase inhibitors," *Pharmacol. Rev. 11* (1959), 567-688.
31. J. Pal, "Physostigmin ein Gegengift des Curare," *Zbl. Physiol. 14* (1900), 255-258.
32. H. Roeber, Über die Wirkungen des Calabarextractes auf Herz und Rückenmark (Berlin, Dissertation 1868). "Safe use of pesticides in public health." Sixteenth Report of the WHO Expert Committe on Insecticides. WHO Technical Report Series (1967), 356.
33. E. Harnack and L. Witkowski, "Pharmakologische Untersuchungen über das Physostigmin und Calabarin." *Arch. Exp. Path. Pharmacol. 5* (1871), 401-454.
34. E. H. Greenhow, "Report–Clinical Society of London–Calabar Bean (Traumatic Tetanus)," *Brit. Med. J.* (1869), 234, 235.

35. W. Munro, "Suggestions as to the use of Calabar bean in Cholera Asiatica," *Edinb. Med. J. 17* (1871), 327-329.
36. S. W. D. Williams, "The *Physostigma venenosum* in epilepsy and progressive paralysis of the insane," *Practitioner, London 8* (1872), 75-81.
37. S. Ringer and W. Murrell, "On the use of Physostigma in some nervous affections," *The Lancet* (1877), 912-914, 950-952.
38. Leibholz, "Zwei Physostigminvergiftungen," *Vjschr. gerichtl. Med. Dritte Folge 3* (1892), 284-287.
39. G. F. Reynalds, "Case of poisoning by Calabar bean," *J. Trop. Med. Hyg.* (London, 1899), 206-207.
40. J. Cameron, "The recent cases of poisoning by Calabar bean (Reported by Mr. J. H. Evans, House-Surgeon)," *Med. Tms. Gaz.* (London, 1864), 406-410.
41. Kleinwächter, "Beobachtung über die Wirkung des Calabar-Extracts gegen Atropinvergiftung," *Berl. Klin. Wschr.* (1864), 369-371.
42. Lingen, "Vergiftung durch Calabar-Bohne," *St. Petersburger Medizin. Zschr.* (1864), 244-246.
43. J. Jobst and O. Hesse, "Über die Bohne von Calabar," *Ann. Chemie Pharmacie 129, 130* (1864), 115-121.
44a. A. A. Vée, *Récherches chimiques et physiologiques sur la fève du Calabar* (Paris, Thèse, 1865). See also 44b.
44b. A. Vée and M. Leven, "De l'alcaloide de la fève du Calabar et expériences physiologiques avec ce même alcaloide," *C. R. de la Soc. de Biol.* (1864), 160-172.
45. A. Ehrenberg, "Über Alkaloide der Calabarbohnen," *Verh. Vers. dtsch. Ntf. u. Ärzte. II* (1893), 102-103. See also *Chem. Zentr. 2* (1894), 439.
46. C. F. Boehringer & Söhne, *Pharmaceut. Post.* (1888), 663-664.
47. W. Eber, "Ein neues Physostigminderivat und seine pharmacologische Bedeutung," *Berl. thierärztl. Wschr.* (1888), 41-49, 57-68.
48. Ogiu, "Über das Isophysostigmin," *Apoth. Ztg. 19* (1904), 891.
49. A. H. Salway, "Chemical Examination of Calabar beans," *J. Chem. Soc. 99* (1911), 2148-2159.
50a. M. Polonovski and C. Nitzberg, "Étude sur les alcaloides de la fève de Calabar. (II) La Génésérine, nouvel alcaloid de la fève," *Bull. Soc. Chim. Fr. 17* (1915), 244-256.
50b. M. Polonovski and C. Nitzberg, "Étude sur les alcaloides de la fève de Calabar (IV). Synthèse partielle de l'ésérine et de la génésérine," *Bull. Soc. chim. Fr. 19* (1916), 27-37.
51. A. H. Salway, *Pharm. J. 87* (1912), 719.
52. S. Ellis, "Studies on physostigmine and related substances. IV. Chemical studies on physostigmine breakdown products and related epinephrine derivatives," *J. Pharmacol. 79* (1943), 364-372.
53. A. Petit and M. Polonovski, "Étude sur l'ésérine," *Bull. Soc. chim. Pr. 9* (1893), 1008-1018.
54. M. Polonovski, "Étude sur les alcaloides de la fève de Calabar (I). Esérine," *Bull. Soc. chim. Fr. 17* (1915), 235-256.
55. L. Marion, "The indole alkaloids," in R. H. F. Manske, ed. (New York, Academic Press, 1952), vol. II.
56a. M. Polonovski, "Action de CH^3I sur les bases de la série de l'ésérine et de la génésérine," *Bull. Soc. chim. Fr. 23* (1918), 335-361.

56b. M. Polonovski, "Étude sur les alcaloides de la fève de Calabar (VII). Dégradation par iodomethylations successives des noyaux de l'ésérine et de la génésérine." No. 46 *Bull. Soc. chim. Fr. 23* (1918), 335-361.
56c. M. Polonovski, "Chimieorganique—Di iodomethylates dans la série de l'ésérine," *C. R. Acad. Sci.* (Paris) *176* (1923), 1813-1815.
57. F. E. King and R. Robinson, "Experiments on the synthesis of physostigmine (eserine) Part XI. The later phases of the synthetical investigations," *J. Chem. Soc. 2* (1935), 755-759.
58. E. Stedman and G. Barger, "Physostigmine (Eserine), Part III." *J. Chem. Soc. 128* (1925), 247-258.
59. P. L. Julian and J. Pikl, "Studies in indole series; V. Complete synthesis of physostigmine (eserine)," *J. Amer. Chem. Soc. 51* (1935), 755-757.
60. J. A. Aeschliman and M. Reinert, "The pharmacological action of some analogues of physostigmine," *J. Pharmacol. 43* (1931), 413-444.
61. R. Hunt and M. Taveau, "On the physiological action of certain choline derivatives and new methods for detecting choline," *Brit. Med. J. 2* (1906), 1788-1789.
62. J. H. Gaddum, "Anticholinesterases—the history of work on anticholinesterases," *Chem. and Ind.* (1954), 266-268.
63. F. Lembeck and W. Giere, *Otto Loewi—Ein Lebensbild in Dokumenten* (Berlin, Heidelberg, New York, Springer Verlag, 1968).
64. H. H. Dale, *Adventures in Physiology with Excursions in Autopharmacology* (London, The Wellcome Trust, 1965).
65. W. Feldberg, "Henry Hallett Dale, 1875-1968," *Brit. J. Pharmacol. 35* (1969), 1-9.
66. G. B. Koelle, sub. ed., "Cholinesterases and anticholinesterase agents," in *Handb. d. experim. Pharmakol.*, vol. 15 (Berlin, Springer Verlag, 1963).
67. H. K. Anderson, "The paralysis of involuntary muscle. Part III. On the action of pilocarpine, physostigmine, and atropine upon the paralysed iris," *J. Physiol. 33* (1905-6), 414-438.
68. W. E. Dixon, "On the mode of action of drugs," *Med. Mag.* (London) *16* (1907), 454-457.
69. O. Loewi and G. Mansfeld, "Über den Wirkungsmodus des Physostigmins (III Mitteilung)," *Arch. exp. Path. Pharmak. 62* (1911), 180-185.
70. O. Loewi, "Untersuchungen zur Physiologie und Pharmakologie des Herzvagus. III. Mitteilung: Vaguserregbarkeit und Vagusgifte," *Arch. f. Pharmakologie 70-71* (1912-1913), 351-368.
71. H. H. Dale, "The action of certain esters and ethers of choline and their relation to muscarine," *J. Pharmacol. 6* (1914), 147-190.
72. R. Hunt, "Vasodilator reactions. I," *Amer. J. Physiol. 45* (1918), 197-230.
73. R. Hunt, "Vasodilator reactions. II," *Amer. J. Physiol. 45* (1918), 231-267.
74. R. Hunt, "Some physiological actions of the homocholins and some of their derivatives," *J. Pharmacol. 6* (1915), 477-525.
75. H. Fühner, "Der toxicologische Nachweis des Physostigmins," *Biochem. Z.* (Berlin) *92* (1918), 347-354.
76. W. Feldberg, "Letter to the author," April 26, 1968.
77. O. Loewi, "An autobiographic sketch," *Perspec. Biol. Med.* (1960), 3-25.
78. O. Loewi, "Über humorale Uebertragbarkeit der Herznervenwirkung. (I. Mitteilung)," *Pflügers Arch. ges. Physiol. 189* (1921), 239-242.
79. O. Loewi, "Über humorale Uebertragbarkeit der Herznervenwirkung. (II. Mitteilung)," *Pflügers Arch. ges. Physiol. 193* (1921), 201-213.

80. O. Loewi and E. Navrathil, "Über humorale Uebertragbarkeit der Herznervenwirkung (XI. Mitteilung. Über den Mechanismus der Vaguswirkung von Physostigmin und Ergotamin)," *Pflügers Arch. ges. Physiol. 214* (1926), 689-696.
81. H. H. Dale, "Otto Loewi 1873-1961," *Biographical Memoirs of Fellows of the Royal Society 8* (1962), 67-89.
82. O. Galehr and F. Plattner, "Über das Schicksal des Acetylcholins im Blute," *Arch. ges. Physiol. 218* (1927), 488-505.
83. E. Engelhart and O. Loewi, "Fermentative Azetylcholinspaltung im Blut und ihre Hemmung durch Physostigmin," *Arch. exp. Path. Pharmak. 150* (1930), 1-13.
84. C. Matthes, "Action of blood on acetylcholine," *J. Physiol. 70* (1930), 338-348.
85. E. Stedman and E. Stedman, "Studies on relationship between chemical constitution and physiological action; inhibitory action of certain synthetic methanes on activity of liver esterase," *Biochem. J. 25* (1931), 1147-1167.
86. E. Stedman, E. Stedman, and L. H. Easson, "Cholinesterase enzyme present in blood serum of horse," *Biochem. J. 26* (1932), 2056-2066.
87a. W. Leuzinger, A. L. Baker, and E. Cauvin, "Acetylcholinesterase, II. Crystallization, absorption-spectra, isoionic point," *Proc. Nat. Acad. Sci.* (Washington, D.C.) *59* (1968), 620-623.
87b. E. Stedman, "Studies on the relationship between chemical constitution and physiological action. I. Position of isomerism in relation to miotic activity of some synthetic urethanes," *Biochem. J. 20* (1926), 719-734.
87c. J. E. Casida, "Mode of action of carbanates," *Ann. Rev. Entomol. 8* (1963), 39-58.
88. F. Hobbiger, "Anticholinesterases. The *in vitro* inhibition of cholinesterases by physostigmine, neostigmine and other related (non-phosphorus) compounds," *Chem. and Ind.* (1954), 415-418.
89. I. Wilson, M. Hatch, and S. Ginsburg, "Carbamylation of acetylcholinesterase," *J. Biol. Chem. 235* (1960), 2312-2315.
90. I. B. Wilson, M. A. Harrison, and S. Ginsburg, "Carbamyl derivatives of acetylcholinesterase," *J. Biol. Chem. 236* (1961), 1498-1500.
91. R. D. O'Brien, "The mechanism of inhibition of cholinesterase by carbamates," *Exp. Med. Surg.,* Suppl. Issue (1965), Brooklyn Medical Press, Inc.
92. H. H. Dale and J. H. Gaddum, "Reactions of denervated voluntary muscle and their bearing on the mode of action of parasympathetic and related nerves," *J. Physiol.* (London) *70* (1930), 109-144.
93. U. S. von Euler and J. H. Gaddum, "Unidentified depressor substance in certain tissue extracts," *J. Physiol.* (London) *72* (1931), 74-87.
94. H. H. Dale and H. W. Dudley, "The presence of histamine and acetylcholine in the spleen of the ox and the horse," *J. Physiol.* (London) *68* (1929), 97-123.
95. W. Feldberg and J. H. Gaddum, "The chemical transmitter at synapses in a sympathetic ganglion," *J. Physiol.* (London) *81* (1934), 305-319.
96. A. W. Kibjakow, "Über humoral Übertragung der Erregung von einem Neuron auf das Andere," *Pflüger's Arch. ges. Physiol. 232* (1933), 432-443.
97. A. W. Kibjakow, "Über humorale Übertragung der Erregung von einem Neuron Ganglien," *Proc. XV. Internat. Physiol. Congr.* (Leningrad-Moscow, 1935).
98. C. H. Chang and J. H. Gaddum, "Cholineesters in tissue extracts," *J. Physiol.* (London) *79* (1933), 255-285.
99. R. J. Banister, V. P. Whittaker, and S. Wijesundera, "The occurrence of homo-

logues of acetylcholine in ox spleen," *J. Physiol.* (London) *121* (1953), 55-71.
100. H. H. Dale, W. Feldberg, and M. Vogt, "Release of acetylcholine at voluntary motor nerve ending," *J. Physiol.* (London) *86* (1936), 353-380.
101. A. Schweitzer, E. Stedman, and S. Wright, "Central action of anticholinesterases," *J. Physiol.* (London) *96* (1939), 302-336.
102. C. G. Hammar, I. Hanin, B. Holmstedt, R. J. Kitz, D. J. Jenden, and B. Karlén, "Identification of acetylcholine in fresh rat brain by combined gas chromatopraphy-mass spectrometry," *Nature 220* (1968), 915-917.
103. A. R. Goni, *Myasthenia gravis,* trans. Georgianna Simmons Gittiner (Baltimore, Md., Williams & Wilkins, 1946).
104. F. Jolly, "Über myasthenia gravis pseudoparalytica," *Berl. klin. Wschr. 32* (1895), 1-7.
105. F. Jolly, "Pseudoparalysis myasthenica," *Neurol. Zbl. 14* (1895), 34-36.
106. Th. Kirchhoff, ed., *Deutsche Irrenärzte. Jolly, Friedrich 1844-1904* (Berlin, Julius Springer Verlag, 1924).
107. A. Fick and R. Böhm, "The effect of veratrine on the muscle fibres," *Verh. phys. med. Ges. Würzb.* (1872).
108. Walker-Overend, "On the influence of curare and veratrine on striated muscle," *Arch. exp. Path. Pharmak. 26* (1889).
109. T. W. Eden, "On the effect of proveratrine," *Pharmacological Institute 29* (18 ?), Leipzig.
110. H. R. Viets, "The miracle at St. Alfege's," *Med. Hist. 9* (1965), 184-185.
111. M. B. Walker, "Treatment of myasthenia gravis with physostigmine," *Lancet 226* (1934), 1200-1201.
112. M. B. Walker, "Case showing the effect of prostigmine on myasthenia gravis," *Proc. Roy. Soc. Med. 28* (1934-35), 759-761.
113. L. Remen, "Zur Pathogenese und Therapie der myasthenia gravis pseudoparalytica," *Dtsch. Z. Nervenheilk. 128* (1932), 66-78.
114. H. R. Viets and R. S. Schwab, "Prostigmine in the diagnosis of myasthenia gravis," *New Engl. Med. J. 213* (1935), 1280-1283.
115. H. W. Whipple, "Myasthenia gravis," *Ann. N. Y. Acad. Sci. 135* (1966).
116. W. W. Hofmann, "Newer drugs for myasthenia gravis. A microphysiologic study of effects," *J. Pharmacol. 160* (1968), 349-359.
117. A. Albert, *Selective Toxicity* (London, Methuen, 1968).
118a. H. Gysin, "Un Nouveau Group de substances à activité insecticide." *3e Congrès International de Phytopharmacie* (Paris, September, 1952).
118b. H. Gysin, "Über einige neue Insektizide," *Chimia 8* (1954), 205-210, 221-228.
119. A. Buxtorf and M. Spindler, eds., *Fifteen Years of Geigy Pest Control* (Basel, Switz. Buchdruckerei Karl Werner AG, 1954).
120. "Symposium on Carbamate Insecticides," *J. Agric. Food Chem. 13* (1965), no. 3.
121. R. L. Metcalf et al., "Carbamate insecticides—Where are they headed?" *Pharm. Chem. 130* (1967), 45-58.
122. "The Safe Use of Pesticides," WHO Technical Report Series, 1967.
123. A. Heyndrickx, "Fatal and chronic intoxications by man due to insecticides and phytopharmaceuticals in the Belgian Congo and Ruanda-Urundi," *Mededelingen van de Landbouwhogeschool en de Opzockingsstations van de Staat te Gent 25* (1960), 1542-1548.

124. M. D. Mashkovski and R. P. Kruglikova-Lvova, "Pharmacology of a new alkaloid in Russian," *Farmak. Toks. 14* (1951), 27-30.

125a. R. L. Irwin and H. J. Smith, "A: cholinesterase inhibition by galanthamine and lycoramine," *Biochem. J. 3* (1960), 147-148.

125b. R. L. Irwin and H. J. Smith, "B: The activity of galanthamine and related compounds on muscle," *Arch. int. Pharmacodyn. 127* (1960), 314-330.

125c. R. L. Irwin, H. J. Smith, and M. Hein, "The activity of certain lycoramine derivatives on muscle," *J. Pharmacol. 134* (1961), 53-59.

125d. R. L. Irwin and M. M. Hein, "The inhibition of rat brain cholinesterase after administration of the dimethyl-carbamates of deoxy-demethyl-lycoramine, neostigmine or physostigmine," *J. Pharmacol. 136* (1962), 20-25.

126. *Atlas of Medicinal Plants of the USSR* (in Russian). Moscow, Government Pharmaceutical Editions, 1962.

Index